2

QUADERNI

SELECTIONS

R. Asgari
School of Physics
Institute for Research
in Fundamental Sciences
(IPM)
Tehran 19395-5531, Iran

J. Capps
Department of Physics
Carnegie Mellon University
Pittsburgh, PA 15213

S. Chesi
Beijing Computational
Science Research Center
Beijing 100084, China

M. Daniels
Department of Physics
Carnegie Mellon University
Pittsburgh, PA 15213

M. V. Fistul
Theoretische Physik III
Ruhr-University Bochum
D-4081 Bochum, Germany
and
Theoretical Physics
and Quantum Technologies
Department
National University
of Science and Technology
MISIS
Moscow, Russia

J. K. Jain
Department of Physics
The Pennsylvania State
University
University Park
PA 16802, USA

Y. Lyanda-Geller
Department of Physics
Purdue University
West Lafayette
IN 47907, USA

D. C. Marinescu
Department of Physics
and Astronomy
Clemson University
Clemson
South Carolina 29634 USA

P. Muzikar
Department of Physics
Purdue University
West Lafayette
IN 47907, USA

F. Pellegrini
SISSA
Via Bonomea 265
34136 Trieste, Italia

V. Pellegrini
Istituto Italiano di Tecnologia,
Graphene Labs,
Via Morego, 30
I-16163, Genova, Italia

M. Polini
Istituto Italiano di Tecnologia,
Graphene Labs,
Via Morego, 30
I-16163, Genova, Italia
and
NEST, Scuola Normale
Superiore,
Piazza dei Cavalieri, 7
56126, Pisa, Italia

L. P. Rokhinson
Department of Physics
Purdue University
West Lafayette
IN 47907, USA

G. Santoro
SISSA
Via Bonomea, 265
34136 Trieste, Italia

G. Simion
Department of Physics
Purdue University
West Lafayette
IN 47907, USA

C. E. Sosolik
Department of Physics
and Astronomy
Clemson University
Clemson
South Carolina 29634, USA

E. Tosatti
SISSA
Via Bonomea, 265
34136 Trieste, Italia

G. Vignale
Department of Physics
and Astronomy
University of Missouri
Columbia
Missouri 65211, USA

S. Yarlagadda
CMP Div.
Saha Institute
of Nuclear Physics
Kolkata, India

No-nonsense Physicist
An overview of Gabriele Giuliani's work and life

No-nonsense Physicist

An overview of Gabriele Giuliani's work and life

edited by
Marco Polini, Giovanni Vignale,
Vittorio Pellegrini and Jainendra K. Jain

EDIZIONI
DELLA
NORMALE

ISBN: 978-88-7642-535-6
eISBN 978-88-7642-536-3

Figure 1. Gabriele (left) and Giovanni Vignale (right) promoting the sales of their newly published book at the Cambridge University Press boot at the 2005 American Physical Society March Meeting in Los Angeles (USA).

Figure 2. Gabriele with a big smile on his face at the Montreal "Grand Prix du Canada" in 2010.

Figure 3. Gabriele (left) and Giovanni Vignale (right) at the 1988 American Physical Society March Meeting in New Orleans (USA).

Figure 4. Gabriele was a passionate supported of F.C. internazionale, one of the two soccer teams based in the city of Milano (Italy). The picture was taken in the summer of 2009.

Figure 5. Gabriele (right) with his (in)famous red minivan and his Formula Ford car, which he used to call "his black girlfriend" (circa 2010).

Contents

Preface

Gabriele Francesco Giuliani, Professor of Physics at Purdue University (West Lafayette, Indiana, USA), died on November 22, 2012 (Thanksgiving Day) after a 12-year-long battle with a rare, slow-growing, relentless form of cancer. His passion for life, his taste for simple and earthy pleasures, and his no-nonsense attitude in science as in all other aspects of life, remained with him until the last minute. This volume is as much an attempt to pay tribute to his scientific contributions as to record the many facets of his personality: scientist, educator, author, sportsman, coach, polemist, family man, and provocateur. The gallery of photographs collected in the following pages will give an idea of the range of his interests. The volume opens with a series of personal recollections of Gabriele by his immediate family and friends. This is followed by a set of scientific papers by former colleagues, collaborators and students of Gabriele. These papers, especially written for this volume, report original research on topics to which Gabriele made important contributions, and often are inspired precisely by those contributions. A selection of reprints of Gabriele's most important scientific papers concludes the volume.

Born in Ascoli Piceno, Italy, on April 13, 1953, Giuliani was educated at the University of Pisa where he graduated cum laude in 1976 under the guidance of Professor Mario Tosi. He continued his studies at the Scuola Normale Superiore in Pisa and was a researcher in Rome and in Trieste, where he worked with Professor Erio Tosatti. In 1979 he met Professor Albert Overhauser, who was to be the decisive influence in his career. Fascinated by the physics of broken symmetry phases in simple metals, he joined Overhauser at Purdue University in the study of collective modes of charge-density waves, the so-called "phasons" and "amplitons". He eventually became a member of the physics faculty at Purdue in 1984 –but not before completing an extremely fruitful postdoctoral experience at Brown University with Professor John Quinn. It was during this period that some of his best known contributions sprang to life, such as the calculation of plasmon dispersions in semiconductor superlattices, the discovery of the singular behavior of the quasiparticle linewidth as a function of temperature in a two-dimensional electron gas, and the prediction of a ferromagnetic phase transition in the two-dimensional electron gas in the quantum Hall regime (now experimentally observed).

Gabriele Giuliani's field of research was the theoretical study of the properties of low-dimensional electronic systems, particularly those that are controlled by electron-electron interactions. His enthusiasm for the theory of the interacting Electron Gas earned him the nickname of "EG" since the Trieste days. Many of his contributions are widely known and some of them are featured in textbooks. Besides the already mentioned works, these include an elegant analysis of the role of impurities in the integer quantum Hall effect, an experimentally confirmed theory of the effect of a magnetic field on the critical current of a layered superconductor, and numerous contributions to the foundations of the theory of Fermi liquids, most recently in the presence of spin-orbit interactions. Among his legacies is a monograph, co-authored with Giovanni Vignale, on the "Quantum theory of the electron liquid" (Cambridge University Press, Cambridge, 2005), which has become a standard reference for beginning students and advanced researchers.

Giuliani was known in physics circles for his flamboyant personality, his quick sense of humor, and his unremitting critical eye. Shunning the superficial and the fashionable he always strove for genuine accomplishment and complete intellectual honesty. His criticism could be abrasive, but never intentionally so. His ability to entertain and provoke with humorous word play was unmatched. He never lost the purity and the enthusiasm of his happy childhood in Ascoli Piceno. An avid soccer player and sports critic to his last days, he successfully coached soccer teams of all age groups at Purdue and in the Lafayette area. Other interests of his were Hammond-based blues, wildlife (he boasted to have once met a grizzly bear at Yellowstone), and all kinds of "italica", ranging from spaghetti alla carbonara "better than sex", to espresso brews, to Italian politics which he followed with a mixture of wit, concern, and shame. In the early 2000, after surviving against all odds a first major operation to remove a large liposarcoma, he decided to act on one of the great passions of his youth: auto racing. Equipped with an old van Diemen, which he used to call "his black girlfriend", he experienced the adrenaline rush of Formula Ford racing. During this period he also participated as a volunteer race marshal in several racing events. Above all, Gabriele Giuliani loved his family, his mentors, and his students, several of which are now professors in various countries. With his mentors he always enjoyed a relation based on admiration, respect and love, and to his students he tried to give back what his mentors had given him. A few weeks before dying, musing on the twelve years that followed his first encounter with cancer, he told his wife: "If somebody had offered me this twelve years ago, I would've signed on the dotted line in a minute to be able to live as well as I did for these twelve years". He died as he had lived, joyously, defiantly, and deeply engaged in his own life.

This volume is dedicated to Gabriele's wife, Pamela Wilhelm-Giuliani, his children Daniela, Adriana and Giuseppe and his siblings, Alessandro and Carla Cutolo. But, more broadly, it is dedicated to the whole scientific community of which Gabriele was part, in the firm belief that the fruits of his scientific ingenuity will outlive him.

ACKNOWLEDGEMENTS. The Editors of this Volume wish to thank Luisa Ferrini and the whole team of the "Edizioni della Normale" at the Scuola Normale Superiore (Pisa, Italy) for their interest in this Volume.

Curriculum vitae

EDUCATION:

Undergraduate: *Laurea* in Physics, 110/110 *cum laude*),
University of Pisa, Italy, 1976;
Doctoral: *Perfezionamento*, 70/70 *cum laude*,
Scuola Normale Superiore, Pisa, Italy, 1983.

EMPLOYMENT:

1995-2012: Professor of Physics, Purdue University
1990-1995: Associate Professor of Physics, Purdue University
1984-1990: Assistant Professor, Purdue University
1983-1984: Assistant Professor (Research), Brown University
1982-1983: Postdoctoral Research Associate, Brown University
1979-1981: Postdoctoral Research Associate, Purdue University
1976-1979: *Ricercatore*, Scuola Normale Superiore, Pisa, Italy.

PROFESSIONAL ASSOCIATIONS:

Member of the American Physical Society since 1979.
Fellow of the American Physical Society since 2006.

Research Journal Publications:

PUBLICATIONS:

BOOK: **Quantum Theory of the Electron Liquid**
Gabriele F. Giuliani and Giovanni Vignale
Cambridge University Press, March 2005

1. *Excitonic instability in a quasi-one-dimensional electron gas*, Nuovo Cimento Lett. **16**, 385 (1976); with E. Tosatti and M. P. Tosi.
2. *Electronic response and longitudinal phonons of a charge-density-wave distorted linear chain*, Nuovo Cimento **47B**, 135 (1978); with E. Tosatti.

3. *Quasi-one-dimensional excitonic insulator*, Journal of Phys. C: Solid State Physics **12**, 2769 (1979); with E. Tosatti and M. P. Tosi.
4. *Longitudinal phonon spectrum of incommensurate one-dimensional charge-density-waves*, *Lecture Notes in Physics* n. 95: Proceedings of the International Conference on One-Dimensional Conductors I, Dubrovnik, Yugoslavia, 1978; ed. S. Barisic et al., (Springer Verlag, Berlin, 1979), v. I, pp. 191-198; with E. Tosatti.
5. *Wave-vector orientation of a charge-density-wave in Potassium*, Phys. Rev. B **20**, 1328-1331 (1979); with A. W. Overhauser.
6. *Theory of transverse phasons in potassium*, Phys. Rev. B **21**, 5577 (1980); with A. W. Overhauser.
7. *Charge-density-wave satellites intensity in potassium*, Phys. Rev. B **22**, 3639 (1980); with A. W. Overhauser.
8. *Observation of phasons in metallic rubidium*, Phys. Rev. Lett. **45**, 1335 (1980); with A. W. Overhauser.
9. *Structure factor of a charge-density-wave*, Phys. Rev. B **23**, 3737 (1981); with A. W. Overhauser.
10. *Microscopic theory of phase and amplitude instabilities of an incommensurate charge-density-wave*, Phys. Rev. B **26**, 1660 (1982); with A. W. Overhauser.
11. *Spin response of a charge-density-wave*, Phys. Rev. B **26**, 1671 (1982); with A. W. Overhauser.
12. *Lifetime of a quasiparticle in a two-dimensional electron gas*, Phys. Rev. B **26**, 4421 (1982); with J. J. Quinn.
13. *Quantization of the Hall conductance in a two-dimensional electron gas*, Phys. Rev. B **28**, 2969 (1983); with J. J. Quinn and S. C. Ying.
14. *Charge density excitations at the surface of a semiconductor superlattice: a new type of surface polariton*, Phys. Rev. Lett. **51**, 919 (1983); with J. J. Quinn.
15. *Acoustic surface plasmons in type-II semiconductor superlattices*, Phys. Rev. B **28**, 6144 (1983); with G. Qin and J. J. Quinn.
16. *Coulomb inelastic lifetime of a quasiparticle in a two-dimensional electron gas*, Surface Sci. **142**, 48 (1994); with J. J. Quinn.
17. *Intrasubband plasma modes of a semi-infinite superlattice: a new type of surface wave*, Surface Sci. **142**, 433 (1984); with G. Qin and J. J. Quinn.
18. *Surface and bulk plasmon-polaritons in periodic metallic heterostructures*, Journal de Physique Colloque **45**, C5 285 (1984); with J. J. Quinn and R. F. Wallis.
19. *Effects of diffusion on the plasma oscillations of a two-dimensional electron gas*, Phys. Rev. B **29**, 2321 (1984); with J. J. Quinn.

20. *Existence of acoustic surface plasmons in semiconducting superlattices*, Proceedings of the 17th International Conference on the Physics of Semiconductors, J. D. Chady and W. A. Harrison editors, (Springer-Verlag, New York, 1985), pp. 511-513; with G. Qin and J. J. Quinn.
21. *Breakdown of the random phase approximation in the anomalous quantum Hall effect regime*, Phys. Rev. B **31**, 3451 (1985); with J. J. Quinn.
22. *Triplet exciton and ferromagnetic instability of a two-dimensional electron gas in a large magnetic field with filling factor* $\nu = 2$, Solid State Commun. **54**, 1013 (1985); with J. J. Quinn.
23. *Spin-polarization instability in a tilted magnetic field of a two-dimensional electron gas with filled Landau levels*, Phys. Rev. B **31**, 6228 (1985); with J. J. Quinn.
24. *Theory of surface plasmon polaritons in truncated superlattices*, Surface Sci. **166**, 45 (1986); with R. Szenics, R. F. Wallis and J. J. Quinn.
25. *Plasmon bands in periodic conducting heterostructures*, Phys. Rev. B **33**, 1405 (1986); with G. Eliasson, J. J. Quinn and R. F. Wallis.
26. *Magnetic instabilities of a two-dimensional electron gas in a large magnetic field*, Surface Sci. **170**, 316 (1986); with J. J. Quinn.
27. *Plasmons in semiconducting superlattices with complex unit cell*, Phys. Rev. B **33**, 8390 (1986); with R. A. Mayanovic and J. J. Quinn.
28. *Elementary excitations at the surface of a semiconductor superlattice and their coupling to external probes*, Physica Scripta **36**, 946 (1987); with P. Hawrylak and J. J. Quinn.
29. *Theory of surface magnetoplasmon-polaritons in truncated superlattices*, Phys. Rev. B **36**, 1218 (1987); with R. Szenics, R. F. Wallis and J. J. Quinn.
30. *Acoustic plasmons in a conducting double layer heterostructure*, Phys. Rev. B **37**, 937 (1988); with G. E. Santoro.
31. *Plasmon mechanism for superconductivity in semiconducting heterostructures: the effect of acoustic plasmons*, Surface Sci. **196**, 476 (1988).
32. *Exact limits of the many-body local fields in a two-dimensional electron gas*, Phys. Rev. B **37**, 4813 (1988); with G. E. Santoro.
33. *Raman scattering from plasma excitations in a conducting double layer*, Phys. Rev. B **37**, 8443 (1988); with G. E. Santoro.
34. *Frequency dependence of the electron self-energy in a two-dimensional electron gas*, Solid State Commun. **67**, 681 (1988); with G. E. Santoro.
35. *Many-body effective mass and anomalous g-factor in inversion layers*, Phys. Rev. B **38**, 10966 (1988); with S. Yarlagadda.

36. *Quasiparticle energy and interaction in a degenerate Fermi liquid*, Solid State Commun. **69**, 677 (1989); with S. Yarlagadda.
37. *Exact asymptotic behavior of the charge and spin susceptibilities in an interacting Fermi liquid*, Phys. Rev. B **39**, 3386 (1989); with S. Yarlagadda.
38. *Electron self-energy in two dimensions*, Physical Review B **39**, 12818 (1989); with G. E. Santoro.
39. *Spin susceptibility in a two-dimensional electron gas*, Phys. Rev. B **40**, 5432 (1989); with S. Yarlagadda.
40. *Many-body local fields and quasiparticle renormalization effects in two-dimensional electronic systems*, Surface Sci. **229**, 410 (1990); with S. Yarlagadda.
41. *Perturbation theory of the Anderson model*, Solid State Commun. **76**, 1177 (1990); with G. E. Santoro.
42. *Impurity spin susceptibility of the Anderson model: a perturbative approach*, Phys. Rev. B **44**, 2209 (1991); with G. E. Santoro.
43. *Light metal hydrides and superconductivity*, Electric Power Research Institute, **EL-7507**, Project 7911-3, 1992; with A. W. Overhauser, E. Negishi, R. D. Hong, and Z. Owczarczyk.
44. *Critical current of a Josephson junction in layered superconductors*, Phys. Rev. B **47**, 11341 (1993); with M. V. Fistul.
45. *Two-impurity Anderson model: some exact results within Fermi liquid theory*, Modern Physics Letters B **8**, 367 (1994); with G. E. Santoro.
46. *Two-Impurity Anderson model: results of a perturbative expansion in* U, Phys. Rev. B **49**, 6746-6762 (1994); with G. E. Santoro.
47. *Quasiparticle pseudo Hamiltonian for an infinitesimally polarized Fermi liquid*, Phys. Rev. B **49**, 7887-7897 (1994); with S. Yarlagadda.
48. *Many-body local fields and Fermi liquid parameters in a quasi-two-dimensional electron liquid*, Phys. Rev. B **49**, 14188 (1994); with S. Yarlagadda.
49. *Landau theory of the Fermi liquids and the integration-over-the-coupling-constant algorithm*, Phys. Rev. B **49**, 14172 (1994); with S. Yarlagadda.
50. *Critical current of a lateral Josephson junction for layered superconductors*, Phys. Rev. B **50**, 7026 (1994); with M. V. Fistul.
51. *Magnetic field dependence of the critical current of a layered superconductor*, Physica C **230**, 9 (1994); with M. V. Fistul.
52. *Theory of finite size effects and vortex penetration in small Josephson junctions*, Phys. Rev. B **51**, 1090 (1995); with M. V. Fistul.
53. *Magneto-transport behavior of polycrystalline* $YBa_2Cu_3O_7$*: a possible role for surface barriers*, Phys. Rev. B **52**, 747 (1995); with Shi Li, M. V. Fistul, J. Deak, P. Metcalf, M. McElfresh and D. L. Kaiser.

54. *Force balance equation for a pinned lattice of Abrikosov vortices*, Phys. Rev. B **54**, 15468 (1996); with K. Bark.
55. *Generalized Eck peak in inhomogeneous Josephson junctions*, Physica C **273**, 309 (1997); with M. V. Fistul.
56. *Effects of intrinsic inelastic scattering on the critical current of a Josephson junction*, Europhys. Lett. **39**, 317 (1997); with S. Ranjan and M. V. Fistul.
57. *Current-voltage characteristic of a Josephson junction with randomly distributed Abrikosov vortices*, Phys. Rev. B **56**, 788 (1997); with M. V. Fistul.
58. *Effect of randomly distributed anisotropic vortices on the critical current of a layered superconductor*, Physica C **289**, 291 (1997); with M. V. Fistul.
59. *Critical current of a long Josephson junction in the presence of a perturbing Abrikosov vortex*, Phys. Rev. B **58**, 9343 (1998); with M. V. Fistul.
60. *Abrikosov vortices in long Josephson junctions*, Phys. Rev. B **58**, 9348 (1998); with M. V. Fistul.
61. *Screened interaction and self-energy in an infinitesimally polarized electron gas via the Kukkonen-Overhauser method*, Phys. Rev. B **61**, 12556 (2000); with S. Yarlagadda.
62. *Localized modes in finite and phase-inhomogeneous Josephson tunnel junctions*, Phys. Rev. B **61**, 12285 (2000); with A. K. Setty.
63. *Spin Instabilities in semiconductor superlattices*, Phys. Rev. B **61**, 7245 (2000); with D. C. Marinescu and J. J. Quinn.
64. *Magnetic phase diagram of a semiconductor superlattice at $\nu = 2$*, Physica E **6**, 807 (2000); with D. C. Marinescu and J. J. Quinn.
65. *Analytical expressions for the charge-charge local-field factor and the exchange-correlation kernel of a two-dimensional electron gas*, Phys. Rev. **B 64**, 153101 (2001); with B. Davoudi, M. Polini and M. P. Tosi.
66. *Analytical expressions for the spin-spin local-field factor and the spin-antisymmetric exchange-correlation kernel of a two-dimensional electron gas*, Phys. Rev. B **64**, 233110 (2001); with B. Davoudi, M. Polini and M. P. Tosi.
67. *Quasiparticle Lifetime in a bilayer system*, Physica E **12**, 331 (2001); with D. C. Marinescu and J. J. Quinn.
68. *Tunneling between dissimilar quantum wells*, Phys. Rev. B **65**, 045325 (2002); with D. C. Marinescu and J. J. Quinn.
69. *Friedel oscillations in a two-dimensional Fermi liquid*, Solid State Commun. **127**, 789 (2003); with G. Simion.

70. *Quasiparticle self-energy and many-body effective mass enhancement in a two-dimensional electron liquid*, Phys. Rev. B **71**, 045323 (2005); with R. Asgari, B. Davoudi, M. Polini, M. P. Tosi and G. Vignale.
71. *Many-body effective mass enhancement in a two-dimensional electron liquid*, Proceedings of CMT28; with R. Asgari, B. Davoudi, M. Polini, M. P. Tosi and G. Vignale.
72. *On the RKKY range function of a one dimensional non interacting electron gas*, Phys. Rev. B **72**, 033411 (2005); with G. Vignale and T. Datta.
73. *On the two dimensional electron liquid in the presence of spin-orbit coupling*, Proceedings of "Highlights in the Quantum Theory of Condensed Matter" 2004; with S. Chesi.
74. *Friedel Oscillations in a Fermi Liquid*, Phys. Rev. B **72**, 045127 (2005); with G. E. Simion.
75. *Correlation energy in a spin-polarized two-dimensional electron liquid in the high-density limit*, Phys. Rev. B **75**, 153306 (2007); with S. Chesi.
76. *Exchange energy and generalized polarization in the presence of spin-orbit coupling in two dimensions*, Phys. Rev. B **75**, 155305 (2007); with S. Chesi.
77. *Many-body local fields theory of quasiparticle properties in a three-dimensional electron liquid*, Phys. Rev. B **77**, 035131 (2008); with G. E. Simion.
78. *Absence of certain exchange driven instabilities of an electron gas at high densities*, Phys. Rev. B **78**, 075110 (2008); with G. Vignale.
79. *Many-body local fields theory of quasiparticle properties in a three-dimensional electron liquid*, Phys. Rev. B **77**, 035131 (2008); with G. E. Simion.
80. *Two exact properties of the perturbative expansion for the two-dimensional electron liquid with Rashba or Dresselhaus spin-orbit coupling*, Phys. Rev. B **83**, 235308 (2011); with S. Chesi.
81. *High-density limit of the two-dimensional electron liquid with Rashba spin-orbit coupling*, Phys. Rev. B **83**, 235309 (2011); with S. Chesi.
82. *Anomalous spin-resolved point-contact transmission of holes due to cubic Rashba spin-orbit coupling*, Phys. Rev. Lett. **106**, 236601 (2011); with S. Chesi, L. P. Rokhinson, L. N. Pfeiffer, and K. W. West.
83. *Spin density waves in a semiconductor superlattice in a tilted magnetic field*, Phys. Rev. B **84**, 205321 (2011); with Liqiu Zheng and D. C. Marinescu.

Personal recollections

Daniela Giuliani

Adriana Giuliani

My father was a bear. This aspect changed from perspective.

I saw a grizzly, bristled and growling, wild and protective. His mentor Tosatti saw a teddy bear. His students saw a majestic and noble beast.

There are lots of bears. Honey-loving Pooh bears. Lame-joke Yogi bears. Coke-drinking polar bears. Yellowstone-cabin soap bears. My father was each and all of these.

He was a bear.

It is only fitting, for he loved bears.

When we would go hiking, he would seek them out. He would stand one hill away, binoculars in hand, inching forward towards the animal he was. He would stride, camcorder in hand, towards a bear cowering on the side of the road. Park rangers yelled at him to get back. He would yell back in uncomprehending Italian.

Bears can be protective or solitary. Prone to tremendous wrath and wide smiles. Harmless unless provoked or hungry.

For a long time, I only saw one face of the bear. I knew him always and only as my father, far replaced from the cub he invariably was or the teddy bear he could seem. I never knew him as an erudite physicist nor fervent calciatore nor unabashed racer nor stimulating professor. He was my father. He was fierce, he was strong, he was unrelenting.

He expected and insisted upon only the best from his children, even if this expectation was limited in scope to lawyer or physician. He was smartly and often scathingly sarcastic. He stubbornly and bravely held a moral standard and true compassion which too many surrender. It was under his order that I was raised, tossed between his math lessons and his Befana masquerades. Mostly, I feared him and in that fear, I respected him.

He loved- infuriatingly, passionately, insistently- he loved. “I love you guys”, he breathed on his deathbed, “so much”.

He loved life. He loved speeding up the top of a mountain so he could look down. He loved cooking with the colors of the Italian flag. He loved swimming past the rocks, to the furthest proper boundaries and beyond. He loved Christmas ornaments that glowed and whistling the tune of "The Good, the Bad, and the Ugly". When he smiled, it was broad and wrinkled the edges of his eyes. When he laughed, it was loud and emanated from deep in the throat. His spirit burst forth and embedded itself in those moments.

He was free, unabashed, daring. He spoke with words that were only ever his own. He drove alone, pedal pushed down, speeding his racecar around the next corner. He flung curses at whoever needed them the most. He climbed onto rocks where he would fall.

When he fell, and fall he would, he always stood back up. He always defied the odds, beating them down with what only was pure will. When he died, and die he would, it was only with the consent of that same will. The odds never stood a chance against this bear of a man, with daunting blue eyes and thick mane of black hair.

I never saw a mark or wrinkle or gray hair to show what he had overcome. For my twenty long years, he was my same father. Always protective and guarded, always holding us up and telling us to be better.

He put his shoulder around me as I was shaking in an ambulance, moments from a scare with a car crash, and spoke in low tones "It's okay, baby. It's okay". He sent me a jubilant text when I completed my studies in Guatemala, calling me *gringa* and signing *el papa'* and again mentioning that I am *muy blanca*. "Adriana, Adriana mia", he called me, half-mumbling the words.

"Be good", he always managed to say to me, as I kissed a grizzled cheek and said goodbye. Eventually, the goodbyes came to an end.

It is only after his death, in my own adventures in Italy, in my moments with my Nonna, in my sifting through old letters, in my meeting older friends, that I have come to know the other faces of the bear. I learned about the cigarette burns administered by his officemate in Trieste, about his mandate on the cleanliness of his carpet in Providence, about his insistence on getting water from his mother "per bere" instead of "per piacere". More stories trickle down to me, unveiling more and more the character of a man I only knew as my father. More stories I still hold, in the recesses of my mind, of the bear I saw in my eye. This is the best proof that these memories and this spirit lives on.

A bear is a beast not to be trifled with and never to be forgotten.

February 7, 2014

Adriana Giuliani

Giuseppe Giuliani

When thoughts concerning my father come to mind, I always find myself lost. Boys look up to their fathers. I must look up to what my father was and what he stood for. Who was he? This is the hardest question which I must approach everyday. Gabriele Giuliani constantly drove himself to wrench the most out of life. He seemed to be able to accomplish anything. Starting from the day he was born on the kitchen table of his parents' home in Ascoli Piceno, he pushed himself into becoming Professor Giuliani, a great soccer player, a passionate musician, and a courageous racecar driver. Most everyone knew him from one of these very different and discrete worlds, but I knew him as the man that raised me. Gabriele was very set in his ways; he would never compromise what he truly believed in for anything. This uncompromising nature was what I thought was most unique about him. My father took many actions that were considered unusual or unaccepted because he believed whole-heartedly that what he did was right, not that he would profit from it. Many people today do not make decisions based on a moral basis but on a dormant nihilism that lies in their instinctive nature. Though simple, I always admired this in him because it made him more trustworthy in serious situations. In my eyes, this along with his intelligence, passion for knowledge and drive to push life to its limits made him who he was. I do not claim to understand the full impact my father had on the academic world, the football world and the racing world; I hope that through life and my studies I will discover more and more as to who my father was and what I have to look up to. I loved my father, and I hope that what whatever I become would have made him proud.

April 1, 2014

Giuseppe Giuliani

Memories of Gabriele

Paul Muzikar

It is not an easy task to capture Gabriele's unique personality by putting pen to paper. The excitement and liveliness he brought to everything he did made a deep mark on those who knew him. In a profound way, he showed us all the value of living one's life with attention and with spirit. In his remarks at Gabriele's memorial service, his friend Nick Giordano laid particular stress on Gabriele's passion; this emphasis on passion is a wise guide to follow in our thoughts of Gabriele. In the following I will offer a few memories.

One absolutely crucial aspect of Gabriele's years at Purdue was his special connection with Al Overhauser. Gabriele had been a postdoc with Al, and then, after working with John Quinn, returned to Purdue as a colleague. They differed in many significant ways: conservative vs. socialist, devout Catholic vs. skeptic. Yet they had a deep respect for each other, and their mutual affection was wonderful to see. Their special bond permitted them to have passionate discussions (*i.e.* arguments) concerning all aspects of physics. An unwitting observer of such a discussion might easily misunderstand what was going on.

In a sense their strengths were complementary. Al was a tremendously creative scientist; he was a constant source of ideas, and was not afraid of voicing even somewhat unconventional ones. Gabriele's clear, incisive, critical intelligence served as a sounding board, and their mutual trust and affection allowed for an exuberant and lively exchange.

Gabriele and Al shared a passion for condensed matter physics, believing it to be a sublimely beautiful subject. Gabriele transmitted this outlook to his students and colleagues, and as the years went on, assumed a leading role in the condensed matter group at Purdue.

In his teaching, Gabriele was demanding; he wanted his students to really understand the physics. But he was very generous with his time and effort in helping even an unskilled student, if he felt the student was willing to work. Someone not willing to tackle physics with effort and passion did not receive as much sympathy.

Besides family and physics, Gabriele's chief passion was soccer. I think in his heart he believed that if physics had not worked out, he could have had another life as a player in Serie A. He played soccer with passion, and taught others with passion. More than once, the two of us took an afternoon break from physics, went to my house, and watched a Champion's League game. His comments always gave me a mini-clinic on the subtleties of the game.

His devotion to the Italian national team was intense. If the team lost a critical game, it would be a bad mistake to make jokes about the defeat after the game. A recovery period of perhaps a week was needed. It took a lot of effort to convince him to grudgingly admit that a team not in Serie A was very good.

Yellowstone National Park was another of Gabriele's great passions. His view, often strongly expressed, was that visiting Yellowstone far outranked almost any other trip one could make. Perhaps as a European, he avoided the tendency of Americans to take Yellowstone for granted. He was quite proud of his somewhat dangerous encounter there with a grizzly bear. Gabriele's enthusiasm was the direct cause of the first trip my wife and I made to Yellowstone, and I owe him for this.

Gabriele held strong views about food and cooking, and it must be admitted that his skill in the kitchen entitled him to those views. Countless friends have eaten some of their best meals with Gabriele and his family. On the other hand, it was somewhat risky to invite him over for dinner. If Gabriele felt the food being served was perhaps not the best, it was essentially impossible for him to offer false praise to his hosts; food was too important to permit any diplomatic 'white lies'. The converse of this was that any praise from Gabriele had a deep meaning for the cook.

In all these endeavors (physics, soccer, cooking) the crucial point was to approach the job with care, flair, and effort. He held himself, and all of us, to a high standard in which laziness and inattention were the enemies. Of course, this attitude could make Gabriele a 'pain in the neck' at times, on the soccer pitch, in the kitchen, or in the physics building.

The reader may notice that, as promised in the first paragraph, the word "passion" has perhaps been overused in the preceding recollections. But I feel that no correction or revision is needed. Gabriele's passion, ultimately a form of love, coupled with his keen and quick mind, made him the remarkable man we all miss so much.

Scientific contributions

Introduction to electronic and optical properties of two-dimensional molybdenum disulfide systems

Reza Asgari

Abstract. Two-dimensional (2D) nanomaterials have attracted increasing attention because of their unusual physical and chemical properties. Among these 2D nanomaterials, the monolayers of layered transition metal dichalcogenides exhibit intriguing electronic and optical properties. In this chapter, therefore, the electronic and optical properties of monolayer MoS_2 are briefly reviewed. We present a model Hamiltonian within tight-binding theory, some transport properties like the charge compressibility in the mean-field approximation, plasmon modes in the Random-Phase Approximation and intrinsic optical properties of monolayer MoS_2. Finally, we briefly discuss many-body ground-state of the system and its quantum phase transition in physical parametric space within Hartree-Fock theory.

1. introduction

Two-dimensional (2D) materials have been one of the most interesting subjects in condensed matter physics for potential applications due to the wealth of unusual physical phenomena that occur when charge, spin and heat transport are confined to a 2D plane [1]. These materials can be mainly classified in different classes which can be prepared as a single atom thick layer namely, layered van der Waals materials, layered ionic solids, surface growth of monolayer materials, 2D topological insulator solids and finally 2D artificial systems and they exhibit novel correlated electronic phenomena ranging from high-temperature superconductivity, quantum valley or spin Hall effect to other enormously rich physics phenomena. 2D materials can be mostly exfoliated into individual thin layers from stacks of strongly bonded layers with weak interlayer interaction and a famous example is graphene, hexagonal boron nitride [2] and black phosphorus [3].

The 2D transition metal dichalcogenides, on the other hand, exhibit novel correlated electronic phenomena ranging from insulator to superconductor show a wide range of electronic, optical, mechanical, chemical, and thermal properties [4]. In contrast to the zero-gap graphene, the 2D transition metal dichalcogenides possess sizable bandgaps, very important to field-emission transistors and optoelectronic devices. They are a class of materials with the formula XY_2, where X is a transition metal element from group IV (Ti, Zr, Hf, and so on), group V (for instance V, Nb, or Ta) or group VI (Mo, W, and so on), and Y is a chalcogen (S, Se, or Te). These materials have crystal structures consisting of weakly coupled layers YXY, where a X-atom layer is enclosed within two Y layers and the atoms in layers are hexagonally packed. Adjacent layers are weakly held together by van der Walls interaction to form the bulk crystal in a variety of polytypes, which vary in stacking orders and metal atom coordination. The overall symmetry of transition metal dichalcogenides is hexagonal or rhombohedral, and the metal atoms have octahedral or trigonal prismatic coordination.

The monolayer MoS_2 has recently attracted great interest because of its potential applications in 2D nanodevices [5, 6], owing to the structural stability and lake of dangling bands, although it had been obtained and studied in the several decades ago [7]. The monolayer MoS_2 is a direct gap semiconductor with a bandgap of 1.8 eV [5], and can be easily synthesized by using scotch tap or lithium-based intercalation [5–8]. The mobility of the monolayer MoS_2 can be at least 217 $cm^2V^{-1}s^{-1}$ at room temperature using hafnium oxide as a gate dielectric, and the monolayer transistor shows the room temperature current on/off ratios of 10^8 and ultralow standby power dissipation [5]. Recently, the MoS_2 nanoribbons have been obtained by using electrochemical method [9]. The experimental achievements triggered the theoretical interests on the physical and chemical properties of monolayer MoS_2 nanostructures to reveal the origins of the observed electrical, optical, mechanical, and magnetic properties, and guide the design of novel MoS_2-based devices. The layered structure of MoS_2 is formed by graphene-like hexagonal arrangement of Mo and S atoms bonded together to give S-Mo-S sandwiches. The S-Mo-S units are stacked on top of each other and are held together by weak noncovalent interactions. In this arrangement, in each layer Mo atom is covalently bonded to six sulfur atoms, whereas each sulfur atom is connected to three Mo atoms. The structure is uniquely determined by the hexagonal lattice constant a, the out-of-plane lattice constant c, and the internal displacement parameter z. The experimental lattice constants and the internal displacement parameter were determined as $a = 3.16$ Å, $c = 12.58$ Å, and $z = 0.12$ Å.

Optical spectroscopy, on the other hand, is a broad field and useful to explore the electronic properties of solids. Optical properties can be tuned by varying the Fermi energy or the electronic band structure of 2D systems. Recently, developed 2D systems such as gapped graphene [10], thin film of the topological insulator [11,12], and monolayer of transition metal dichalcogenides [4] provide the electronic structures with direct band gap signatures. The optical response of semiconductors with direct band gap is strong and easy to explore experimentally since photons with energy greater than the energy gap can be absorbed or omitted. The thin film of the topological insulator, on the other hand, has been fabricated experimentally by using Sb_2Te_3 slab [13] and has been shown that a direct band gap can be formed owing to the hybridization of top and bottom surface states. Furthermore, a non-trivial quantum spin Hall phase has been realized experimentally which was predicted previously in this system [14–16]. Although pristine graphene and surface states of the topological insulator reveal massless Dirac fermion physics , by opening an energy gap they become formed as massive Dirac fermions. The thin film of the topological insulator and monolayer transition metal dichacogenides can be described by a modified-Dirac Hamiltonian. The optical response in monolayer MoS_2 to increases in comparison with its bulk and multilayer structures [17–21] since, as we mentioned before, a monolayer of the molybdenum disulfide is a direct band gap semiconductor [5], however its multilayer and bulk show indirect band gap [4].

One of the main properties of MoS_2 is a circular dichroism aspect responding to a circular polarized light where the left or right handed polarization of the light couples only to the K or K' valley and it provides an opportunity to induce a valley polarized excitation which can profoundly be of interest in the application for valleytronics [22–24]. Another peculiarity of MoS_2 is the coupled spin-valley in the electronic structure which is owing to the strong spin-orbit coupling originating from the existence of a heavy transition metal in the lattice structure and the broken inversion symmetry too [25]. These two aspects are captured in a minima massive Dirac-like Hamiltonian introduced by Xiao et al. [25] However it has been shown , based on the tight-binding [26,27] and $k.p$ method [28], that other terms like an effective mass asymmetry, a trigonal warping, and a diagonal quadratic term might be included in the massive Dirac-like Hamiltonian. The effect of the diagonal quadratic term is very important, for instance, if the system is exposed by a perpendicular magnetic field, it will induce a valley degeneracy breaking term [26]. The optical properties of MoS_2 have been evaluated by *ab-initio* calculations [29] and studied theoretically based on the simplified massive Dirac-like model Hamiltonian [30], which is by itself valid only near the main absorbtion edge. A part of the model Hamiltonian which describes the dynamic of massive Dirac fermions are known in graphene committee to have an optical response quite different from that of a standard 2D electron gas. Thus it would be worthwhile to generalize the optical properties of such systems by using the modified-Dirac fermion model Hamiltonian [31].

ACKNOWLEDGEMENTS. R. A. would like to thank H. Rostami with whom some part of this work on MoS_2 have been done. This chapter is dedicated to the memory of late Prof. Gabriele F. Giuliani, distinguished theoretical condensed matter physicist best known for his contribution to the correlations of two-dimensional quantum liquid systems and passed away on 22nd of November, 2012.

2. Theory and method

2.1. Electronic band structure and low-energy Hamiltonian

The applications of MoS_2 monolayer in electronic and optoelectronic devices are directly dependent on its electronic properties, such as band structure and density of states. The band structure of bulk MoS_2 calculated from first-principles show indirect-semiconducting behavior with a bandgap of 1.2 eV, which originates from transition from the top of valence band situated at Γ to the bottom of conduction band halfway between Γ and K high symmetry points [32]. The optical direct bandgap is situated at the K point. As the number of layers decreases, the indirect bandgap increases. In the monolayer, the material changes into a 2D direct bandgap semiconductor with a gap of 1.8 eV. At the same time, the optical direct gap at the K point stays almost unchanged and close to the value of the optical direct bandgap at the K point of a bulk system. The change in the band structure with layer number in MoS_2 is due to quantum confinement and the resulting change in hybridization between p_z orbitals on S atoms and d orbitals on Mo atoms. The states near Γ point are due to combination of the anti-bonding p_z orbitals on the S atoms and d orbitals on Mo atoms and have a strong interlayer coupling effect.

The calculated bandgap of MoS_2 monolayer varies from 1.6 to 1.9 eV due to different approximations for the exchange and correlation functionals [33]. The theoretical results based on the Perdew-Burke-Ernzerhof functional form of the generalized gradient approximation showed that the bandgap of MoS_2 monolayer is about 1.9 eV [32], which agrees with the experimental data observed from photoluminescence [5]. It is known, however, that DFT always underestimates the bandgaps of materials owing to the calculated unreliable exited states. Furthermore, the strong exciton binding due to the weak screening compared to bulk cases in low-dimensional systems may affect the bandgap. Therefore, the good bandgap agreement between theoretical and experimental results for MoS_2 monolayer may be a mere coincidence. The GW approach is expected to yield more accurate gaps and predicted that the gap of

MoS_2 monolayer is 2.7-2.9 eV due to the effect of confinement and environment on the electronic structure and exciton binding energy [34]. It was argued that the experimentally observed gap was optical gap and the exciton binding energy was about 0.8-1.0 eV [34]. The fundamental bandgap of MoS_2 monolayer, therefore, is about 2.8 eV. The direct experimental confirmation on the prediction is still not available, and the issue is open to question.

Moreover, the inversion symmetry is broken in monolayer MoS_2 similar to gapped graphene however it is preserved in gapless graphene. In a monolayer MoS_2, the sulfur atoms will be transformed into an empty site and therefore, the transition from bulk MoS_2 to a monolayer MoS_2 removes inversion symmetry and breaks down the Kramers degeneracy in most of the Brillouin zone. We can say that the dispersion relation or Hamiltonian has time-reversal symmetry if the band energy satisfies $E(\mathbf{k}, \tau, s) = E(-\mathbf{k}, -\tau, -s)$ and it will has inversion symmetry if $E(\mathbf{k}, \tau, s) = E(-\mathbf{k}, -\tau, s)$ where s and τ indicate the spin and valley indexes, respectively. In a system where the inversion symmetry is broken, the valley degree of freedom can be distinguished. Thus, there is an intrinsic valley dependent Berry phase effect that can result in a valley Hall transport with carriers in different valleys turning into opposite directions transverse to an in-plane electric field.

The band structures and bandgaps of monolayer MoS_2 are very sensitive to the external strain [35]. Compared to that of graphene, a much smaller amount of strain is required to vary the bandgap of MoS_2. The mechanical strains reduced the bandgap of semiconducting MoS_2 causing a direct-to- indirect bandgap and a semiconductor-to-metal transition. These transitions, however, significantly depend on the type of applied force. In addition, the results demonstrate that the homogeneous biaxial tensile strain of around 10% leads to semiconductor-to-metal transition in all semiconducting MoS_2. With the applied strain influencing the band gap, the effective mass of carriers is also changed. By considering the accurate bandgap of MoS_2, the required strain for the transition should be increased.

One of simplifications over the first-principle calculations is the tight-binding Hamiltonian. In this regard, it is assume that the orbitals that are very similar at atomic scales, can be used as a basic for expanding the wave function. In order to construct a tight-binding Hamiltonian for the system, the knowledge of the lattice symmetry is needed. The symmetry space group of MoS_2 is D^1_{3h} which contains the discrete symmetries C_3 (trigonal rotation), σ_v (reflection by the yz plane), σ_h (reflection by the xy plane) and any of their products [25]. Besides the symmetry of the lattice, it is essential to consider the local atomic orbitals symmetries. The trigonal prismatic symmetry dictates that the d and p orbitals split into three and two groups, respectively, $\{d_{z^2}\}, \{d_{x^2-y^2}, d_{xy}\}, \{d_{xz}, d_{yz}\}$ and $\{p_x, p_y\}$, $\{p_z\}$. The reflection symmetry along the z direction allows the coupling of Mo d_{xz}, d_{yz} orbitals with only the p_z orbital of the S atom, whose contribution at the valence band maximum (VBM) and the conduction band minimum (CBM) located at the symmetry points is negligible according to first principle calculations [36]. Therefore the conduction band minimum is mainly formed from Mo d_{z^2} orbitals and the valence band maximum is constructed from the Mo $\{d_{x^2-y^2}, d_{xy}\}$ orbitals with mixing from S $\{p_x, p_y\}$ (Refs. [36,37]) in both cases.

We thus constructed [26] the tight-binding Hamiltonian for MoS_2 by using symmetry adapted states and assuming nearest neighbor hopping terms;

$$\hat{H}_{TB} = \sum_{i\mu\nu} \left\{ \epsilon^a_{\mu\nu} a^\dagger_{i\mu} a_{i\nu} + \epsilon^b_{\mu\nu} b^\dagger_{i\mu} b_{i\nu} + \epsilon^{b'}_{\mu\nu} b'^\dagger_{i\mu} b'_{i\nu} \right\} + \sum_{\langle ij\rangle,\mu\nu} t_{ij,\mu\nu} a^\dagger_{i\mu} (b_{i\nu} + b'_{i\nu}) + H.c. \tag{2.1}$$

Here a and $b(b')$ indicate the Mo and S atoms in the up (down) layer, respectively. The indices μ and ν show the orbital degrees of freedom labelled as $\{1, 2, 3\} \equiv \{d_{z^2}, d_{x^2-y^2} + id_{xy}, d_{x^2-y^2} - id_{xy}\}$ and $\{1', 2'\} \equiv \{p_x + ip_y, p_x - ip_y\}$ for Mo and S atoms, subsequently. Therefore the matrices ϵ^a, $\epsilon^b(\epsilon^{b'})$, and $t_{ij,\mu\nu} =< a, i, \mu|H|b, j, \nu >$ are responsible for the on-site energies of Mo and S atoms, and hopping between different neighboring sites in the space of different orbitals, respectively. We do need to take into account the overlap integrals, $\mathcal{S}$, defined similar to the hopping terms of the Hamiltonian with elements $s_{ij,\mu\nu} =< a, i, \mu|b, j, \nu >$.

Due to the trigonal rotational symmetry of the Hamiltonian, the symmetry properties of the lattice the number of independent variables are reduced [26]. Furthermore, a good approximation is provided by the Slater-Koster method [38] in which all of the hopping and overlap integrals are written as a linear combinations of the hopping integrals $V_{pd\sigma}$, $V_{pd\pi}$ and overlap integrals $S_{pd\sigma}$, and $S_{pd\pi}$ where $V_{pd\sigma} =< \mathbf{R}', p, \sigma|H|\mathbf{R}, d, \sigma >$ and $S_{pd\sigma} =< \mathbf{R}', p, \sigma|\mathbf{R}, d, \sigma >$, for instance.

One of the ways in which MoS_2 differs from graphene is the presence of strong spin-orbit interactions. The coupling between the spin and angular momentum of electrons creates an internal magnetic field that can break down Kramers degeneracy in systems without inversion symmetry, such as, for example, zinc blende semiconductors as the twofold degeneracy throughout the Brillouin zone is no longer required. To complete our effective Hamiltonian, we need to add the spin-orbit interaction (SOI) in the model which causes spin-valley coupling in the valence band. The large SOI in MoS_2 can be approximately understood by intra atomic contribution $H_{SO} = \xi(r)\mathbf{S}.\mathbf{L}$. The spinvalley coupling at the valence-band edges suppresses spin and valley relaxation, as flip of each index alone is forbidden by the valley-contrasting spin splitting. As a result, a change in valence carrier spin state and hence valley in k-space will be far less likely in monolayer MoS_2. We only consider only the most important contribution of the Mo atoms which gives rise to the spin-orbit coupling term $H_{SO}^{Mo} = \lambda\,\mathrm{diag}\{0, s, -s\}$ in the basis of states $\{1, 2, 3\}$ where λ is the spin-orbit coupling and $s = \pm$.

Generally, our tight-binding model leads to seven bands for each spin component, however, in the absence of external bias *i.e.* $U^c = 0$ the symmetry between top and bottom S sublayers reduces the number of bands to five. Two of them correspond to the conduction and valence bands, from which we calculate the effective electron and hole masses, energy gap, and valence band edge. Moreover, since the conduction band minimum mostly comes from d orbitals [36], we assume 10% mixing with p orbitals for the conduction band. This assumption is in good agreement with the result reported in Ref. [39]. This provides us with five equations for seven unknown parameters based on the values obtained from *ab initio* calculations and experimental measurements. Furthermore, it is reasonable to consider $s_{\mu\nu}/t_{\mu\nu} = 0.1\ \mathrm{eV}^{-1}$ which reduces the number of unknown parameters to five. We consider the energy gap $\Delta = 1.9$ eV, spin-orbit coupling $\lambda = 80$ meV, effective electron and hole masses $m_e = 0.37m_0$ and $m_h = -0.44m_0$ (m_0 is the free electron mass) [40], and $E_{VBM} = -5.73$ eV [41]. Eventually, all parameters can be fixed and we then obtain [26] the on-site energies $A_1 = -1.45$ eV, $A_2 = -5.8$ eV, $B = 5.53$ eV and hopping integrals $e^{i\pi/6}t_{11} = 0.82$ eV, $e^{-i\pi/6}t_{21} = -1.0$ eV, and $e^{-i\pi/2}t_{22} = 0.51$ eV.

Now, we present an effective low-energy two-band continuum Hamiltonian governing the conduction and valence bands around the K and K' points, by exploiting the Löwdin partitioning method [42]. After straightforward but lengthy algebra, the final result for the two-band Hamiltonian describing the conduction and valence bands near K or K' reads,

$$H_{\tau s} = \frac{\Delta}{2}\sigma_z + \lambda\tau s\frac{1-\sigma_z}{2} + t_0 a_0 \mathbf{q}\cdot\boldsymbol{\sigma}_\tau + \frac{\hbar^2|\mathbf{q}|^2}{4m_0}(\alpha + \beta\sigma_z) + t_1 a_0^2 \mathbf{q}\cdot\boldsymbol{\sigma}_\tau^*\sigma_x\mathbf{q}\cdot\boldsymbol{\sigma}_\tau^* \qquad (2.2)$$

for spin $s = \pm$ and valley $\tau = \pm$, with Pauli matrices $\boldsymbol{\sigma}_\tau = (\tau\sigma_x, \sigma_y)$ and momentum $\mathbf{q} = (q_x, q_y)$. The numeric values of the two-band model parameters are $t_0 = 1.68$ eV, $t_1 = 0.1$ eV, $\alpha = 0.43$, and $\beta = 2.21$. Notice that $\alpha = m_0/m_+$ and $\beta = m_0/m_- - 4m_0v^2/(\Delta - \lambda)$ where $m_\pm = m_e m_h/(m_h \pm m_e)$ and $v = t_0 a_0/\hbar$. Moreover, a quadratic correction $\delta\lambda \approx (0.03\ \text{eV})(a_0|\mathbf{q}|)^2$ arises to the spin orbit coupling due to folding down of the five-band model to a two-band one. This correction is estimated by using the effective masses of two spin-split valence band branches as $m_h(\uparrow) = -0.44m_0$ and $m_h(\downarrow) = -0.46m_0$ at the K point. Notice that the correction term can be safely ignored in the validity range of the effective low-energy two-band model, $(a_0|\mathbf{q}| \ll 1)$.

The Hamiltonian differs from that introduced by Xiao *et al.* [25] because of the second order terms in q. The diagonal q^2 terms, which contribute in the energy, to the same way as does the first order off-diagonal term, are responsible for the difference between electron and hole masses recently reported by using *ab initio* calculations [40]. Moreover, the last term leads to anisotropic q^3 corrections to the energy which contribute to the trigonal warping effect. Importantly, α vanishes for the case that $m_e = -m_h$, however β remains a constant. Basically, there is the possibility to have a cubic off-diagonal term in the low-energy Hamiltonian which in the calculation of the eigenvalues of the Hamiltonian are multiplied with the off-diagonal q terms and eventually contributes at the same order as the diagonal q^2 terms. Since that term is very small, we thus ignore the q^3 off-diagonal term.

Figure 1 shows the band structure of MoS_2 consisting of five bands for each spin in the absence of an external field. Two of them are spin polarized (dot-dashed and dashed lines) and the others are spin degenerate (solid lines). We note that due to the limitations of our model, the high energy bands may not be comparable with those of first principal calculations in a quantitative manner. Figure 1(b) shows a comparison between our results and those calculated by density functional theory [40] indicating that our theory is in good agreement with density functional theory results close to the **K** point up to a high particle (hole or electron) density 10^{14}cm^{-2} (the Fermi energy is $E_F - E_{\text{CBM}} \simeq 0.2$ eV). Nevertheless our effective model Hamiltonian does not provide a good description of the physics around the Γ point where other orbitals like p_z must be considered in order to describe the electronic dispersion [39].

We further investigate the band structure close to the valence and conduction bands and our numerical results are shown, via contour plots which show the isoenergy lines, in Figure 2. A strong anisotropy of the constant energy lines can be seen around K points in the valence band, due to the trigonal warping, while in the conduction band all lines are al-

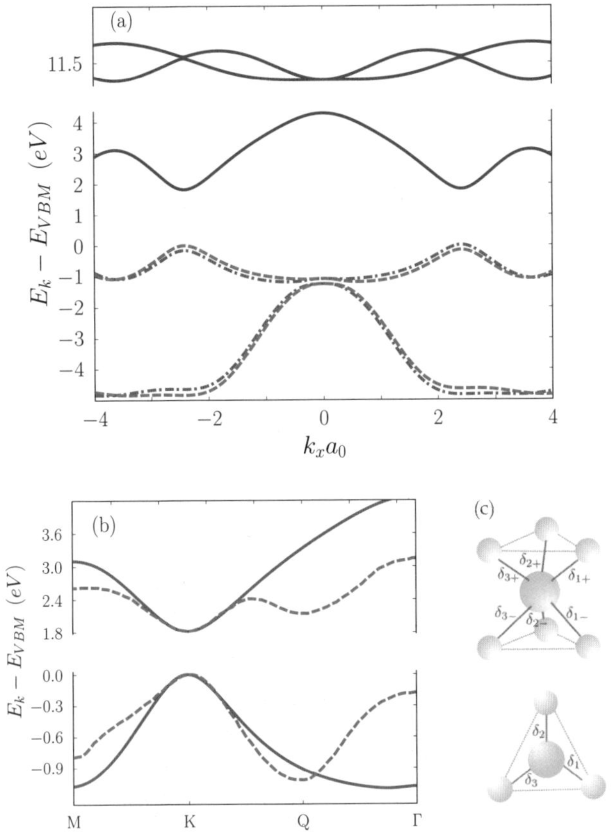

Figure 1. (Color online) (a) Band structure of MoS_2 consisting of five bands in which two are spin split in the valence band. The dot-dashed line refers to one spin and the dashed line denotes another spin component. Solid lines refer to the spin degenerate band. (b) A comparison between the band structure calculated by the present theory (solid lines) and the results calculated in Ref. [40] (dashed lines) based on density functional theory. Notice that our theory works quite well around the **K** point for the particle (hole or electron) density less than 10^{14}cm^{-2} ($E_F - E_{CBM} \simeq 0.2$ eV). Here, $a_0 = a\cos\theta$ where a is the length of Mo-S bond and θ is the angle between the bond and the xy plane. (c) Side and top views of the lattice structure are seen where Mo atom (larger green sphere) is surrounded by six S atoms.

most isotropic; the warping is due to the difference of the orbital structure of the conduction and valence bands.

To study the interplay of spin and valley physics, we introduce, by ignoring trigonal warping, the effect of a time reversal symmetry breaking

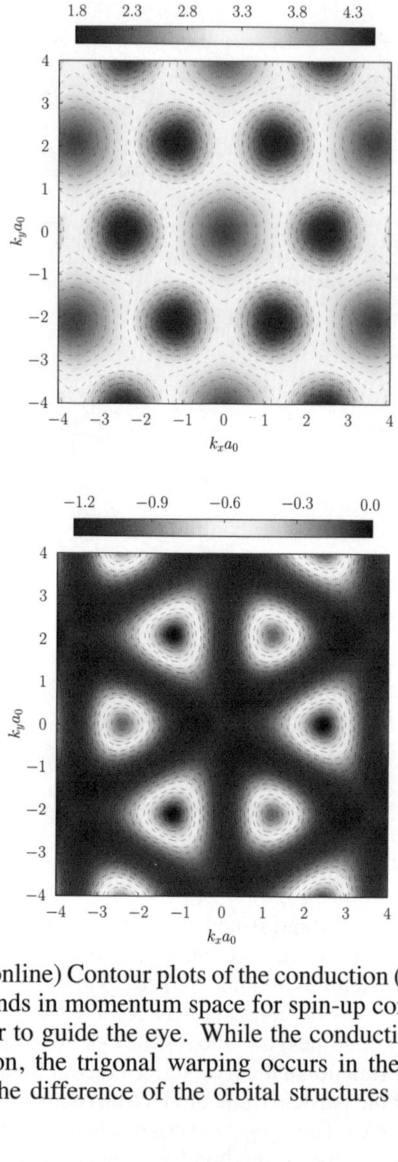

Figure 2. (Color online) Contour plots of the conduction (top panel) and valence (bottom panel) bands in momentum space for spin-up component together with isoenergy lines for to guide the eye. While the conduction band shows almost isotropic dispersion, the trigonal warping occurs in the valence band around K points due to the difference of the orbital structures of the conduction and valance bands.

term by applying a perpendicular magnetic field, leading to the appearance of Landau levels (LLs) as follows,

$$E^{\pm}_{n,\tau s} = \pm\sqrt{\left[\frac{\Delta - \lambda\tau s}{2} + \hbar\omega_c\left(\beta n - \frac{\alpha\tau}{2}\right)\right]^2 + 2\left(\frac{t_0 a_0}{l_B}\right)^2 n} + \frac{\lambda\tau s}{2} + \hbar\omega_c\left(\alpha n - \frac{\beta\tau}{2}\right), \quad n = 0, 1, \ldots \tag{2.3}$$

where $\omega_c = eB/2m_0$ and $l_B = \sqrt{\hbar/(eB)}$ are the cyclotron frequency and magnetic length, respectively. It should be noticed that the trigonal warping term, t_1, leads to a second order perturbation correction in the Landau level energy and accordingly its effect on the Landau levels is very weak. In contrast to Ref. [60], we see an additional valley degeneracy breaking term which is the reminiscent of the Zeeman-like coupling for valleys. For more details of the derivation of the integral quantum Hall effect, which was first suggested by Laughlin in 1981, see Ref. [43]. Figure 3 shows the band structure in the presence of the perpendicular magnetic field calculated by the full tight-binding Hamiltonian. It should be noticed that the edge band indicated in the figure might not be reliable owing to the absence of the symmetries at the edges and the contribution of other d-orbitals which are not included in the tight-binding Hamiltonian. The conduction band LLs are valley polarized and the valence band LLs are both valley and spin polarized although we have not yet considered the usual Zeeman interaction for spins. In particular, the $n = 0$ LLs, $E^+_{0,\tau s} = [\Delta - \hbar\omega_c\tau(\beta+\alpha)]/2$ and $E^-_{0,\tau s} = \lambda\tau s - [\Delta + \hbar\omega_c\tau(\beta-\alpha)]/2$, depend on the magnetic field strength in opposite ways for the two valleys. More intriguingly, the splittings of LLs in the conduction and valence bands $\delta E^+ \approx 5.4\hbar\omega_c$ and $\delta E^- \approx 4.6\hbar\omega_c$, differ from each other due to the difference of m_e and $-m_h$. The full numerical results also confirm our results obtained by the low-energy Hamiltonian and shown in Figure 3.

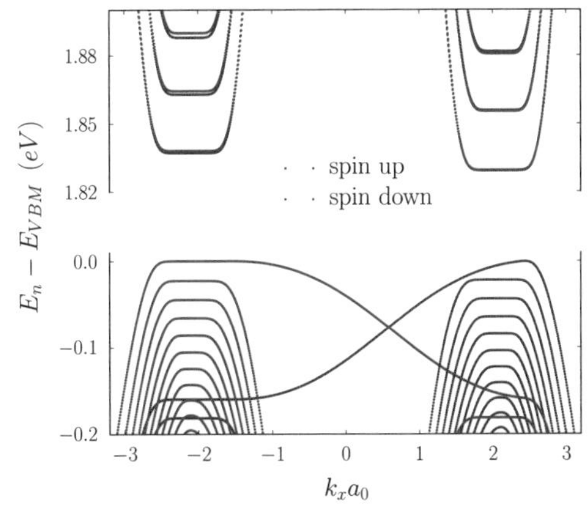

Figure 3. (Color online) Band structure of a zigzag nano-ribbon MoS_2 in the presence of the magnetic field with width $W = 149a_0$ and the absence of the Zeeman term. Notice that the edge band indicated in the figure might not be fully correct due to the absence of the symmetries at the edges and the contribution of other d-orbitals which are not included in the tight-binding Hamiltonian. The magnetic length is $l_B = W/10$.

2.2. Plasmons in MoS_2

In the following, we want to investigate the plasmon spectrum and the screening behavior. For this purpose, we need to calculate the dielectric function, restricting ourselves to Random Phase Approximation [44] in order to account for electron-electron interactions, given by

$$\varepsilon(q,\omega) = 1 - V(q)\chi(q,\omega) \tag{2.4}$$

where $V(q) = e^2/2\pi\epsilon_0 q$ is the Fourier transform of the Coulomb potential in two dimensions, ϵ_0 the background dielectric constant. The free charge-charge response function given by a two-dimensional integral in momentum space

$$\chi_0(q,\omega) = \sum_{s,\tau,\sigma,\sigma'=\pm} \int \frac{d^2k}{(2\pi)^2} | < \chi_\sigma^{\tau s}(\mathbf{k}) | \chi_{\sigma'}^{\tau s}(\mathbf{k}+\mathbf{q}) > |^2 \times \frac{f[\varepsilon_\sigma^{\tau s}(\mathbf{k})] - f[\varepsilon_{\sigma'}^{\tau s}(\mathbf{k}+\mathbf{q})]}{\omega - \varepsilon_{\sigma'}^{\tau s}(\mathbf{k}+\mathbf{q}) + \varepsilon_\sigma^{\tau s}(\mathbf{k}) + i0} \tag{2.5}$$

where $\varepsilon_\sigma^{\tau s}(\mathbf{k})$ and $\chi_\sigma^{\tau s}(\mathbf{k})$ are the energies and eigenstates of the low-energy Hamiltonian mentioned in the previous subsection, for a given valley τ, spin s, and pseudospin σ. The collective mode [45] is calculating by evaluating $\varepsilon(q,\omega_q) = 0$. The knowledge of the appropriate limiting behavior of the Lindhard function will be crucial in analyzing the excitation spectrum of systems that support different types of collective modes [46]. It is important mentioning that the plasmon energy in monolayer MoS_2 is of the form $\omega_q \sim \sqrt{nq}$ at the long-wavelength behavior as in a conventional two-dimensional electron gas [45].

Due to the large value of the band gap, the interband part of the electron-hole continuum MoS_2 is energetically very high and, subsequently, the plasmon dispersion enters the intraband the electron-hole continuum area [45]. This is quite different compared to graphene where due to the singularity of the free polarizability at the boundary, damping can only be caused by interband transitions. The damping process and the plasmon life time are open questions in monolayer MoS_2.

2.3. Ground-state energy and quantum magnetic phase transition in MoS_2

To study the effect of electron-electron interactions, we use a model which includes both intravalley (long-range) and intervalley (short-range) interactions as introduced by Roldan et al. [47]. We [48] consider the interaction of quasiparticles by using the leading diagram approximation, which is just the exchange interaction. In this regard, the interacting Hamiltonian in the mean-field approximation reads as

$$
\begin{aligned}
\mathcal{V}|_{\text{intra}} &= \frac{1}{2S}\sum_{kk',\tau ss',\alpha\beta}\sum_{q\neq 0} v_q \psi^\dagger_{k-q,\tau s,\alpha}\psi^\dagger_{k'+q,\tau s',\beta}\psi_{k',\tau s',\beta}\psi_{k,\tau s,\alpha} \\
\mathcal{V}|_{\text{inter}} &= \frac{U}{2S}\sum_{kk'q,\tau s,\alpha\beta} \psi^\dagger_{k-q,\tau s,\alpha}\psi^\dagger_{k'+q,\bar{\tau}\bar{s},\beta}\psi_{k',\bar{\tau}\bar{s},\beta}\psi_{k,\tau s,\alpha}
\end{aligned}
\tag{2.6}
$$

where $\bar{s}$ means $-s$, for instance. To account partially the screening effect and importantly to avoid any divergence within mean-field theory in the system with long-range Coulomb interaction, we use an interaction potential including Thomas-Fermi screening

$$
v_q = \frac{2\pi e^2}{\epsilon_0(q_{\text{TF}} + |\mathbf{q}|)} \tag{2.7}
$$

where ϵ_0 is the effective dielectric constant and $q_{\text{TF}} = 2\pi e^2 \mathcal{D}(\epsilon_{\text{F}})/\epsilon_0$ is the Thomas-Fermi screening wave vector in which $\mathcal{D}(\epsilon_{\text{F}}) = (g/2\pi)(kdk/d\varepsilon)$ is the density of states at the Fermi energy, *i.e.* $k = k_{\text{F}}$. Here g indicates the degeneracy of each energy level. $U = U_{4d} \times S$ in which U_{4d} is the Hubbard repulsion coefficient which mostly comes from $4d$ orbitals of Mo atoms [47] and $S = 3\sqrt{3}/2a_0^2$ is the unit cell area.

The Hamiltonian consists of a momentum-dependent pseudospin effective magnetic field that acts in the direction of momentum k. The band eignestates in the positive and negative energy bands have their pseudospins either align or oppose to the direction of the momentum [49–51]. Therefore, the mean-field Hamiltonian can be simplified as

$$
\begin{aligned}
\mathcal{H}_{MF} = \mathcal{H}_0 &- \frac{1}{S}\sum_{kk',\tau s,\alpha\beta} \psi^\dagger_{k,\tau s,\alpha} v_{k-k'}\rho_{\alpha\beta}(k',\tau s)\psi_{k,\tau s,\beta} \\
&+ \frac{U}{S}\sum_{kk',\tau s,\alpha} \text{trace}[\rho(k',\bar{\tau}\bar{s})]\psi^\dagger_{k,\tau s,\alpha}\psi_{k,\tau s,\alpha}
\end{aligned}
\tag{2.8}
$$

where a density matrix is defined as

$$
\rho_{\alpha\beta}(k,\tau s) = \langle\psi_0|\psi^\dagger_{k,\tau s,\beta}\psi_{k,\tau s,\alpha}|\psi_0\rangle \tag{2.9}
$$

The space in which the Hamiltonian is diagonalized is based on electron (c_k) and hole(v_k) operators with $(c^\dagger_{k,\tau s}\ v^\dagger_{k,\tau s}) = (a^\dagger_{k,\tau s}\ b^\dagger_{k,\tau s})\mathcal{U}$, where $\mathcal{U}$ is the unitary matrix which diagonalizes the single particle Hamiltonian given as $\mathcal{U} = (|\psi_+\rangle, |\psi_-\rangle)$. By noticing that $\langle\psi_0|c^\dagger_{k,\tau s}c_{k,\tau s}|\psi_0\rangle = n^c_{k,\tau s}$,

$\langle\psi_0|v^\dagger_{k,\tau s}v_{k,\tau s}|\psi_0\rangle = n^v_{k,\tau s}$ and $\langle\psi_0|c^\dagger_{k,\tau s}v_{k,\tau s}|\psi_0\rangle = \langle\psi_0|v^\dagger_{k,\tau s}c_{k,\tau s}|\psi_0\rangle = 0$, it would be easy to find that

$$\begin{aligned}
\rho_{aa}(k,\tau s) &= (t_0a_0)^2k^2\left(\frac{n^c_{k,\tau s}}{(t_0a_0)^2k^2+D_+^2}+\frac{n^v_{k,\tau s}}{(t_0a_0)^2k^2+D_-^2}\right)\\
\rho_{bb}(k,\tau s) &= \left(\frac{D_+^2n^c_{k,\tau s}}{(t_0a_0)^2k^2+D_+^2}+\frac{D_-^2n^v_{k,\tau s}}{(t_0a_0)^2k^2+D_-^2}\right)\\
\rho_{ab}(k,\tau s) &= \rho^*_{ba}(k,\tau s)\\
&= -(t_0a_0)\tau k e^{-i\tau\phi}\left(\frac{D_+n^c_{k,\tau s}}{(t_0a_0)^2k^2+D_+^2}+\frac{D_-n^v_{k,\tau s}}{(t_0a_0)^2k^2+D_-^2}\right)
\end{aligned} \tag{2.10}$$

where

$$\begin{aligned}
D_\pm &= \frac{\Delta}{2}+\frac{\hbar^2k^2}{4m_0}(\alpha+\beta)-E_\pm\\
E_\pm &= \pm\sqrt{\left(\frac{\Delta-\lambda\tau s}{2}+\frac{\hbar^2k^2}{4m_0}\beta\right)^2+(t_0a_0)^2k^2}+\frac{1}{2}\lambda\tau s+\frac{\hbar^2k^2}{4m_0}\alpha.
\end{aligned} \tag{2.11}$$

Consequently, the mean-field Hamiltonian (more details of analytical calculations will be present elsewhere [48]) can be written as

$$\begin{aligned}
\mathcal{H}_{HF} &= B_0^{\tau s}(\mathbf{k})\sigma_0+\mathbf{B}^{\tau s}(\mathbf{k})\cdot\boldsymbol{\sigma}_\tau\\
B_0^{\tau s}(\mathbf{k}) &= \frac{1}{2}\lambda\tau s+\frac{\hbar^2k^2}{4m_0}\alpha-\frac{1}{2}\int\frac{d^2k'}{(2\pi)^2}v_{k-k'}\{n^c_{k',\tau s}+n^v_{k',\tau s}\}\\
&\quad +U\int\frac{d^2k'}{(2\pi)^2}\{n^c_{k'\bar\tau\bar s}+n^v_{k'\bar\tau\bar s}\}\\
B_z^{\tau s}(\mathbf{k}) &= \frac{\Delta-\lambda\tau s}{2}+\frac{\hbar^2k^2}{4m_0}\beta\\
&\quad -\frac{1}{2}\int\frac{d^2k'}{(2\pi)^2}v_{k-k'}\left\{\frac{(t_0a_0)^2k'^2-D_+^2}{(t_0a_0)^2k'^2+D_+^2}n^c_{k',\tau s}\right.\\
&\quad \left.+\frac{(t_0a_0)^2k'^2-D_-^2}{(t_0a_0)^2k'^2+D_-^2}n^v_{k',\tau s}\right\}\\
B_x^{\tau s}(\mathbf{k})-iB_y^{\tau s}(\mathbf{k}) &= (t_0a_0)k(\cos\phi-i\sin\phi)\\
&\quad +\int\frac{d^2k'}{(2\pi)^2}v_{k-k'}\left\{\frac{(t_0a_0)k'D_+}{(t_0a_0)^2k'^2+D_+^2}n^c_{k',\tau s}\right.\\
&\quad \left.+\frac{(t_0a_0)k'D_-}{(t_0a_0)^2k'^2+D_-^2}n^v_{k',\tau s}\right\}(\cos\phi'-i\sin\phi')
\end{aligned} \tag{2.12}$$

where $n^{c,v}_{k,\tau s} = \Theta(\varepsilon_{\rm F} - \varepsilon^{c,v}_{k,\tau s})$ is the Fermi distribution at zero temperature. It must be noted that instead of following a self-consistent procedure to find particle distribution function, we just use its noninteracting from. Moreover, to investigate the magnetic phase of the ground state in the Hartree-Fock approximation, we use Stoner exchange model in which it is assumed that the system is partially spin polarized. The spin polarization rate and the total charge density are $\zeta = (n_\uparrow - n_\downarrow)/n$ and $n = n_\uparrow + n_\downarrow$, respectively. Notice that $k_{\rm Fs}$ is the wave vector of two spin components of spin at each valleys where for the electron doped case they are the same but in a hope doped case, they differ from each other and the Fermi wave vector is $k_{\rm Fs} = k_{\rm F}(1 + s\zeta)^{1/2}$ in electron doped case. The Fermi wave vector given by $k_{\rm F} = \sqrt{4\pi n/g}$ where g stands for the degeneracy of the band structure which is equal to 4 in the later case. The total energy per particle including the kinetic and the exchange terms for electron doped case, reads as

$$\begin{aligned}
\varepsilon_{\rm tot}(n, \zeta, \epsilon_0, U) &= \frac{E_\uparrow + E_\downarrow}{N_\uparrow + N_\downarrow} \\
E_s &= \sum_{k\tau} \varepsilon^c_{k\tau s} n^c_{k\tau s} = \frac{S}{(2\pi)^2} \sum_\tau \int \varepsilon^c_{k\tau s} n^c_{k\tau s} d^2k \\
&= \frac{S}{2\pi} \sum_\tau \int \varepsilon^c_{k\tau s} n^c_{k\tau s} k dk = \frac{S}{2\pi} \sum_\tau \int_0^{k_{\rm Fs}} \varepsilon^c_{k\tau s} k dk \\
N_s &= \sum_{k\tau} n^c_{k\tau s} = \frac{S}{2\pi} \sum_\tau \int n^c_{k\tau s} k dk = \frac{S}{2\pi} k^2_{\rm Fs}
\end{aligned} \tag{2.13}$$

where the total energy of occupied state in the valence band is considered as the vacuum energy and we ignore its contribution in the energy per particle. At zero temperature and electron doped case we thus have

$$\begin{aligned}
\varepsilon_{\rm tot}(n, \zeta, \epsilon_0, U) &= \frac{\sum_{\tau s} \int_0^{k_{\rm Fs}} \varepsilon^c_{k\tau s} k dk}{k^2_{\rm F\uparrow} + k^2_{\rm F\downarrow}} \\
&= \frac{1}{2k^2_{\rm F}} \sum_{\tau s} \int_0^{k_{\rm Fs}} [|\mathbf{B}^{\tau s}(k)| + B_0^{\tau s}(k)] k dk
\end{aligned} \tag{2.14}$$

and in addition, for the case of the hole dope system, the ground state energy can be written as [48]

$$\begin{aligned}
\varepsilon_{\rm tot}(n, \zeta, \epsilon_0, U) &= -\frac{\sum_{\tau s} \int_0^{k_{\rm Fs}} \varepsilon^v_{k\tau s} k dk}{k^2_{\rm F\uparrow} + k^2_{\rm F\downarrow}} \\
&= \frac{1}{2k^2_{\rm F}} \sum_{\tau s} \int_0^{k_{\rm Fs}} [|\mathbf{B}^{\tau s}(k)| - B_0^{\tau s}(k)] k dk.
\end{aligned} \tag{2.15}$$

Finally, the critical density [52] in which the paramagnetic-to-ferromagnetic Bloch phase transition [53] occurs can be obtained by following criteria $\varepsilon_{tot}(n_{cr}, 1, \epsilon_0, U) = \varepsilon_{tot}(n_{cr}, 0, \epsilon_0, U)$. Efforts to observe the ferromagnetic phase predicted by Bloch have likewise been frustrated by the difficulty of achieving low values of the charge density. The closest thing to an experimental observation of this transition has come so far from experiments in the 2D electron gas at high magnetic field. Under appropriate conditions the magnetic field suppresses not only the kinetic energy, but also the correlation energy. This leaves the exchange energy master of the field, and leads to a ferromagnetic transition [54]. Here, we show that the such a transition takes place for a hole doped system rather than an electron doped case in the absence of the magnetic field.

We introduce an ultraviolet cutoff, k_c for which the low-energy Hamiltonian is valid. A typical value for the cutoff is $1/a_0$, however we choose $k_c = 0.5/a_0$ to be more precise based on the comparison between the electron dispersion relation calculated by our Hamiltonian and the results obtained by *ab initio* band structure [40].

In Figure 4(a), we report Hartree-Fock theory results for the inverse thermodynamic density of states $\partial\mu/\partial n$. The decrease in $\partial\mu/\partial n$ with density is a consequence of the difference between hyperbolic and parabolic dispersion. We see that $\partial\mu/\partial n$ is positive and enhanced by exchange interactions over the density range covered in this plot. In Ref. [55] a nonmonotonic behavior was also found for a bilayer graphene system within the Hartee-Fock approximation and the change in sign of the inverse thermodynamic density of states predicted in very low density.

To calculate the magnetic phase transition, we investigate the condition for which $\varepsilon_{tot}(n, 1, \epsilon_0, U) = \varepsilon_{tot}(n, 0, \epsilon_0, U)$ is satisfied by giving n, u and ϵ_0 parameters. Figure 4(b) shows the magnetic phase diagram at given the charge density. The critical value of the intervalley interaction in which the phase transition is occurred is plotted as a function of the dielectric constant for for both electron and hole cases. The results suggest that the system with hole charge carrier can easily go to the ferromagnetic phase in comparison with a situation in which the charge carrier is the electron.

It turns out that the Bloch transition, a ferromagnetic ground-state, is quantitatively incorrect in the Hartee-Fock approximation. In order to obtain accurate ground-state energy, a renormalized Hamiltonian for low-energy excitations [56] theory so that to derive the expression for the interaction function in a paramagnetic system and the knowledge of the energy functional appropriate to an infinitesimally polarized electron liquid is needed.

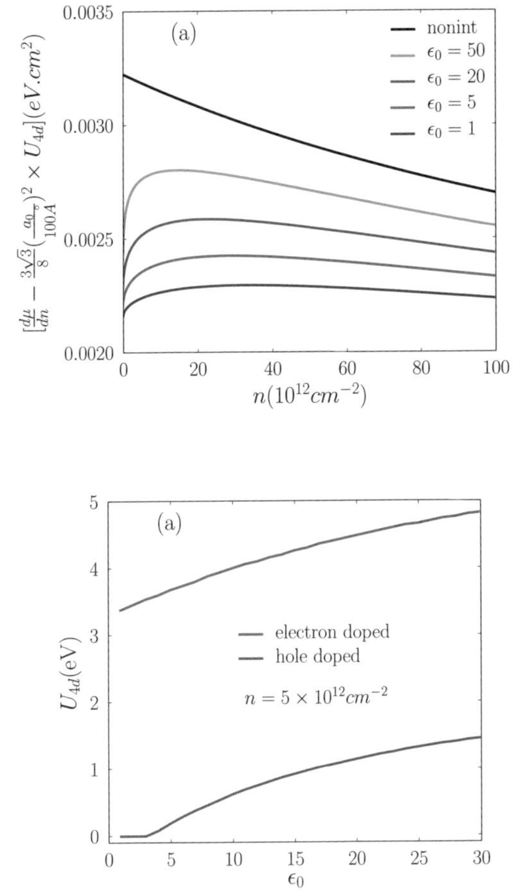

Figure 4. (color online) (a) Charge compressibility, defined by $(n^2\kappa)^{-1} = \partial\mu/\partial n$ where μ is the chemical potential for the electron doped case as a function of the charge density for varies dielectric constant. (b) Magnetic phase diagram in the parameters space where each curve indicates the part of the parameter space in which ground-state is paramagnetic (ferromagnetic)

2.4. Optical properties

The optical absorbtion spectrum of bulk MoS_2 shows two main peaks corresponding to exciton bands, the so called *A* and *B* excitons [5]. They are the direct band gap transitions at the *K* point of the Brillouin zone between the maxima of split valance bands and the minimum of the conduction band. *A* peak centered about 1.9 eV while *B* peak centered at

2.05 eV. Photoluminescence spectra at various temperature [57] show that the two peaks red-shift and broaden with the increased temperature. Furthermore, experiments [18,58] reported observation and electrostatic tunability of charging effects in positively charge X^+, neutral X^0 and negatively charged X^- excitons in field effect transistors via photoluminescence and by using high quality monolayer MoS_2. The conversion from X^0 to trion can be made by absorbing one electron or hole. By setting a gate voltage to be negative, experimental sample is p-doped favoring excitons to form lower energy bound complexes with free holes. As the gate voltage decreases, more holes are injected into the sample and all X^0 turn into X^- to form a positively charge hole-trion gas with positive the gate voltage a similar situation occurs with free electrons to form an electron-trion gas.

In this subsection, we would like to consider the intrinsic optical properties of MoS_2 with no exciton physics. To do so, the modified Dirac Hamiltonian, which describes the physics of monolayer MoS_2 around K point, can be used to describe optical conductivity of the system [31]. By having the Hamiltonian and calculating velocity operators along the x and y directions $\hbar v_x = \frac{\partial H}{\partial q_x}$ and $\hbar v_y = \frac{\partial H}{\partial q_y}$, the intrinsic optical conductivity can be calculated by using the Kubo formula [61–63] in a clean sample and it is given by

$$\begin{aligned}
\sigma_{xy}(\omega) &= -i\frac{e^2}{2\pi h}\int d^2q\,\frac{f(\varepsilon_c)-f(\varepsilon_v)}{\varepsilon_c-\varepsilon_v}\left\{\frac{\langle\psi_c|\hbar v_x|\psi_v\rangle\langle\psi_v|\hbar v_y|\psi_c\rangle}{\hbar\omega+\varepsilon_c-\varepsilon_v+i0^+}\right.\\
&\qquad\left.+\frac{\langle\psi_v|\hbar v_x|\psi_c\rangle\langle\psi_c|\hbar v_y|\psi_v\rangle}{\hbar\omega+\varepsilon_v-\varepsilon_c+i0^+}\right\}\\
\sigma_{xx}(\omega) &= -i\frac{e^2}{2\pi h}\int d^2q\,\frac{f(\varepsilon_c)-f(\varepsilon_v)}{\varepsilon_c-\varepsilon_v}\left\{\frac{\langle\psi_c|\hbar v_x|\psi_v\rangle\langle\psi_v|\hbar v_x|\psi_c\rangle}{\hbar\omega+\varepsilon_c-\varepsilon_v+i0^+}\right.\\
&\qquad\left.+\frac{\langle\psi_v|\hbar v_x|\psi_c\rangle\langle\psi_c|\hbar v_x|\psi_v\rangle}{\hbar\omega+\varepsilon_v-\varepsilon_c+i0^+}\right\}
\end{aligned} \tag{2.16}$$

where $f(\omega)$ is the Fermi distribution function. We include only the interband transitions and the contribution of the intraband transitions, which leads to the fact that the Drude-like term, is no longer relevant here since the momentum relaxation time is assumed to be infinite. This approximation is valid at low-temperature and a clean sample where defect, impurity, and phonon scattering mechanisms are ignorable. We also do not consider the bound state of exciton in the systems. After straightforward calculations, the real and imaginary parts of diagonal and off-diagonal components of the conductivity tensor at $\tau = +$ are given by [31]

$$\sigma_{xy}^{\Re,\tau s}(\omega) = \frac{2e^2}{h}\mathbb{P}\int dq(f(\varepsilon_c) - f(\varepsilon_v))$$

$$\times \left\{ \frac{\tau(\Delta'_{\tau s}q - \beta' q^3)}{\sqrt{(\Delta'_{\tau s} + \beta' q^2)^2 + q^2}[4((\Delta'_{\tau s} + \beta' q^2)^2 + q^2) - (\hbar\omega/t_0)^2]} \right\}$$

$$\sigma_{xy}^{\Im,\tau s}(\omega) = \frac{\pi e^2}{2h}\int dq(f(\varepsilon_c) - f(\varepsilon_v))$$

$$\times \left\{ \frac{\tau(\Delta'_{\tau s}q - \beta' q^3)}{(\Delta'_{\tau s} + \beta' q^2)^2 + q^2} \right\} \delta(\hbar\omega/t_0 - 2\sqrt{(\Delta'_{\tau s} + \beta' q^2)^2 + q^2})$$

$$\sigma_{xx}^{\Im,\tau s}(\omega) = -\frac{2e^2}{h}\hbar\omega\mathbb{P}\int dq(f(\varepsilon_c) - f(\varepsilon_v))$$

$$\times \left\{ \frac{q}{\sqrt{(\Delta'_{\tau s} + \beta' q^2)^2 + q^2}[4((\Delta'_{\tau s} + \beta' q^2)^2 + q^2) - (\hbar\omega/t_0)^2]} \right.$$

$$\left. \frac{q^3[\frac{1}{2} + 2\beta'\Delta'_{\tau s}]}{((\Delta'_{\tau s} + \beta' q^2)^2 + q^2)^{3/2}[4((\Delta'_{\tau s} + \beta' q^2)^2 + q^2) - (\hbar\omega/t_0)^2]} \right\}$$

$$\sigma_{xx}^{\Re,\tau s}(\omega) = -\frac{\pi e^2}{2h}\int dq(f(\varepsilon_c) - f(\varepsilon_v))$$

$$\times \left\{ \frac{q}{\sqrt{(\Delta'_{\tau s} + \beta' q^2)^2 + q^2}} - \frac{q^3[\frac{1}{2} + 2\beta'\Delta'_{\tau s}]}{((\Delta'_{\tau s} + \beta' q^2)^2 + q^2)^{3/2}} \right\} \delta\Big(\hbar\omega/t_0$$

$$- 2\sqrt{(\Delta'_{\tau s} + \beta' q^2)^2 + q^2}\Big) \tag{2.17}$$

where $\Delta'_{\tau s} = (\Delta - \lambda\tau s)/2t_0$, $\alpha' = b\alpha/t_0$, $\beta' = b\beta/t_0$, $\sigma_{xy}^{\tau,s} = \sigma_{xy}^{\Re,\tau s} + i\sigma_{xy}^{\Im,\tau s}$, $b = \hbar^2/4m_0a_0^2$ and $\sigma_{xx}^{\tau,s} = \sigma_{xx}^{\Re,\tau s} + i\sigma_{xx}^{\Im,\tau s}$.

$\Re$ and $\Im$ refer to the real and imaginary parts of σ and $\mathbb{P}$ denotes the principle value. It is worthwhile mentioning that the conductivity for MoS_2 for $\tau = -$ can be found by implementing $p_x \rightarrow -p_x$ and $\lambda \rightarrow -\lambda$. Using these transformations, the velocity matrix elements around the K' point can be calculated by taking the complex conjugation of the corresponding results for the $\tau = +$ case.

Noticeably, the spin and valley *transverse* ac-conductivity are given by

$$\sigma^s_{xy} = \frac{\hbar}{2e}\sum_\tau [\sigma^{\tau,\uparrow}_{xy} - \sigma^{\tau,\downarrow}_{xy}]$$
$$\sigma^v_{xy} = \frac{1}{e}\sum_s [\sigma^{K,s}_{xy} - \sigma^{K',s}_{xy}] \tag{2.18}$$

and for the *longitudinal* ac-conductivity case, an electric field can only induce a charge current and corresponding conductivity is given as

$$\sigma_{xx} = \sum_\tau [\sigma^{\tau,\uparrow}_{xx} + \sigma^{\tau,\downarrow}_{xx}]. \tag{2.19}$$

2.5. Intrinsic dc-conductivity

To find the static conductivity in a clean sample, we set $\omega = 0$ and thus the interband longitudinal conductivity vanishes. Consequently, we calculate only the transverse conductivity in this case. At zero temperature, the Fermi distribution function is given by a step function, *i. e.* $f(\varepsilon_{c,v}) = \Theta(\varepsilon_F - \varepsilon_{c,v})$. Most of the interesting transport properties of MoS_2 originates from its spin splitting band structure for the hole doped case. Therefore, for the later case, when the upper spin-split band contributes to the Fermi level state, the dc-conductivity is given by

$$\sigma^{K\uparrow}_{xy} = -\sigma^{K'\downarrow}_{xy} = -\frac{e^2}{2h}\int_{q_F}^{q_c} \frac{(\Delta'_{K\uparrow} q - \beta' q^3)dq}{((\Delta'_{K\uparrow} + \beta' q^2)^2 + q^2)^{\frac{3}{2}}}$$
$$= -\frac{e^2}{2h}\mathcal{C}^{K\uparrow} + \frac{e^2}{2h}\frac{2\mu + 2b(\alpha-\beta)q_F^2}{\Delta - \lambda + 2\mu + 2b\alpha q_F^2} \tag{2.20}$$

and for the spin-down component we thus have

$$\sigma^{K\downarrow}_{xy} = -\sigma^{K'\uparrow}_{xy} = -\frac{e^2}{2h}\int_0^{q_c} \frac{(\Delta'_{K\downarrow} q - \beta' q^3)dq}{((\Delta'_{K\downarrow} + \beta' q^2)^2 + q^2)^{\frac{3}{2}}}$$
$$= -\frac{e^2}{2h}\mathcal{C}^{K\downarrow} \tag{2.21}$$

where q_c is the ultra violate cutoff and $\mu/t_0 = \sqrt{(\Delta'_{K\uparrow} + \beta' q_F^2)^2 + q_F^2} - \Delta'_{K\uparrow} - \alpha' q_F^2$ stands for the chemical potential and it is easy to show that $\mathcal{C}^{Ks} = \mathrm{sgn}(\Delta - \lambda s) - \mathrm{sgn}(\beta)$ at large cutoff values. In a precise definition,

$\mathcal{C}^{Ks}$ terms are the Chern numbers for each spin and valley degrees of freedom and the total Chern number is zero owing to the time reversal symmetry. Intriguingly, the quadratic term in Eq. (3), β, leads to a new topological characteristic. When $\beta\Delta > 0$, with $\Delta > \lambda$, system has a trivial phase with no edge mode closing the energy gap however for the case that $\beta\Delta < 0$, the topological phase of the system is a non-trivial with edge modes closing the energy gap. In the case of the MoS_2, the tight binding model [26,59] predicts the trivial phase ($\beta > 0$) with $\mathcal{C}^{Ks} = 0$. However, a non-trivial phase is expected by Refs. [27,28] (where $\beta < 0$) which leads to $\mathcal{C}^{Ks} = 2$. In other words, the term proportional to β has a topological meaning in Z_2 symmetry invariant like the ultra-thin film topological insulator system [14] and the sign of β plays important role.

The transverse intrinsic dc-conductivity for the hole doped MoS_2 case, is given by

$$\sigma^{s}_{xy} = \frac{\hbar}{e}[\sigma^{K\uparrow}_{xy} - \sigma^{K\downarrow}_{xy}] = \frac{e}{2\pi}\frac{\mu + b(\alpha-\beta)q_{\mathrm{F}}^2}{\Delta - \lambda + 2\mu + 2b\alpha q_{\mathrm{F}}^2}$$
$$\sigma^{v}_{xy} = \frac{2}{e}[\sigma^{K\uparrow}_{xy} + \sigma^{K\downarrow}_{xy}] = -\frac{e}{h}\mathcal{C}^{K} + \frac{2}{\hbar}\sigma^{s}_{xy} \tag{2.22}$$

where, at large cutoff, $\mathcal{C}^{K} = [\mathrm{sign}(\Delta-\lambda) + \mathrm{sign}(\Delta+\lambda)]/2 - \mathrm{sign}(\beta)$ stands for the valley Chern number and it equals to zero or 2 corresponding to the non-trivial or trivial band structure, respectively. It should be noted that in the absence of the diagonal quadratic term, the non-zero valley Chern number at zero doping predicts a valley Hall conductivity, which is proportional to sign(Δ). Therefore, the exitance of edge states, which can carry the valley current, is anticipated. However, Z_2 symmetry prevents the edge modes from existing. Since the Z_2 topological invariant is zero when the gap is caused only the inversion symmetry breaking [65], thus the topology of the band structure is trivial and there are no edge states to carry the valley current when the chemical potential is located inside the energy gap. Therefore, we can ignore the valley Chern number in σ^{v}_{xy} and thus the results are consistent with those results reported by Xiao *el al.* [25] at a low doping rate where $\mu \ll \Delta - \lambda$.

2.6. Intrinsic dynamical conductivity

In this subsection, we analytically calculate the dynamical conductivity of the modified-Dirac Hamiltonian. Using the two-band Hamiltonian, including the quadratic term in momentum, the optical Hall conductivity for each spin and valley components are given by

$$\begin{aligned}
\sigma_{xy}^{\Re,\tau s}(\omega) &= \tau\frac{e^2}{h}[G_{\tau s}(\omega,q_{\rm F}) - G_{\tau s}(\omega,q_c)] \\
\sigma_{xy}^{\Im,\tau s}(\omega) &= \tau\frac{\pi e^2}{2h}\frac{\Delta'_{\tau s} - \beta' q_{0,\tau s}^2}{\hbar\omega' n(\omega')} \\
&\quad \times [\Theta(2\varepsilon'_{\rm F} - \lambda'\tau s - 2\alpha' q_{0,\tau s}^2 - \hbar\omega') - (\omega' \to -\omega')] \\
&\quad \times \Theta(n(\omega') - (1 + 2\beta'\Delta'_{\tau s}))
\end{aligned} \tag{2.23}$$

where $\Re$ and $\Im$ indicate to the real and imaginary parts, respectively and $G_{\tau s}(\omega, q)$ reads as below

$$\begin{aligned}
G_{\tau s}(\omega,q) = {}& \frac{\Delta'_{\tau s}}{\hbar\omega' n(\omega')}\ln\left|\frac{\hbar\omega'\frac{m(q)}{n(\omega')} - 2\sqrt{(\Delta'_{\tau s}+\beta' q^2)^2 + q^2}}{\hbar\omega'\frac{m(q)}{n(\omega')} + 2\sqrt{(\Delta'_{\tau s}+\beta' q^2)^2 + q^2}}\right| \\
&+ \frac{1}{4\beta'\hbar\omega' n(\omega')}\ln\left|\frac{\hbar\omega'\frac{m(q)}{n(\omega')} - 2\sqrt{(\Delta'_{\tau s}+\beta' q^2)^2 + q^2}}{\hbar\omega'\frac{m(q)}{n(\omega')} + 2\sqrt{(\Delta'_{\tau s}+\beta' q^2)^2 + q^2}}\right| \\
&- \frac{1}{4\beta'\hbar\omega'}\ln\left|\frac{\hbar\omega' - 2\sqrt{(\Delta'_{\tau s}+\beta' q^2)^2 + q^2}}{\hbar\omega' + 2\sqrt{(\Delta'_{\tau s}+\beta' q^2)^2 + q^2}}\right|
\end{aligned}$$

where $m(q) = 1 + 2\beta'\Delta'_{\tau s} + 2\beta'^2 q^2, n(\omega') = \sqrt{1 + 4\beta'\Delta'_{\tau s} + \beta'^2(\hbar\omega')^2}$, $\hbar\omega' = \hbar\omega/t_0$, $\varepsilon'_{\rm F} = \varepsilon_{\rm F}/t_0$ and $\lambda' = \lambda/t_0$. The value of $q_{0,\tau s}$ can be evaluated from $m(q_{0,\tau s}) = n(\omega')$. Note that q_c, the ultra violate cutoff, is assumed to be equal to $1/a_0$. Some special attentions might be taken for the situation in which there is no intersection between the Fermi energy and the band energy, for instance in a low doping hole case of the MoS_2 in which the Fermi energy lies in the spin-orbit splitting interval. In this case, the Fermi wave vector ($q_{\rm F}$, which has no contribution to the Fermi level) vanishes.

The quadratic terms can also affect profoundly on the longitudinal dynamical conductivity which plays main role in the optical response when the time reversal symmetry is preserved. In this case, one can find

$$\begin{aligned}
\sigma_{xx}^{\Re,\tau s}(\omega) &= -\frac{\pi e^2}{4h}\frac{1}{n(\omega')}\left(1 - \frac{1+4\beta'\Delta'_{\tau s}}{2}\left(\frac{2q_{0,\tau s}}{\hbar\omega'}\right)^2\right) \\
&\quad \times [\Theta(2\varepsilon'_{\rm F} - \lambda'\tau s - 2\alpha' q_{0,\tau s}^2 - \hbar\omega') - (\omega' \to -\omega')] \\
&\quad \times \Theta(n(\omega') - (1+2\beta'\Delta'_{\tau s})) \\
\sigma_{xx}^{\Im,\tau s}(\omega) &= -\frac{e^2}{h}[H_{\tau s}(\omega,q_{\rm F}) - H_{\tau s}(\omega,q_c)]
\end{aligned} \tag{2.24}$$

where $H_{\tau s}(\omega, q)$ is given by

$$H_{\tau s}(\omega, q) = \frac{(1 + 2\beta'\Delta'_{\tau s})m(q) - (1 + 4\beta'\Delta'_{\tau s})}{2\beta'^2\hbar\omega'\sqrt{(\Delta'_{\tau s} + \beta' q^2)^2 + q^2}}$$
$$+ \frac{1 + 4\beta'\Delta'_{\tau s}}{2\beta'^2(\hbar\omega')^2} \ln\left|\frac{\frac{\hbar\omega'}{2} - \sqrt{(\Delta'_{\tau s} + \beta' q^2)^2 + q^2}}{\frac{\hbar\omega'}{2} + \sqrt{(\Delta'_{\tau s} + \beta' q^2)^2 + q^2}}\right|$$
$$+ \frac{(1 + 2\beta'\Delta'_{\tau s})(1 + 4\beta'\Delta'_{\tau s}) + \beta'^2(\hbar\omega')^2}{2\beta'^2(\hbar\omega')^2 n(\omega')}$$
$$\times \ln\left|\frac{\frac{\hbar\omega'}{2}\frac{m(q)}{n(\omega')} - \sqrt{(\Delta'_{\tau s} + \beta' q^2)^2 + q^2}}{\frac{\hbar\omega'}{2}\frac{m(q)}{n(\omega')} + \sqrt{(\Delta'_{\tau s} + \beta' q^2)^2 + q^2}}\right|. \tag{2.25}$$

Notice that, dropping the λ, α and β terms, it gives rise to the optical conductivity of gapped graphene and the result is in good agreement with the universal conductivity of graphene [64] for $\Delta = \lambda = \alpha = \beta = 0$.

The real part of the optical Hall conductivity for two set of parameters are illustrated in Figure 5 where top and bottom panels indicate electron and hole doped systems, respectively. The effect of the mass asymmetry between the effective masses of the electron and hole (α) bands is neglected and it will be discussed later. It is clear that the quadratic term, β, causes a reduction of the intensity of the optical Hall conductivity with no changing of the position of peaks for both electron and hole doped cases. The position of peaks in the real part of Hall conductivity is given by $\hbar\omega = \sqrt{(\Delta - \lambda\tau s)^2 + 4t_0^2 {q_{Fs}}^2}$ for $\beta = 0$ case and $\hbar\omega' m(q_{Fs}) n(\omega')^{-1} - 2\sqrt{(\Delta'_{\tau s} + \beta' {q_{Fs}}^2)^2 + {q_{Fs}}^2} = 0$ and $\hbar\omega' - 2\sqrt{(\Delta'_{\tau s} + \beta' {q_{Fs}}^2)^2 + {q_{Fs}}^2} = 0$ for each spin component with corresponding Fermi wave vector q_{Fs} and for the case that $\beta \neq 0$. Surprisingly, the last two equations for the later case are simultaneously fulfilled the equation $m(q_{Fs}) = n(\omega')$ in frequency. In the energy range shown in the figures, the numerical value of the peak position for both cases are approximately equal and it indicates that the position of peaks and steplike configuration don't change due to the β term in a certain Fermi energy. It should be noticed that the intensity of the real part of σ_{xx} decreases with the quadratic term. Consequently, it indicates that the effective mass approximation of the Hamiltonian for the MoS_2 is not completely valid because two sets of parameters with the same effective masses are showing distinct results.

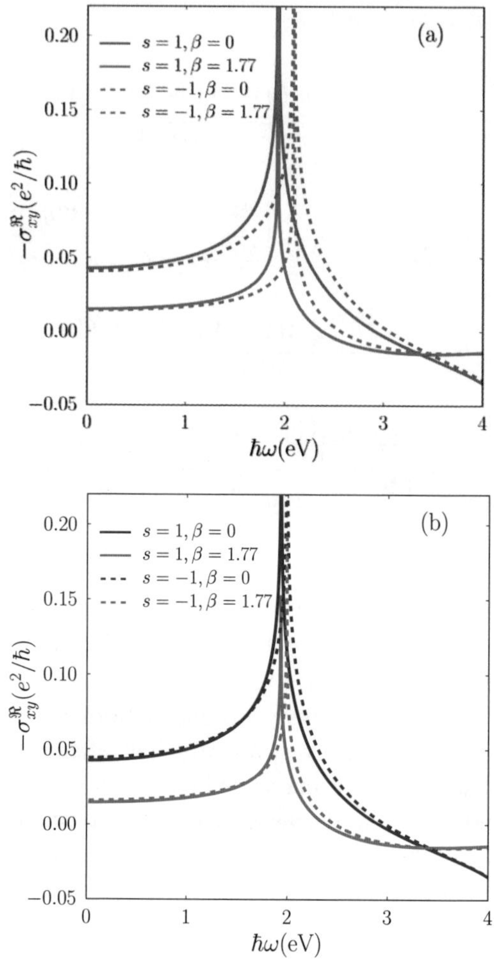

Figure 5. (Color online). Real part of the Hall conductivity (in units of $e^2/\hbar$) for (a) electron with $\varepsilon_F = 1$ eV and (b) hole with $\varepsilon_F = -1$ eV $+ \lambda$ doped cases as a function of photon energy (in units of eV) around the K point. Electron and hole masses are set to be $0.5m_0$ and for two set of parameters, $\beta = 0$, $t_0 = 2.02$ eV and $\beta = 1.77$, $t_0 = 1.51$.

2.7. Circular dichroism and optical transmittance

One of the main optical properties of the monolayer transition metal dichalcogenide system is the circular dichroism when it is exposed by a circularly polarized light in which left- or right-handed light can be absorbed only by K or K' valley for the sake of angular momentum conservations and time-reversal symmetry and it makes the material promising for the valleytronic field. This effect originates from the broken inversion symmetry and it can be understood by calculating the interband optical selection rule $\mathcal{P}_{\pm} = m_0 \langle \psi_c | v_x \pm i v_y | \psi_v \rangle$ for incident right-(+) and left-(-)handed light. The photoluminescence probability for the modified Dirac fermion Hamiltonian is

$$|\mathcal{P}_{\pm}| = \frac{m_0 t_0 a_0}{\hbar} \left(1 \pm \tau \frac{\frac{\Delta - \lambda s}{2} - \frac{\hbar^2}{2m_0} \beta q^2}{\sqrt{(\frac{\Delta - \lambda s}{2})^2 + t_0^2 a_0^2 q^2}} \right) \tag{2.26}$$

where $q^2 = q_x^2 + q_y^2$. Notice that the mass asymmetry term, α, has no effect on the optical selection rule. The selection rule can simply prove the circular dichroism in the MoS_2. Another approach which helps us to understand this effect is to calculate the optical conductivity around the K point of two kinds of light polarizations as $\sigma_{\pm} = \sum_s \{\sigma_{xx}^{Ks} \pm \sigma_{xy}^{Ks}\}$ which has been calculated by using the Dirac-like model [24, 30] and now, we modify that by using the modified-Dirac Hamiltonian. Our results [31] show that $\Re e[\sigma_-]$ is large and comparable in size for either spin up or down while $\Re e[\sigma_+]$ is small in comparison. The valley around the K point can couple only to the left-handed light and this effect is washed up by increasing the frequency of the light and the result is in good agreement with recent experimental measurements [17]. The direct gap transition at the two degenerate valleys, together with this valley-contrasting selection rule, suggest that one can optically generate and detect valley polarizations in this class of materials.

In many semiconductor structures, the circular polarization of luminescence from circularly polarized excitation originates from electron or hole spin polarization. But in monolayer MoS_2, the optical selection rule originates from orbital magnetic moments at K valley independent of electron spin. Actually, there was no experimental noticeable difference between the photoluminescence polarization at zero field and in an in-plane finite magnetic field. Therefore, based on the analysis of inversion symmetry breaking, the helicity of the luminescence should exactly follow that of the excitation light. In other words, the right-handed circularly polarized excitation generates right-handed luminescence, and the left-handed circularly polarized excitation generates left-handed luminescence.

Furthermore, the optical transmittance is an important physical quantity and it can be evaluated stemming from the conductivity. The optical transmittance of a free standing thin film exposed by a linear polarized light is given by [66]

$$T(\omega) = \frac{1}{2} \left\{ \left| \frac{2}{2 + Z_0 \sigma_+(\omega)} \right|^2 + \left| \frac{2}{2 + Z_0 \sigma_-(\omega)} \right|^2 \right\} \tag{2.27}$$

where $Z_0 = 376.73\Omega$ and $\sigma_{\pm}(\omega) = \sigma_{xx}(\omega) \pm i\sigma_{xy}$ are the vacuum impedance and the optical conductivity of the thin film, respectively. For the MoS_2 case, the total Hall conductivity in the presence of the time reversal symmetry is zero and the total longitudinal conductivity is given by $\sigma_{xx} = 2(\sigma_{xx}^{K\uparrow} + \sigma_{xx}^{K\downarrow})$. The optical transmittance of the multilayer of MoS_2 systems has been recently measured [59] and it is about 94.5% for each layer in the optical frequency range. The optical transmittance of the MoS_2 is displayed in Figure 6 for both electron and hole doped cases using the numerical value defined as set_0. The result shows that the optical transmittance is about 98% for the frequency range in which both spin components are active for giving response to the incident light. Importantly, for the electron dope case, there are two minimums with distance about 0.16 $\mathrm{eV}/\hbar$ in frequency which mostly indicates the spin-orbit splitting (2λ) in the valence band. The optical transmittance for electron doped case is about 98% in all frequency range. Moreover, for the hole dope case, the optical transmittance changes by tuning doping rate. Interestingly, at $\mu = -0.942$ eV the difference between the position of peaks of two spin components, is approximately zero and consequently, the total optical conductivity enhances in this resonating doping rate which has significant effect on the optical transmittance of the system where the transmittance decreases and particularly reaches to a value less than 90% at the resonance frequency. Our numerical calculations show that the hole doped MoS_2 is darker than the electron doped one specially close to the resonance frequency. Furthermore, this feature provides an opportunity with measuring the spin-orbit coupling by an optical transmittance measurement.

3. Summary

This chapter only reviews our recent activity regarding the electronic and optical properties of monolayer MoS_2. The study of two-dimensional crystal structures, especially transition-metal dichalcogenides, is an active and rapidly growing field of research, which is driven, on one hand, by application-oriented investigation of transport and optoelectronic devices, and on the other hand, by basic research efforts to study material properties and novel effects related to the peculiar band structure.

The monolayer MoS_2 and nanoribbon MoS_2 offer many opportunities for the investigation of fundamental phenomena and their practical applications. The basic electronic structure properties of exfoliated monolayer MoS_2 is now well understood. MoS_2 undergoes an indirect to direct band gap transition when is thinned down to one layer. The inversion symmetry is not present for odd MoS_2 layers. The loss of the inver-

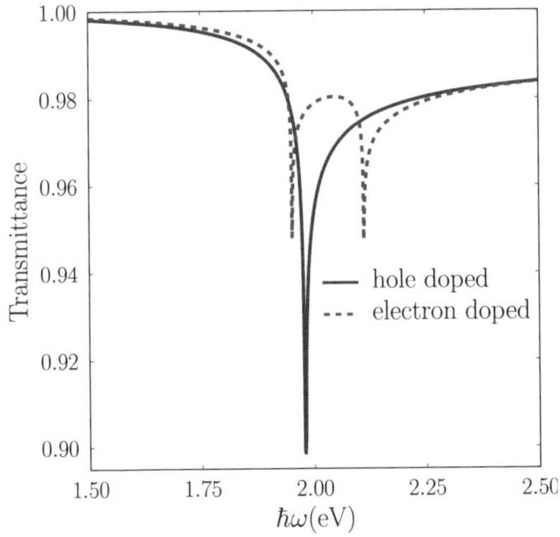

Figure 6. (Color online) Optical transmittance in a finite frequency for the electron ($\varepsilon_F = 1$ eV) and hole ($\varepsilon_F = -1$ eV $+ \lambda$) doped cases including mass asymmetry.

sion symmetry in conjunction with strong spin-orbital coupling leads to a set of unique optical selection rules that couple spin and valley degrees of freedom. Their versatile and tunable properties make them applicable from energy storage and membrane to nanodevices (electronics, optoelectronics, and spintronics). Some key features of MoS2 monolayer and nanoribbon, including bandgap, model Hamiltonian, plasmon modes, ground-state phase transition, charge compressibility and optical properties are highlighted. Especially, we have analytically calculated the intrinsic conductivity of the electronic systems which govern a modified-Dirac Hamiltonian by using the Kubo formula. The theoretical studies showed that their applications in electronics, spintronics, and valleytronics could be achievable by controlling the doping and functionalization. There are many open questions regarding the correlation many-body effects, transport properties, spin and charge relaxation times in monolayer MoS_2 systems. It would be interesting to consider hybrid systems, for example, an assembly of two different metal dichalcogenide sheets. For these heterosystems, the band lineups and deformation potentials will need to be established

References

[1] M. Xu, T. Liang, M. Shi and H. Chen, *Chemical Reviwes*, **113** (2013), 3766.

[2] A. K. Geim and I. V. Grigorrieva, Nature (London) **499** (2013), 419.

[3] E. S. Reich, Nature **506** (2014), 19; L. Li, Y. Yu, G. J. Ye, Q. Ge, X. Ou, H. Wu, D. Feng, X. H. Chen and Y. Zhang, Nature Nanotechnology (2014).

[4] Q. H. Wang, K. Kalantar-Zadeh, A. Kis, J. N. Coleman and M. S. Strano, Nat. Nanotechnol. **7** (2012), 699.

[5] K. F. Mak, C. Lee, J. Hone, J. Shan and T. F. Heinz, Phys. Rev. Lett. **105** (2010), 136805.

[6] B. Radisavljevic, A. Radenovic, J. Brivio, V. Giacometti and A. Kis, Nature Nanotechnol. **6** (2011), 147.

[7] S. Banerjee, W. Richardson, J. Coleman and A. Chatterjee, Electron Dev. Lett. **8** (1987), 347; D. Yang and R. F. Frindt, J. Appl. Phys. **79** (1996), 2376; R. F. Frindt, J. Appl. Phys. **37** (1966), 1928.

[8] Z. M. Wang, "MoS_2, Materials, Physics and Devices", Springer International Publishing, Switzerland, 2014.

[9] Q. Li, J. T. Newberg, E. Walter, J. Hemminger and R. Penner, Nano Lett. **4** (2004), 277.

[10] A. H. Castro Neto, F. Guinea, N. M. R. Peres, K. S. Novoselov and A. K. Geim, Rev. Mod. Phys. **81** (2009), 109.

[11] M. Z. Hasan and C. L. Kane, Rev. Mod. Phys. **82** (2010), 3045.

[12] Shun-Qing Shen, "Topological Insulator: Dirac Equation in Condesed Matters", Springer, 2012.

[13] T. Zhang, J. Ha, N. Levy, Y. Kuk and J. Stroscio, Phys. Rev. Lett. **111** (2013), 056803.

[14] H.-Z. Lu, W.-Y. Shan, W. Yao, Q. Niu and S.-Q. Shen, Phys. Rev. B **81** (2010), 115407.

[15] H. Li, L. Sheng, D. N. Sheng and D. Y. Xing, Phys. Rev. B **82** (2010), 165104.

[16] H. Li, L. Sheng and D. Y. Xing, Phys. Rev. B **85** (2012), 045118.

[17] K. F. Mak, K. He, J. Shan and T. F. Heinz, Nat. Nanotechnol. **7** (2012), 494.

[18] K. F. Mak, K. He, C. Lee, G. H. Lee, J. Hone, T. F. Heinz and J. Shan, Nat. Mat. **12** (2013), 207.

[19] H. Zeng, J. Dai, W. Yao, D. Xiao and X. Cui, Nat. Nanotechnol. **7** (2012), 490.

[20] T. Cao, G. Wang, W. Han, H. Ye, C. Zhu, J. Shi, Q. Niu, P. Tan, E. Wang, B. Liu and J. Feng, Nature Commun. **3** (2012), 887.

[21] S. Wu, J. S. Ross, G. B. Liu, G. Aivazian, A. Jones, Z. Fei, W. Zhu, D. Xiao, W. Yao, D. Cobden and X. Xu, Nat. Phys. **9** (2013), 149.
[22] A. Rycerz, J. Tworzydlo and C. W. J. Beenakker, Nat. Phys. **3** (2007), 172.
[23] D. Xiao, W. Yao and Q. Niu, Phys. Rev. Lett. **99** (2007), 236809.
[24] W. Yao, D. Xiao and Q. Niu, Phys. Rev. B **77** (2008), 235406.
[25] Di Xiao, Gui-Bin Liu, W. Feng, X. Xu and W. Yao, Phys. Rev. Lett. **108** (2012), 196802.
[26] H. Rostami, A. G. Moghaddam and R. Asgari, Phys. Rev. B **88** (2013), 085440.
[27] G. -B. Liu, W. -Y. Shan, Y. Yao, W. Yao and D. Xiao, Phys. Rev. B **88** (2013), 085433.
[28] A. Kormanyos, V. Zolyomi, N. D. Drummond, P. Rakyta, G. Burkard and V. I. Fal'ko, Phys. Rev. B **88** (2013), 045416.
[29] A. Carvalho, R. M. Ribeiro and A. H. Castro Neto, Phys. Rev. B **88** (2013), 115205.
[30] Zhou Li and J. P. Carbotte, Phys. Rev. B **86** (2012), 205425.
[31] H. Rostami and R. Asgari, Phys. Rev. B, **89** (2014), 115413.
[32] A. Kuc, N. Zibouche and T. Heine, Phys. Rev. B **83** (2011), 245213.
[33] P. L. Liao and E. A. Carter, Chem. Soc. Rev. **42** (2013), 2401; H. Jiang, J. Phys. Chem. C **116** (2012), 7664; Y. Ping, D. Rocca and G. Galli, Chem. Soc. Rev. **42** (2013), 2437; H. S. S. Matte, A. Gomathi, A. K. Manna, D. J. Late, R. Dutta, S. K. Pati, C. N. R. Rao, Chem. Int. Ed. **49** (2010), 4059; E. S. Kadantsev and P. Hawrylak, Solid State Commun. **152** (2012), 909.
[34] Y. Ding, Y. L. Wang, J. Ni, L. Shi, S. Q. Shi and W. H. Tang, Physica B **406** (2011), 2254; T. Cheiwchanchamnangij and W. R. L. Lambrecht, Phys. Rev. B **85** (2012), 205302; H. Komsa and A. V. Krasheninnikov, Phys. Rev. B **86** (2012), 241201(R).
[35] P. Johari and V. B. Shenoy, ACS Nano **6** (2012), 5449; Q. Yue, J. Kang, Z. Z. Shao, X. A. Zhang, S. L. Chang, G. Wang, S. Q. Qin and J. B. Li, Phys. Lett. A **376** (2012), 1166.
[36] E. S. Kadantsev and P. Hawrylak, Solid State Commun. **152** (2012), 909; H. Shi, H. Pan, Y.-W. Zhang and B. I. Yakobson, Phys. Rev. B **87** (2013), 155304.
[37] J. Kang, J. Kang, S. Tongay, J. Li and J. Wu, Appl. Phys. Lett. **102** (2013), 012111.
[38] J. C. Slater and G. F. Koster, Phys. Rev. **94** (1954), 1498.
[39] E. Cappelluti, R. Roldán, J. A. Silva-Guillén, P. Ordejón and F. Guinea, Phys. Rev. B **88** (2013), 075409.

[40] H. Peelaers and C. G. Van de Walle, Phys. Rev. B **86** (2012), 241401(R).
[41] Y. Li, Y.-L. Li, C. Moyses Araujo, W. Luo and R. Ahuja, arXiv:1211.4052 (2012).
[42] R. Winkler, "Spin Orbit Coupling Effects in Two-Dimensional Electron and Hole Systems", Springer, Berlin, 2003.
[43] F. G. Giuliani, J. J. Quinn and S. C. Ying, Phys. Rev. B **28** (1983), 2969.
[44] G. F. Giuliani and G. Vignale, "Quantum Theory of the Electron Liquid", Cambridge University Press, Cambridge, 2005.
[45] A. Scholz, T. Stauber and J. Schliemann, Phys. Rev. B **88** (2013), 035135.
[46] G. Santoro and G. F. Giuliani, Phys. Rev. B **37** (1988), 937.
[47] R. Roland, E. Cappelluti1 and F. Guinea, Phys. Rev. B **88** (2013), 054515.
[48] H. Rostami and R. Asgari, ArXiv:1412.8134.
[49] H. Rostami and R. Asgari, Phys. Rev. B **86** (2012), 155435.
[50] G. Borghi, M. Polini, R. Asgari and A. H. MacDonaldd, Solid State Commun. **149** (2009), 1117.
[51] Hongki Min *et al.*, Phys. Rev. B **77** (2008), 041407(R).
[52] A. Qauimzadeh and Reza Asgari, Phys. Rev. B **80** (2009), 035429.
[53] F. Bloch, Z. Physik **57** (1929), 545.
[54] G. F. Giuliani and J. J. Quinne, Phys. Rev. B **31** (1985), 6228.
[55] S. V. Kusminskiy, J. Nilsson, D. K. Campbell and A. H. Castro Neto, Phys. Rev. Lett. **100** (2008), 106805.
[56] S. Yarlagadda and G. F. Giuliani, Phys. Rev. B **40**, 5432 (1989), 5432; Phys. Rev. B **49** (1994), 7887; Phys. Rev. B **49** (1994), 14172.
[57] T. Korn, S. Heydrich, M. Hirmer, J. Schmutzler and C. Schuller, Appl. Phys. Lett. **99** (2011), 102109.
[58] Jason S. Ross, Sanfeng Wu, Hongyi Yu, Nirmal J. Ghimire, Aaron M. Jones, Grant Aivazian, Jiaqiang Yan, David G. Mandrus, Di Xiao, Wang Yao and Xiaodong Xu, Nature Communi. **4** (2013), 1474.
[59] A. C.-Gomez, R. Roldán, E. Cappelluti, M. Buscema, F. Guinea, H. S. J. van der Zant and G. A. Steele, Nano Lett., **13** (2013), 5361.
[60] X. Li, F. Zhang and Q. Niu, Phys. Rev. Lett. **110** (2013), 066803.
[61] T. Stauber, N. M. R. Peres and A. K. Geim, Phys. Rev. B **78** (2008), 085432.

[62] WANG-KONG TSE and A. H. MACDONALD, Phys. Rev. B **84** (2011), 205327.
[63] STEVEN G. LOUIE and MARVIN L. COHEN, "Conceptual Foundations of Materials: A Standard Model for Ground- and Excited-State Properties", Elsevier, 2006.
[64] K.ZIGLER, Phys. Rev. B **75** (2007), 233407.
[65] WANG YAO, SHENGYUAN A. YANG and QIAN NIU, Phys. Rev. Lett. **102** (2009), 096801.
[66] A. FERREIRA, J. V.-GOMES, Y. V. BLUDOV, V. PEREIRA, N. M. R. PERES and A. H. CASTRO NETO, Phys. Rev. B **84** (2011), 235410.

Friedel oscillations in a lateral superlattice with spin-orbit interaction

Jeremy Capps, M. Daniels, C. E. Sosolik and D. C. Marinescu

Abstract. We investigate the Friedel oscillations that can be sustained in the presence of the Coulomb interaction in a two-dimensional lateral superlattice (SL) with spin-orbit interaction (SOI) linear in the electron momentum (Rashba). The superlattice is modeled as a periodic array of infinitely attractive quantum wells whose periodicity determines the apparition of energy minibands in the single particle spectrum that are further spin-split by SOI. The Friedel oscillations are obtained from the static real-space density response function $\Delta\nu(\mathbf{r})$ to an external perturbation, evaluated self-consistently within the random-phase approximation of the Coulomb interaction. The interplay in the momentum space between the spin-orbit coupling and periodicity determines the overall characteristics of the density fluctuations. In a singly occupied, chiral-split miniband approximation, the amplitude and phase of the oscillations are studied numerically as functions of several significant parameters of the system such as the miniband width, the strength of the spin-orbit coupling and the superlattice constant.

1. Introduction

As a real-space manifestation of the many-body Coulomb interaction, Friedel oscillations originate in the electron density fluctuations induced by external potentials. They were observed in many STM measurements at surfaces, where steps, impurities, surface dislocations, and point defects can give rise to its characteristic Fermi wavelength-dependent oscillatory signature in tunneling spectroscopy scans [1–8]. On their account, modulation effects have been registered on other phenomena, such as quantum confinement and charge spilling, which appear in the so-called "electronic growth" model for thin metallic films on semiconductors. This model has been used to understand the critical thicknesses observed in the growth modes of Ag and Pb films [9,10]. More recently, it has been proposed that the Friedel oscillation may similarly modulate the growth modes for graphene films on vicinal surfaces [11].

In this paper we are inspired by recent experimental investigations of the vicinal stepped Au (111) [12] surface that probed the physics of a periodic quasi-two dimensional system endowed with spin-orbit interaction (SOI) to study the Friedel oscillations that can be supported in this context. To extend the applicability limit of our results, the theory is formulated for a standard template of a periodic system, a semiconductor lateral superlattice (SL) with SOI, essentially a 2D electron layer patterned by a periodic array of electrostatic gates [13] that is simultaneously characterized by the redistribution in **k** space of the single particle state energies on account of the spin-orbit interaction and by a geometric-real space periodic confinement that introduces a mini-band energy structure. Because the characteristic parameters of this system, such as particle density, miniband width, and periodicity, can be externally controlled, it presents a good test case for theoretical predictions and experimental observation. Results derived in this framework can serve then as guidance for similar problems in metallic surfaces where the particle concentration is fixed.

The considerable interest dedicated to understanding the consequences of the spin-orbit interaction in two-dimensional (2D) semiconductor-based structures is primarily motivated by its potential application to devices where the control of the electron spin is realized by electric fields. While direct resolution of this problem has not been reached, the physics of such systems continues to propose interesting subjects. Originating either in the asymmetry of the quantum well structure (Rashba) [14] or in the inversion asymmetry of the crystal (Dresselhaus) [15], SOI coupling has been found responsible for a plethora of very diverse effects that are generated by the frustrated spin motion. Among these, special situations are encountered when SOI is competing with other dynamic restrictions such as geometrical constraints as it happens in quantum wires or superlattices or the Coulomb interaction that amplifies the role of the spin effects through exchange.

Previously, the effects of the Coulomb repulsion on the physical properties of the homogeneous electron systems with SOI have been explored in great detail both in terms of single particle properties [16] or as collective phenomenology [17–21] in complementary numerical and analytical approaches [22,23]. Moreover, the superposition of SOI and spatial confinement has been shown before to produce specific phenomena that are absent in homogeneous systems such as the induced spin accumulation in the presence of an electric field [24–26] and the enhancement of the excitation frequency of the collective plasma modes [27].

The behavior of Friedel oscillations in 2D homogeneous systems is well understood as a consequence of the non-analyticity of the static polarization function $\Pi(\mathbf{q}, 0)$ whose first order derivative is discontinuous at a wavevector $\mathbf{q}$ equal to the diameter of the Fermi surface [28]. This characteristic is responsible for the r^{-d} decay of the oscillations at large distances in a d-dimensional space. In systems with SOI, several particular features of the electrostatic screening and of the Friedel oscillations have been identified as a consequence of the modified single-particle spectrum. Significant examples are the small-q high-temperature oscillations [29] and the beatings of the Friedel oscillations predicted to appear, under certain circumstances, in the simultaneous presence of the Rashba and Dresselhaus interactions [30].

In this work, we analyze the principal characteristics of the Friedel oscillations in a superlattice with spin-orbit interaction, a system that allows a momentum space interplay between the mini-band distortion introduced by the spin-orbit coupling and the SL periodicity. Following the traditional approach, we evaluate the the static polarization function of the system within the random-phase-approximation (RPA) of the many-body Coulomb interaction and calculate the real space density fluctuations as the Fourier transform of the response function. In a singly occupied, spin-split miniband approximation, the density oscillations are studied numerically as a function of several system parameters, such as the strength of the interaction, the miniband width and the lattice periodicity.

2. System Description

We consider the general model of a lateral semiconductor superlattice obtained by applying a periodic potential along a certain spatial direction - say $\hat{x}$ in an isotropic two dimensional (2D) system [13]. A positive background exists to assure charge neutrality. The electron spin $\boldsymbol{\sigma}$ is coupled to its momentum $\mathbf{p}$ through a Rashba spin-orbit coupling of strength α, described by the Hamiltonian [14],

$$H_{SOI} = \alpha(\boldsymbol{\sigma} \times \mathbf{p}) \ . \tag{2.1}$$

The periodic potential along the x-axis is modulated as a sequence of N infinitely attractive δ-functions equally spaced at distance a, whose role is to produce a localization of the electron along the axis,

$$V(x) = -\lambda \sum_{l} \delta(x - la) \ . \tag{2.2}$$

This potential has the benefit of leading to only one miniband in the presence of weak tunneling, since there is a single bound state in each quantum well, thus allowing for the realization of a good qualitative and approximately quantitative model. The eigenstate and energy spectrum in each isolated well are given, respectively, by $\nu(x) = \sqrt{\kappa} e^{-\kappa|x|}$, and $\epsilon_0 = -\frac{m^*\kappa^2}{2\hbar^2}$, with $\kappa = 2m^*\lambda/\hbar^2$. ($m^*$ is the electron effective mass).

The broadening of the single energy eigenstate into a miniband on account of tunneling in the simultaneous presence of the periodic potential in Equation (2.2) and SOI in Equation (2.1) has been obtained by both numerical [31] and analytical methods [32, 33]. Conceptually, the difference in the two approaches is established by the balance between the magnitude of the two competing effects that determine the exact shape of the spectrum, namely the strength of the tunneling which affects the widening of the single particle levels embodied by the miniband width Δ and the strength of SOI which couples the electron momentum to its spin. In both instances, however, the salient characteristics of the spectrum, *i.e.* degeneracy at $k_x = 0$ and $k_x = \pm\pi/a$, as well as the overall shape of the dispersion curves are similar. Based on these findings, we anticipate that the validity of our results, derived analytically within the approximation of the dominance of the band effects on the spin-orbit coupling will maintain even in strongly SOI-coupled systems when the situation can be reversed.

Following Ref. [32], the single particle states are built from a superposition of the Bloch functions in the absence of the perturbation. An electron of momentum $\mathbf{k} = \{k_x, k_y\}$ and spin state $|\sigma\rangle$ is described by

$$\psi_{k_x,k_y,\sigma} = \frac{1}{\sqrt{L_y}} e^{ik_y y}|\sigma\rangle \frac{R_{k_x}}{\sqrt{N}} \sum_l e^{ik_x la} v(x - la) , \tag{2.3}$$

where $R_{k_x} = \left[1 + 2e^{-\kappa a}(1 + \kappa a) \cos k_x a\right]^{-1/2}$ is the normalization factor. The corresponding energy eigenstates are calculated within the tight binding approximation,

$$\epsilon_{\vec{k},\sigma} = \frac{\hbar^2 k_y^2}{2m^*} + \frac{\Delta}{2}(1 - \cos k_x a) . \tag{2.4}$$

While k_y, the momentum perpendicular on the SL axis, is continuous, k_x, along the axis, is subject to periodic boundary conditions and consequently quantized $k_x = \frac{2\pi}{Na} j$, where $j \in [-N/2, N/2]$. Δ is a function of the single energy level in the quantum well, $\Delta = 8\epsilon_0 e^{-\kappa a}$.

In the presence of the Rashba interaction corresponding to momenta derived from Equation (2.4),

$$\begin{aligned} p_y &= \hbar k_y , \\ p_x &= \frac{m^*}{\hbar} \frac{\partial \epsilon_{k_x,k_y,\sigma}}{\partial k_x} = \left(\frac{m^* a \Delta}{2\hbar}\right) \sin k_x a , \end{aligned} \tag{2.5}$$

the spin-degenerate miniband splits as a function of the spin eigenstates, leading to chiral mini-bands indexed by $\mu = \pm 1$. The single-particle energy is therefore,

$$E_{\mathbf{k},\mu} = \epsilon_{\mathbf{k}} + \mu\sqrt{\left(\hbar k_y\right)^2 + \left(\frac{m^*\Delta a}{2\hbar}\right)^2 \sin^2 k_x a}\ , \tag{2.6}$$

while the chiral state reflects the out-of-phase superposition of the two spinors,

$$\psi_{\mathbf{k},\mu}(x,y) = e^{ik_y y}\zeta_{k_x}(x)\frac{1}{\sqrt{2}}\left[|\uparrow> + \mu e^{i\mu\varphi_k}|\downarrow>\right]\ , \tag{2.7}$$

with the dephasing angle φ_k satisfying,

$$\tan\varphi_k = \frac{\hbar^2 k_y}{m^*\Delta a \sin k_x a}\ . \tag{2.8}$$

In the following considerations, we will assume that the two minibands of opposite chirality are fully occupied. This condition determines the maximum value of the x-axis momentum, $k_{x\mu}^{\max} = \pi/a$ at the edge of the Brillouin zone, the same for both chiralities. For a given total particle density n_0, and implicitly a set Fermi energy E_F, the maximum value of the y-axis wavevector is determined by solutions of the equation $E_F = E_{k_x,k_y^{\max},\mu}$ for each value of k_x. The maximum momentum along the y axis, as function of k_x, is calculated to be

$$\hbar k_{y\mu}^{\max}(k_x a) = \left\{2m^*\left\{E_F - \Delta\sin^2\frac{k_x a}{2} + m^*\alpha^2\right.\right.$$
$$\left.\left. - \alpha\mu\left[2m^*\left(E_F - \Delta\sin^2\frac{k_x a}{2}\right) + \left(\frac{m^* a\Delta}{2\hbar}\right)^2 \sin^2 k_x a + m^{*2}\alpha^2\right]^{1/2}\right\}\right\}^{1/2}. \tag{2.9}$$

The existence of solutions for both values of μ when $k_x \in [-\pi/a, \pi/a]$ requires that the Fermi energy satisfies $E_F > \Delta[1 + \arctan(\alpha m^* a/\hbar)]$, a condition that constrains the relationship between the equilibrium particle density, ν, and the structure parameters of the superlattice, Δ, a and α. At $T = 0K$, when the particle occupation number is represented by the product of two independent Heaviside functions, $n_{\mathbf{k},\mu}^0 = \theta(\frac{\pi}{a} - |k_x|)\theta(k_{y\mu}^{\max} - |k_y|)$, the particle density is given by

$$\nu_0 = \sum_{\mathbf{k},\mu} n_{\mathbf{k},\mu}^0 = \frac{1}{2\pi^2}\int_0^{\pi/a} dk_x\, [k_{y+}^{\max}(k_x a) + k_{y-}^{\max}(k_x a)] \tag{2.10}$$

where the second equality is obtained by transforming the sum into an integral over the momentum states in the usual fashion.

To illustrate the results of Equation (2.10) we introduce the standard system for our simulations, a 2D InAs lateral superlattice (effective mass $m^* = 0.023 m_e$, with m_e the electron mass) with particle density $\nu = 2.5 \times 10^{11}\text{cm}^{-2}$. While the particle density remains constant, the rest of the SL characteristics, *i.e.* miniband width Δ and the SL constant a, are considered the variable parameters of the problem along with the spin-orbit coupling α. The latter is measured in units of 10^{-11}eVm/$\hbar$ that will not be declared henceforth.

In Figure 1 we show the variation of the Fermi energy E_F with the strength of the spin orbit coupling α for a SL with miniband width $\Delta = 20$ meV and three SL constants, $a = 30$ nm, 40 nm and 50 nm. Within the limits of the analytical model used to determine the single particle states, for a given particle density we register a weak evolution of the Fermi energy with α, an outcome consistent with the assumption that the the spin-orbit coupling effect is secondary to the miniband formation in the system. In the following considerations, this SL description will be the template used in all our numerical simulations.

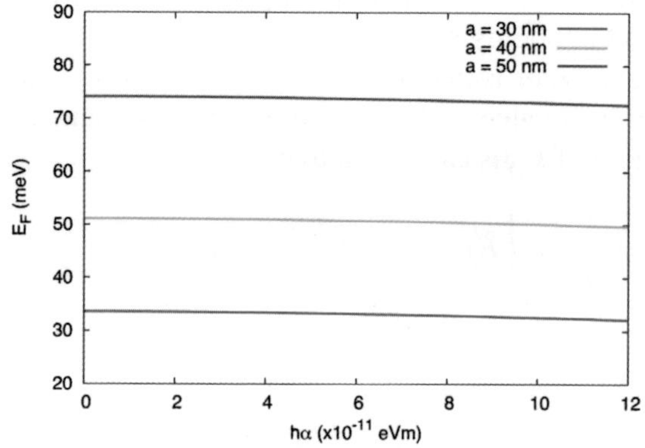

Figure 1. The variation of the Fermi energy in a lateral superlattice with $\Delta = 20$meV as a function of the spin orbit coupling $\alpha = 5$ for $a = 30$nm, $a = 40$nm and $a = 50$nm.

3. The density response function

The linear response of the electron system to a perturbation is described in the Fourier transform space, by a simple proportionality relation that connects the induced density fluctuations self-consistently with the effective potential experienced by the electrons. In the random-phase approximation (RPA) of the Coulomb effects, the effective potential is the superposition between the external potential and the potential associated with the charge fluctuations themselves, leading to the well-known self-consistent equation,

$$\Delta\nu(\mathbf{q},\omega) = P(\mathbf{q},\omega)\left[V_{ex}(\mathbf{q},\omega) + \Delta\nu(\mathbf{q},\omega)\tilde{v}(q)\right] , \qquad (3.1)$$

where $\tilde{v}(q) = \frac{1}{N}\sum_l 2\pi e^2/\sqrt{(q_x + 2\pi l/a)^2 + q_y^2}$ is the Fourier transform of the Coulomb interaction in $2D$ that explicitly incorporates the fact that along the superlattice axis the conservation of the electron momenta in the electrostatic scattering is realized only up to an integer multiple of the reciprocal lattice vector, $2\pi/a$ [27]. The proportionality factor $P(\mathbf{q},\omega)$, the polarization of the electron system, is the Lindhard function written for the single particle energies and states described in Equation (2.6) [18,27]

$$\begin{aligned} P(\mathbf{q},\omega) = \sum_{\mu,\nu}\sum_{k_y,q_y} \frac{n^0_{\mathbf{k}-\mathbf{q}/2,\mu} - n^0_{\mathbf{k}+\mathbf{q}/2,\nu}}{E_{\mathbf{k}-\mathbf{q}/2,\mu} - E_{\mathbf{k}+\mathbf{q}/2,\nu} + \hbar\omega} \\ \times \left|F_{\mu\nu}(k_x,k_y,q_x,q_y)\right|^2 , \end{aligned} \qquad (3.2)$$

with the form factor $F_{\mu\nu}(k_x,k_y,q_x,q_y)$ generated by the overlap of the two different spinors,

$$\left|F_{\mu\nu}(k_x,k_y,q_x,q_y)\right|^2 = \frac{1}{2}\left[1 + \mu\nu\cos(\varphi_{\mathbf{k}-\mathbf{q}/2} - \varphi_{\mathbf{k}+\mathbf{q}/2})\right] . \qquad (3.3)$$

Numerical estimates of the static polarization function presented below are obtained for the InAs SL described above. For a given strength of the SOI coupling and for a given SL periodicity, Equation (2.10) was used to obtain the Fermi energy and the values of the maximum k_y momenta in Equation (2.9). Moreover, we denote by $k^{\max}_{y\mu}$ the absolute maximum value of the electron momentum along the y axis and use it as a scale reference for q_y. In Figure 2 the polarization surfaces are shown for three different values of the SL constant a for the same value of the SOI constant $\alpha = 5$. The polarization values are expressed in terms of the density of states at the Fermi surface, n_0 calculated in the absence of SOI coupling. In general, n_0 is given by

$$\begin{aligned} n_0(\alpha) &= \frac{1}{(2\pi)^2}\sum_{\mathbf{k},\mu}\delta(\epsilon_F - E_{\mathbf{k},\mu}) \\ &= \frac{1}{2\pi^2}\sum_{\mu}\int_0^{\pi/a} dk_x \frac{1}{|\nabla_{k_{y\mu}} E_{\mathbf{k},\mu}|_{k_{y\mu}=k^{\max}_{y\mu}}} . \end{aligned} \qquad (3.4)$$

These pictures reproduce the characteristic behavior of $2D$ systems with SOI, in which the polarization shows an increase in respect to the value at

$\alpha = 0$, on account of the possible transitions between states of opposite chirality. This contribution is magnified by the increase in the SL constant a which enhances the spin-orbit coupling through its effect on the x-axis momentum p_x, Equation (2.5).

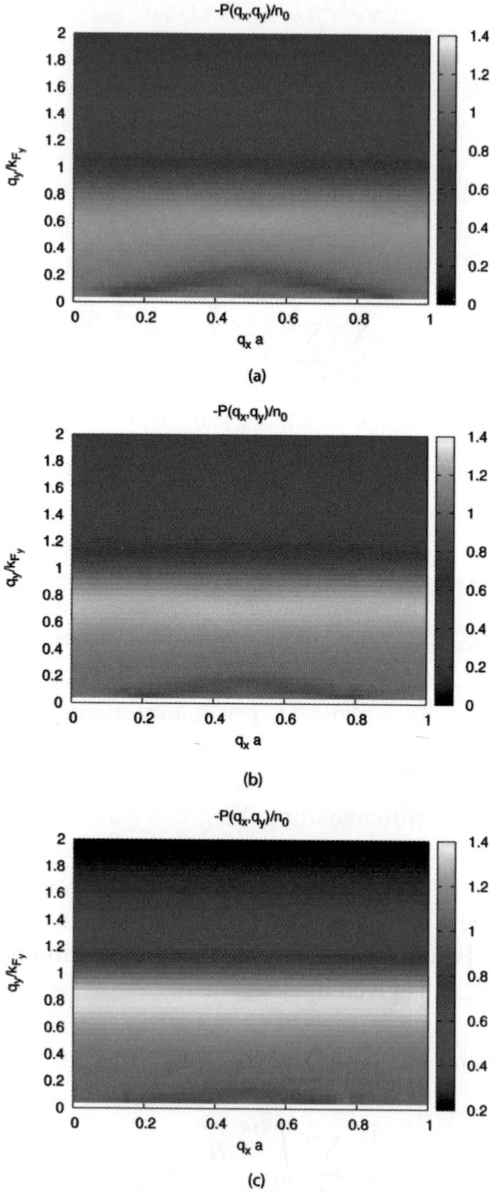

Figure 2. The static polarization function in a SL with $\Delta = 20.0$ meV and $\alpha = 5$ for $a = 30$nm in (a), $a = 40$ nm in (b) and $a = 50$nm in (c).

The static density response function is then evaluated from Equation (3.1)

$$\chi(\mathbf{q},\omega) = \frac{P(\mathbf{q},\omega)}{1-\tilde{v}(q)P(\mathbf{q},\omega)} . \tag{3.5}$$

Using the polarization values obtained before in Figure 2, the density response function is plotted in Figure 3.

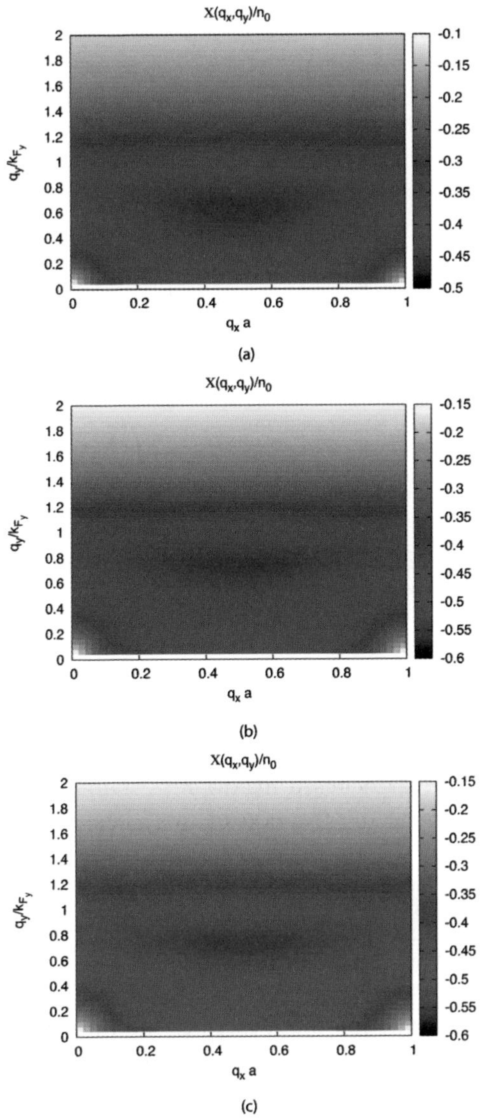

Figure 3. The static density response function in a SL with $\Delta = 20.0$ meV for $\alpha = 5.0$ for $a = 30$nm in (a), $a = 40$ nm in (b) and $a = 50$nm in (c).

4. Friedel oscillations

In the linear regime approximation, the Friedel oscillations result from the density fluctuations registered in response to a perturbing potential V_{ex}. They are given by

$$\Delta\nu(\mathbf{R}) = \frac{1}{N}\sum_{\vec{q}} e^{i\vec{q}\cdot\mathbf{R}}\chi(\mathbf{q})V_{ex}(\mathbf{q}) \,. \tag{4.1}$$

As before, in the periodic system, the sum over $\vec{q}$ has to take into account the fact that the wave vector q_x is defined only up to a multiple of the reciprocal lattice vector, $2\pi l/a$, where l is an integer. In the following considerations we take the perturbing potential to be that of an impurity localized at the origin at the system, $V_{ex} = C\delta(\mathbf{r})$, whose Fourier transform is a constant C. With this choice the quantities involved in Equation (4.1) are periodic with $2\pi/a$ and consequently the sum over q_x can be separated into an integral over the first Brillouin zone and a sum over all its periodic iterations. With this, Equation (4.1) becomes,

$$\Delta\nu(\mathbf{R}) = C\sum_{q_x,q_y}\chi(q_x,q_y)e^{i(q_xR_x+q_yR_y)}\left(\frac{1}{N}\sum_{l} e^{i\frac{2\pi l}{a}R_x}\right) \,. \tag{4.2}$$

The latter sum can be evaluated exactly for a N-well SL,

$$I(R_x) = \frac{1}{N}\sum_{l} e^{i\frac{2\pi l}{a}R_x} = \frac{\sin\frac{(N+1)\pi R_x}{a}}{N\sin\frac{\pi R_x}{a}} \,. \tag{4.3}$$

The function $I(R_x)$ describes an interference term of the single particle states in the SL and reaches a maximum for integer values of the SL constant, $R_x = la$, at the location of the gates. With this, we finally write,

$$\Delta\nu(\mathbf{R}) = \frac{C\,I(R_x)}{4\pi^2}\int_{-\pi}^{\pi}dq_x\int_{-\infty}^{\infty}dq_y\chi(q_x,q_y)e^{i(q_xR_x+q_yR_y)} \,, \tag{4.4}$$

which is the basic equation that describes the Friedel oscillations in the SL. It is easy to see in this configuration that the overall behavior of the density fluctuations is the result of two distinct factors, the interference effects that occur on the account of the geometric periodicity and the real space variation produced by the non-analytical points of the polarization function within the first Brillouin zone.

5. Results and Discussion

Using the same InAs SL template as before, Equation (4.4) is computed numerically to illustrate the behavior of the oscillations induced by an impurity of potential $C = 1\text{meV/m}^2$ located at the $N = 0$ gate, considered the origin of the system.

In Figure 4 we show a representative picture of the interference effect between the oscillatory pattern imposed by the SL periodicity and the density variation determined by the Fourier transform of the density response function within the first Brillouin zone. The interference effect described by Equation (4.3) generates the fast variation of the oscillations with a period proportional to $2a/(N+1)$. They reach significant amplitudes near the position of the gates where the interference factor approaches $(N+1)/N$. $I(R_x)$ modulates the density oscillations that result from the Fourier transform in the first Brillouin zone amplifying the opposite sign density oscillations that occur in the vicinity of $R_x = la$ points. Since this pattern results from the periodicity of the SL, it is reproduced identically in the presence of the spin orbit interaction of any strength. As we show below, the spin orbit interaction changes only the relative amplitude of the oscillations and for this reason, in the following pictures we present only the oscillations that result from the integration of the polarization function over the momentum $\mathbf{q}$ restricted to the first Brillouin zone.

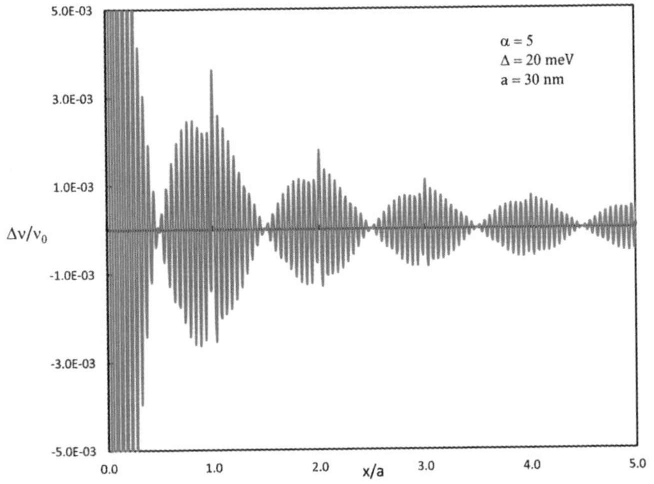

Figure 4. Friedel oscillations induced along the axis of a lateral SL by an impurity located at $x = 0$. The system parameters are $\Delta = 20$ meV, $a = 30$ nm and $\alpha = 5$.

The relative variation of the density oscillations in respect to the equilibrium values is plotted for different values of the SOI coupling strength for the same SL parameters, $\Delta = 20$ meV and $a = 30$ nm in Figure 5. The amplitude of the oscillations decreases compared with the case of $\alpha = 0$, a result of the stronger coupling between the single particle electron states mediated by the spin-orbit interaction. This outcome reproduces the behavior of a 2D homogeneous system, where the amplitude of the Friedel oscillations is known to decrease with α [18]. The density fluctuations are commensurate with the SL period, the zeroes in $\Delta\nu$ being realized at integer and half integer lattice constants. This is a consequence of the periodicity of the polarization function in the momentum space with π/a. The difference in the amplitudes as a function of α decreases with the distance from the impurity.

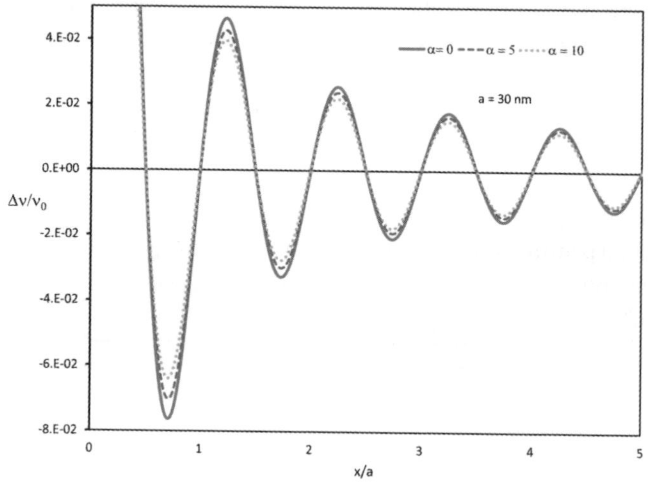

Figure 5. Friedel oscillations induced in a lateral superlattice by an impurity located at $x = 0$ for different SOI coupling values α. The SL parameters are $\Delta = 20$ meV and $a = 30$nm.

In Figure 6 we present the variation of the Friedel oscillations with the SL constant a for the same value of the SOI coupling strength $\alpha = 5$ and miniband width $\Delta = 20$ meV. As the SL constant increases, the amplitude of the oscillations decreases indicating a stronger screening. This feature is a consequence of the Δ dependence of the x-axis momentum involved in the SOI coupling. As before, the periodicity of the polarization in the momentum space is localizes the nodes in the density fluctuations at integer and half-integer lattice constants uniformly.

Further, we plot the oscillations induced along the x axis as a function of the SL miniband width, by comparison with the variation induced

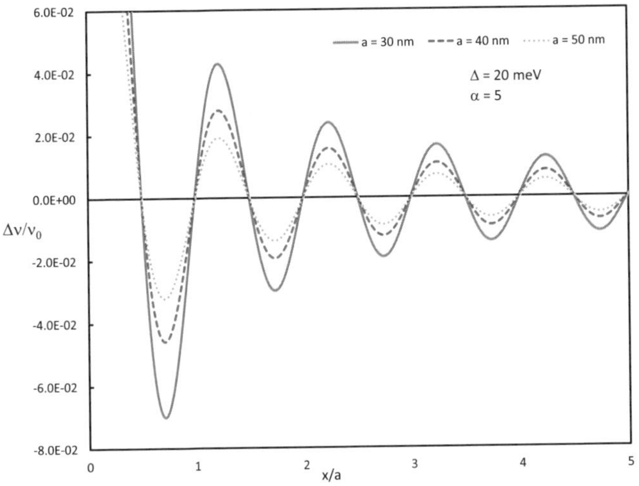

Figure 6. Friedel oscillations induced in a lateral superlattice by an impurity located at $x = 0$ for different SL constants for the same value of the spin-orbit interaction constant, $\alpha = 5$ and $\Delta = 20$ meV.

by SOI, in Figure 7 for the same values of the SL constant. These results indicate a stronger effect of the spin-orbit coupling in enhancing the screening than the miniband width variation.

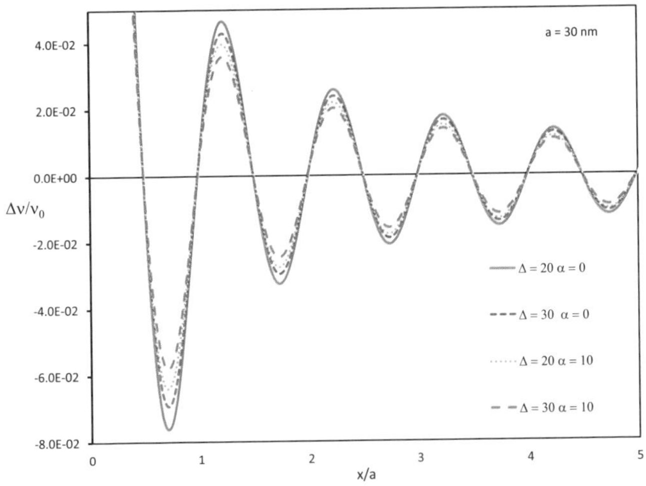

Figure 7. Friedel oscillations induced along the SL axis as a function of the miniband width for $a = 30$ nm and $\alpha = 0, 10$.

These general characteristics also describe the density fluctuations registered along the central y-axis. That spectrum, however, carries the imprint of y maximum momentum being a function of k_x leading to an established pattern of oscillations further from the potential. The variation $\Delta\nu$ is presented in Figure 5 as a function of SOI for a same SL with $\Delta = 20$ nm and $a = 30$nm and for different SL constants at the same value of the SOI coupling, $\alpha = 5$, and $\Delta = 20$ meV in Figure 10. In Figure 9 we present by comparison the change in the amplitude of the oscillations for two miniband widths at two SOI coupling values.

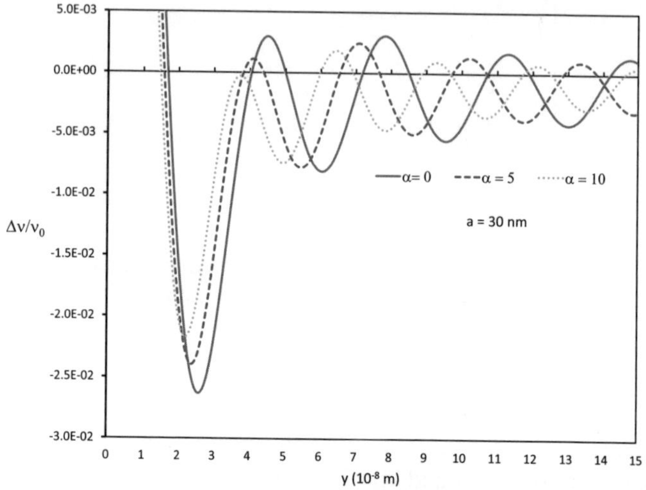

Figure 8. Friedel oscillations induced perpendicular on the SL axis at $x = 0$ as a function of the spin-orbit coupling, for $\Delta = 20$ meV and $a = 30$nm.

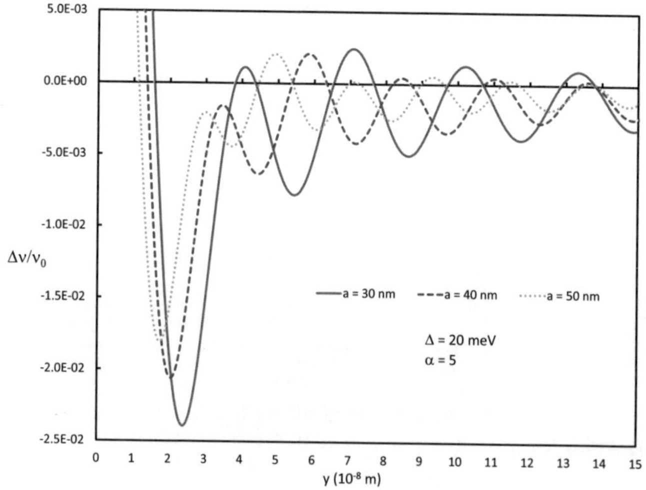

Figure 9. Friedel oscillations induced perpendicular on the SL axis at $x = 0$ as a function of the lattice constant, for $\Delta = 20$ meV and $\alpha = 5$.

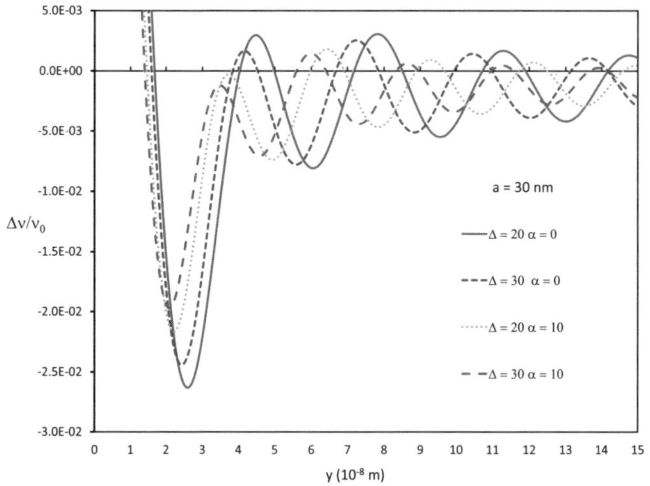

Figure 10. Friedel oscillations induced perpendicular on the SL axis at $x = 0$ as a function of different SL constants and bandwidths.

In conclusion, we analyzed the behavior of the density fluctuations registered in a lateral semiconductor superlattice with spin-orbit interaction in the presence of the Coulomb interaction as a function of the miniband width, the SOI coupling and SL constant. We find that the amplitude of the oscillations, as well as their phase, is affected by the presence of SOI which enhances the coupling between the single particle states indicating a stronger screening.

References

[1] M. F. Crommie, C. P. Lutz and D. M. Eigler, Nature **363** (1995), 524.

[2] M. F. Crommie C. P. Lutz and D. M. Eigler, Science **262** (1993), 218.

[3] D. Fujita, K. Amemiya, T. Yakabe, H. Nejoh, T. Sato and M. Iwatsuki, Phys. Rev. Lett. **78** (1997), 3904.

[4] P. Hofmann, B. G. Briner, M. Doering, H.-P. Rust, E. W. Plummer and A. M. Bradshaw Phys. Rev. Lett. **79** (1997), 265.

[5] M. C. M. M. van der Wielen, A. J. A. van Roij and H. van Kempen, Phys. Rev. Lett. **76** (1996), 1075.

[6] A. Depuydt, C. Van Haesendonck, N. S. Maslova, V. I. Panov, S. V. Savinov and P. I. Arseev, Phys. Rev. B **60** (1999), 2619.

[7] K. Kobayashi, Phys. Rev. B **54** (1996), 17029.

[8] Y. Hasegawa and P. Avouris, Phys. Rev. Lett. **71** (1993), 1071

[9] Z. ZHANG, Q. NIU and C.-K. SHIH, Phys. Rev. Lett. **80** (1998), 5381.
[10] M. M. ÖZER, Y. JIA, B. WU, Z. ZHANG and H. H. WEITERING, Phys. Rev. B **72** (2005), 113409
[11] J. F. WAN and X. Y. KONG, Appl. Phys. Lett. **98** (2011), 013104
[12] J. LOBO-CHECA, F. MEIER, J. H. DIL, T. OKUDA, M. CORSO, V. N. PETROV, M. HENGSBERGER, L. PATTHEY and J. OSTERWALDER, Phys. Rev. Lett. **104** (2010), 187602.
[13] C. ALBRECHT, J. H. SMET, D. WEISS, K. VON KLITZING, R. HENNIG, M. LANGENBUCH, M. SUHRKE, U. RÖSSLER, V. UMANSKY and H. SCHWEIZER, Phys. Rev. Lett. **83** (1999), 2234
[14] Y. A. BYCHKOV and I. E. RASHBA, J. Phys. C **74** (1984), 6039.
[15] G. DRESSELHAUS, Phys. Rev. B **98** (1955), 368.
[16] D. S. SARAGA and D. LOSS, Phys. Rev. B **72** (2005), 195319.
[17] X. F. WANG, Phys. Rev. B **72** (2005), 085317.
[18] M. PLETYUKHOV and V. GRITSEV, Phys. Rev. B **74** (2006), 045307.
[19] S. CHESI and G. F. GIULIANI, Phys. Rev. B **83** (2011), 235308.
[20] S. CHESI and G. F. GIULIANI, Phys. Rev. B **75** (2007), 153306.
[21] J. SCHLIEMANN, Phys. Rev. B **84** (20011), 155201.
[22] A. AMBROSETTI, F. PEDERIVA, E. LIPPARINI and S. GANDOLFI, Phys. Rev. B **80** (2009), 125306.
[23] S. H. ABEDINPOUR, G. VIGNALE and I. V. TOKATLY, Phys. Rev. B **81** (2010), 125123.
[24] M. GOVERNALE and U. ZÜLICKE, Phys. Rev. B **66** (2002), 073311.
[25] I. VURGAFTMAN and J. R. MEYER, Phys. Rev. B **70** (2004), 205319
[26] V. DEMIKHOVSKII and D. KHOMITSKY, JETP Letters **83** (2006), 340.
[27] D. C. MARINESCU and F. LUNG, Phys. Rev. B **82** (2010), 205322.
[28] G. E. SIMION and G. F. GIULIANI, Phys. Rev. B **72** (2005), 045127.
[29] G.-H. CHEN and M. E. RAIKH, Phys. Rev. B **59** (1999), 5090.
[30] S. M. BADALYAN, A. MATOS-ABIAGUE, G. VIGNALE and J. FABIAN, Phys. Rev. B **81** (2010), 205314.
[31] P. FÖLDI, V. SZASZKÓ-BOGÁR and F. M. PEETERS, Phys. Rev. B **82** (2010), 115302.
[32] P. KLEINERT, V. V. BRYKSIN and O. BLEIBAUM, Phys. Rev. B **72** (2005), 195311
[33] D. V. KHOMITSKY, Phys. Rev. B **79** (2009), 205401.

In-plane ferromagnetic instability in a two-dimensional electron liquid in the presence of Rashba spin-orbit coupling

Stefano Chesi and Gabriele F. Giuliani

Abstract. We show that due to the peculiar structure of the non-interacting energy spectrum, the Coulomb interaction leads for all densities to an in-plane ferromagnetic instability in a two-dimensional electron liquid in the presence of sufficiently strong Rashba spin-orbit coupling. This non perturbative phenomenon is characterized by an interesting anisotropic momentum space repopulation and is in nature quite different from the already identified out-of-plane ferromagnetic instability.

The collection of articles in this volume will testify better than me the exuberant and passionate character of Gabriele, as well as the broad range of his interests and scientific contributions. Here I will reproduce (in the next section) an older unpublished manuscript, which originated from the research Prof. Giuliani and I carried out in Purdue in the period from 2002 until 2007, when Gabriele was my PhD advisor and an invaluable example for both scientific and human aspects. The imaginative title of the next Section is borrowed from one of his talks (see Figure 1). The pervasive humor of Gabriele is still at work today: I was reminded of it after searching online without success for the SCEM06 conference cited in his slides. In the spirit of this volume, I hope my introduction will further illustrate the unconventional personality and gifts of Gabriele. Besides mentioning a few memories from when I was a PhD student, I will review some of his scientific ideas from that period and try to connect them with more recent literature, which is obviously missing in the original manuscript.

As a student, as a matter of fact, I regularly approached his office with a feeling of uncertainty. First of all, the electric lights could be seen through the closed door but were generally turned on at any time of the day and night, making it difficult to guess if the office was occupied or not. The light shined through a semi-transparent glass mostly covered by old newspaper clippings (among which a large headline: *Could anyone be worse than Koch? Try Giuliani*). So the worries related to the ongoing research were slightly amplified at the door. After a few moments trying to detect any noise, I would hold on to my notes and knock. During the meetings I would be dragged in a whirlpool of ideas intermixed with a string of provocative remarks, anecdotes, and various considerations on physics and a wide range of other subjects often including soccer (of which Gabriele was a great lover). When I left the office, I was usually quite puzzled on the outcome of the discussion and what to do next. The views of Prof. Giuliani on our ongoing research seemed at first rather paradoxical or far-fetched to my cautious and inexperienced attitude, but they would reveal themselves in due time as useful and deeply true, such that my PhD turned out in the end to be a very productive period of research.

Figure 1. Gabriele loved to wrap physical concepts in colorful terms. In his talks, the topic of this article was introduced as a 'crescent moon' instability (by analogy to by analogy to Figure 4). Other noticeable slides from the same presentation (SCEN06, Pisa) are the *Four pere intermission* (featuring a short video of F. Totti) and *How do we do our calculations? Buy the book!* (obviously referring to [17]).

Some of the ideas he formulated in our discussions have shown in my opinion a remarkable foresight. For example, after we worked out the phase diagram in Figure 3 [1] he liked to mention that the formation of the thin sleeve of spin-polarized states along the dashed curve is very analogous to what happens in the Peierls instability [2]. The dashed curve indicates when the Fermi energy is crossing what he called the 'kissing point' of the two spin bands (see the left panel of Figure 2). In this case, the spontaneous polarization arises by the formation of a gap which removes the degeneracy and leads to a lower total energy of the occupied spin branch. Interestingly, this picture is related to more recent studies of a spontaneous helical nuclear-electronic spin polarization in quantum wires [3]. The formation of these helical states can also be seen (after a gauge transformation generating a Rashba spin-orbit interaction of suitable strength) as due to a similar Peierls-type instability, where the coupled electron-nuclear spin polarization induces a finite gap at the $k = 0$ band crossing of the one-dimensional electron states [4]. The energy gap could be detected in transport and experimental evidence in this direction was recently reported [5]. Returning to two-dimensional systems, another scenario where similar physics plays an important role is provided by pseudospin magnetism in graphene [6].

Beside the FZ ferromagnetic phase shown in Figure 3, in-plane polarized states appear in the complete phase diagram in the Hartree-Fock approximation [7]. Gabriele liked to contrast the 'tilting instability' of spin directions, giving rise to the FZ polarized states, to the 'repopulation instability', a general mechanism giving rise to the in-plane ferromagnetic states. This type of instability is the main topic of the present article. Following his suggestion, the instability could also be studied at large spin-orbit coupling through linear response, from the divergence of the in-plane Pauli spin susceptibility [7]. The instability eventually gives rise to in-plane spin-polarized states with strongly deformed oblong or even 'bean-shaped' occupations, schematically illustrated in the right panel of Figure 4. Recently, the occurrence of these states was proposed in a variety of systems: bilayer graphene [8], electron liquids with short-range interactions [9], and spinor Bose gases [10]. The associated breakdown of Fermi liquid theory was also recently discussed [12,13].

Interestingly, the competition between the FZ phase and in-plane polarized states results in another peculiar feature of the Hartree-Fock phase diagram. When the spin-orbit coupling α approaches zero ($\alpha = 0^+$), the boundary between PM and FZ still occurs at $r_s \simeq 2.01$ as in Figure 3 but a distinct phase boundary between FZ and the in-plane polarized states survives at $r_s = 2.211$ [7]. This behavior is peculiar because at $\alpha = 0$ the magnetic phase is fully isotropic (*i.e.*, there is a single phase boundary at $r_s \simeq 2.01$, the well known Bloch transition [11]) and Gabriele liked particularly this curious 'non-analytic' phase transition with respect to the α, r_s parameters. Although in this case I am not able to point the reader to related literature, I would not be surprised if a similar phenomenon could play an important role in other contexts, given Gabriele's perceptive intuition!

In a similar way as with physics, Gabriele was also a supportive advisor from the personal point of view. While utterly defiant of his disease, he was promptly ready to help us in case of difficulties with his characteristic decisiveness. His sometimes challenging attitude was always directed to stimulate students and co-workers to achieve the best outcomes. Also for this he will be deeply missed.

The crescent moon instability

The problem of a two-dimensional electron liquid in the presence of spin-orbit coupling of the Rashba type is not only of fundamental importance but also of particular technological relevance in view of the considerable recent interest in the possibility of manipulating electronic spins by electric means in modern devices [14–16].

While the effect of the electron interaction in the clean two-dimensional electron liquid is a classic problem studied now for decades [17], the intriguing interplay of many body effects and spin-orbit coupling has only recently began to receive serious attention. The simplest approaches, still not completely characterized, are the random phase approximation (RPA) and the Hartree-Fock (HF) theory. Within the RPA some of the approximate quasiparticle properties were studied in Ref. [18], while the corresponding diagrammatic expansion has been used to extract what amounts to as the exact behavior of the system at high densities in Ref. [19]. Albeit approximate, the HF mean field theory is at the moment the most promising framework to examine the phase diagram. In this respect the behavior of and the observable effects [20] related to the exchange energy in a quantum well were investigated in Ref. [21] for a generalized form of spin-orbit coupling. Furthermore the structure of the HF theory and the peculiar extension of the classic Bloch transition to a homogeneous polarized phase scenario [17] was examined in Ref. [22]. Finally the relevance in this problem of spatially inhomogeneous, charge and spin-density-wave distorted HF states was investigated in Ref. [23].

The purpose of the present paper is to point out the existence in this system of an interesting in-plane ferromagnetic instability of the paramagnetic state that occurs for sufficiently strong Rashba spin-orbit coupling for all densities. The phenomenon acquires particular interest since it is characterized by a peculiar breaking of the rotational symmetry of the momentum space occupation. In this respect it can be seen to be quite different from the already identified out-of-plane ferromagnetic transition. Although we will identify and prove the existence of this non perturbative behavior within the HF theory, the physics underlining the phenomenon is such that it is reasonable to expect that correlation effects will enhance it.

The non interacting problem is defined by the following single-electron Hamiltonian

$$\hat{H}_0 = \frac{\hat{\mathbf{p}}^2}{2m} + \alpha\,(\hat{\sigma}_x \hat{p}_y - \hat{\sigma}_y \hat{p}_x)\,, \tag{1}$$

that describes motion limited to x-y plane and includes a (linear) spin-orbit interaction of the Rashba type [24,25], with α assumed to be positive. The corresponding single-particle eigenfunctions and eigenvalues are given by

$$\varphi_{\mathbf{k},\pm}(\mathbf{r}) = \frac{e^{i\mathbf{k}\cdot\mathbf{r}}}{\sqrt{2L^2}} \begin{pmatrix} \pm 1 \\ i e^{i\phi_{\mathbf{k}}} \end{pmatrix}, \qquad \epsilon_{\mathbf{k}\pm} = \frac{\hbar^2 \mathbf{k}^2}{2m} \mp \alpha \hbar k, \tag{2}$$

where L is the linear size of the system and $\phi_{\mathbf{k}}$ is the angle between the direction of the wave vector and the x-axis. Plots of the non interacting spectrum are provided in Figure 2. It is important to notice that the locus of the points of minimum energy is a circle of radius given by $\frac{m\alpha}{\hbar}$, and that the spinor in $\varphi_{\mathbf{k},+(-)}(\mathbf{r})$ is parallel (antiparallel) to the unit vector $\hat{\phi}_{\mathbf{k}} = -\sin\phi_{\mathbf{k}}\hat{x} + \cos\phi_{\mathbf{k}}\hat{y}$, and therefore can be assigned a positive (negative) chirality.

The corresponding many-body problem is then obtained by accounting for the electronic Coulomb interaction and a suitable homogeneous and rigid neutralizing background [17]. While in the absence of the spin-orbit terms the relevance of the interaction is solely determined by the dimensionless density parameter $r_s^{-1} = \sqrt{\pi a_B^2 n}$ (n being the electron density), here we must also include in our considerations a second dimensionless parameter, *i.e.* $\bar{\alpha} = \frac{\hbar\alpha}{e^2}$: the interplay of these two quantities is responsible for a rich physical scenario.

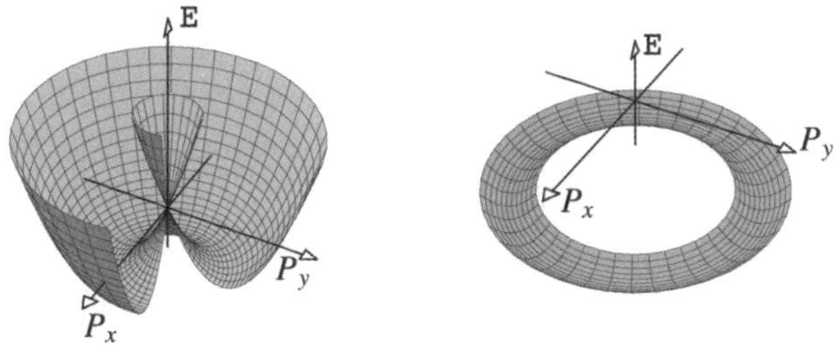

Figure 2. Non interacting single particle spectrum in the presence of linear Rashba spin-orbit interaction. Left: momentum space occupation for small spin-orbit coupling, or large density, with generalized chirality less than one. Right: case of large spin-orbit coupling, or small density, when only the bottom of the lowest band is occupied, a situation in which the generalized chirality is larger than one.

As described in Ref. [22], if one limits the analysis to spatially homogeneous states described by single Slater determinants of plane waves, one finds that the only relevant degree of freedom is the orientation $\hat{s}_{\mathbf{k}}$ of the spin quantization axis of each of the momentum states. As a consequence the HF energy will be in general a functional of $\hat{s}_{\mathbf{k}}$ and the occupation numbers $n_{\mathbf{k}\mu}$. This quantity is readily obtained and is given by:

$$\mathcal{E}[n_{\mathbf{k}\mu}, \hat{s}_{\mathbf{k}}] = \sum_{\mathbf{k};\, \mu=\pm} \left(\frac{\hbar^2 \mathbf{k}^2}{2m} n_{\mathbf{k}\mu} - \hbar \alpha \mu\, k\, \hat{\phi}_{\mathbf{k}} \cdot \hat{s}_{\mathbf{k}}\, n_{\mathbf{k}\mu} \right) - \frac{1}{4L^2} \sum_{\mathbf{k},\mathbf{k}';\, \mu,\mu'=\pm} v_{\mathbf{k}-\mathbf{k}'} \left(1 + \mu\mu'\, \hat{s}_{\mathbf{k}} \cdot \hat{s}_{\mathbf{k}'}\right) n_{\mathbf{k}\mu} n_{\mathbf{k}'\mu'} \,, \qquad (3)$$

where the first line describes the one particle terms, kinetic plus spin-orbit, and the second the exchange energy. For states constructed with symmetric occupation in momentum space the already established phase diagram is depicted in Figure 3 [22]. Neglecting low density phases that are tantamount to a magnetized Wigner crystal, one can identify a paramagnetic chiral phase (PM), that displays a reentrant behavior, and an out-of-plane ferromagnetic chiral phase (FZ) that can be seen as an extension to finite $\bar{\alpha}$ of the classic Bloch instability. Here the latter owes its existence at higher densities to the cusp characterizing the single particle spectrum (see Figure 2), and displays a non trivial spin texture in momentum space [22].

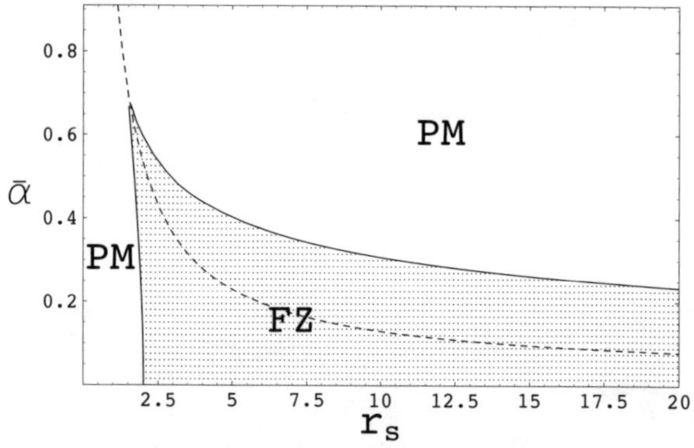

Figure 3. Mean field phase diagram limited to solutions with symmetric momentum space occupation. Within the shaded area the system finds itself in an out-of-plane ferromagnetic chiral phase (FZ). The rest of the phase diagram is occupied by the chiral paramagnetic state (PM). The instability discussed in the text will lead to a modification of this scenario for all densities.

Notice that the FZ phase can persist in a ever shrinking sliver of the plane also at high densities, being located close to the line

$$\bar{\alpha} = \frac{1}{r_s} + \frac{\pi - 1 - 2\mathcal{K}}{2\pi} \,, \qquad (4)$$

where $\mathcal{K} \simeq 0.916$ is the Catalan constant. As it turns out all these states can be elegantly classified by means of one parameter, the generalized chirality χ which is defined in terms of the (dimensionless) radii $\kappa_{0\pm}$ of the circles delimiting the occupied regions in momentum space. When both chiral bands are occupied $\kappa_{0+(-)}$ corresponds to the Fermi radius of the larger (smaller) circle, while when only the lower chiral band is occupied $\kappa_{0+(-)}$ represents the outer (inner) radius of the occupied annulus. In the first case χ coincides with the standard chirality $\chi_0 = \frac{\kappa_{0+}^2 - \kappa_{0-}^2}{\kappa_{0+}^2 + \kappa_{0-}^2}$ while in the second it is larger than one, *i.e.*

$$\chi = \begin{cases} \chi_0 \, , & \text{for } 0 < \chi_0 < 1 \\ \dfrac{\kappa_{0+}^2 + \kappa_{0-}^2}{\kappa_{0+}^2 - \kappa_{0-}^2} \, , & \text{for } \chi_0 = 1 \, , \end{cases} \tag{5}$$

so that (in units of the Fermi wave vector $k_F = \sqrt{2\pi n}$) $\kappa_{0\pm} = \sqrt{|1 \pm \chi|}$. The situation can be readily visualized by inspecting Figures 2 and 4. Notice that both PM and FZ states have a renormalized momentum space occupation.

Consider now the situation at high densities. In this case as $\bar{\alpha}$ exceeds the value of Equation (4) the system appears to settle into a PM state in which only the lower chiral band is occupied and $\chi > 1$. Here, at first sight, it may appear safe to entertain the notion that by increasing the spin-orbit coupling at constant density (and therefore the strength of the non interacting part of the Hamiltonian) one would fall into the familiar paradigm in which the interacting part of the Hamiltonian becomes eventually irrelevant and therefore amenable to perturbative treatment. This is however not the case since, as we will presently show, for sufficiently large $\bar{\alpha}$ the effects of the Coulomb interaction are not perturbative. To demonstrate this effect we will construct a broken symmetry trial state and will show that its energy can be made lower than the corresponding interacting PM. In particular we will consider a state in which the momentum space occupation is repopulated in such a way as to break the circular symmetry in the k_x, k_y space as depicted in Figure 4. To be specific we will construct a Slater determinant with occupation determined by the following Ansatz for the momentum occupation geometry (see Figure 4) [26]

$$\kappa_\pm(\phi) = \kappa_{0\pm} \pm \eta \cos\phi \, , \tag{6}$$

with the azimuthal direction of the spin quantization axis $\hat{s}_{\mathbf{k}} = \hat{\phi}_{\mathbf{k}}$ kept unchanged [27]. Since the distortion is infinitesimal, to lowest order the total energy change will depend on the value of the generalized chirality only. The energy change associated with Equation 6 can be calculated in the limit of large χ. The kinetic plus spin-orbit energy change (in Rydbergs) is given by:

$$\delta\mathcal{E}_0 \simeq \left[\frac{3}{r_s^2} - \frac{\sqrt{2}\bar{\alpha}}{r_s}\left(\sqrt{\chi+1} - \sqrt{\chi-1}\right)\right]\eta^2, \tag{7}$$

while, in the same units, for the exchange energy we find

$$\delta\mathcal{E}_x \simeq -\frac{2\sqrt{2\chi}\ln\chi}{\pi r_s}\eta^2 . \tag{8}$$

Now, since for large $\bar{\alpha}$ we have $\chi \simeq \frac{(r_s\bar{\alpha})^2}{2}$ to leading order these competing energy contributions simplify to

$$\delta\mathcal{E}_0 \simeq \frac{1}{r_s^2}\eta^2 , \qquad \delta\mathcal{E}_x \simeq -\frac{4\bar{\alpha}\ln(\bar{\alpha}r_s)}{\pi}\eta^2 . \tag{9}$$

Thus for sufficiently large $\bar{\alpha}$ the differential instability is established.

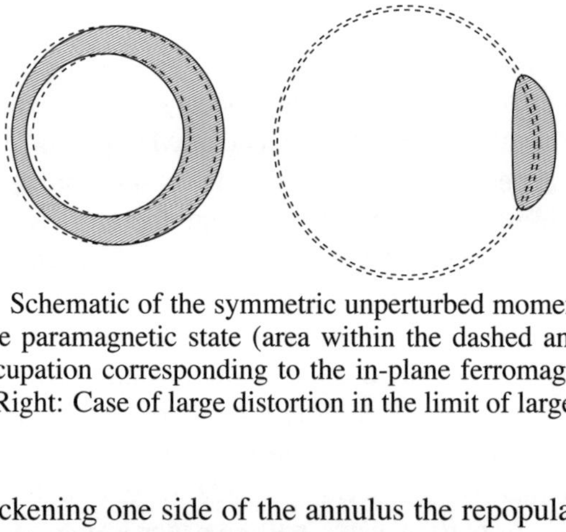

Figure 4. Left: Schematic of the symmetric unperturbed momentum space occupation for the paramagnetic state (area within the dashed annulus) and the asymmetric occupation corresponding to the in-plane ferromagnetic trial state (shaded area). Right: Case of large distortion in the limit of large $\bar{\alpha}$ value.

Clearly by thickening one side of the annulus the repopulation of Equation (6) leads to both an in plane momentum along the x-axis and a polarization along the y-axis. In the same regime of large values of χ and $\bar{\alpha}$ we find $P_x/N = \langle\sum_i \hat{p}_{x,i}\rangle/N \simeq \hbar k_F \chi\eta$ and $S_y/N = \langle\sum_i \hat{\sigma}_{y,i}\rangle/N \simeq \sqrt{\chi}\eta$. On the other hand the two quantities are balanced in such a way as to lead to a vanishing net velocity. In particular $V_x = P_x/mN - \alpha S_y/N = 0$. This clearly minimizes the energy.

Although the resolution of the instability is to be explored we have identified, by a consistency argument, the type of state that eventually takes over in the limit of very large $\bar{\alpha}$ or lower densities. This state corresponds to a fully polarized droplet in momentum space, as illustrated in the second panel of Figure 4. If one assumes that indeed the occupied region is centered about the wave vector $\mathbf{K} = \frac{m\alpha}{\hbar}\hat{x}$ (equal in magnitude to the radius of the occupied annulus, see Figure 4) then in the $r_s\bar{\alpha} \to \infty$ limit, using the fact that the spin quantization axes are asymptotically along $\hat{y}$, the functional (3) simplifies to

$$\mathcal{E} \simeq -\bar{\alpha}^2 + \frac{1}{\pi r_s^2}\int_{\mathcal{D}} k_x^2\, \mathrm{d}\boldsymbol{k} - \frac{\sqrt{2}}{(2\pi)^2 r_s}\int_{\mathcal{D}} \frac{\mathrm{d}\boldsymbol{k}\,\mathrm{d}\boldsymbol{k}'}{|\boldsymbol{k}-\boldsymbol{k}'|}\,, \tag{10}$$

where we have used Rydberg units for the energy and the wave vectors are in units of k_F. Here the integrals are performed over the occupied region $\mathcal{D}$ of extension 2π (which we have folded from $\mathbf{K}$ back to the origin). Equation (10) describes confined classical charges interacting via an hard core potential that forces them to occupy the domain $\mathcal{D}$ and an attractive Coulomb potential in the presence of an additional external parabolic potential along the x direction. The ensuing occupation consists of an oblate region, elongated in the y direction. In the limit of large $\bar{\alpha}$ the actual shape becomes independent of this variable and is solely determined by the density parameter r_s. Since Equation (10) is valid when the linear size (approximately k_F) of the occupied region is small with respect to the radius $m\alpha/\hbar$, it can also be applied in the large r_s limit at constant $\bar{\alpha}$. In this case the $1/r_s^2$ contribution can be neglected so that the consistent HF ground state corresponds to a fully polarized circular droplet of radius $\sqrt{2}k_F$ centered in $\mathbf{K}$. Also in this case the velocity vanishes. The energy of this state can be calculated exactly and it is given by:

$$\mathcal{E}^{(\text{trial})} = -\bar{\alpha}^2 + \frac{2}{r_s^2} - \frac{16}{3\pi r_s}\,. \tag{11}$$

This result can be compared with the energy of the corresponding PM state which is given by

$$\mathcal{E}^{(\text{PM})} \geq -\bar{\alpha}^2 - \frac{1.203}{r_s}\,, \tag{12}$$

where we used $-\bar{\alpha}^2$ as a lower bound for the kinetic and Rashba contributions, and the minimum unpolarized exchange energy [22], which occurs when the generalized chirality is $\chi \simeq 0.9147$. Clearly $\mathcal{E}^{(PM)} \geq \mathcal{E}^{(\text{trial})}$ for $r_s \geq 4.044$, thus establishing the instability of the PM phase also in the low density limit.

It is important to realize that the physical underpinning of this symmetry breaking phenomenon can be attributed to the fact that as $\bar{\alpha}$ is increased, the occupied region in momentum space becomes an annulus of radius $\frac{m\alpha}{\hbar}$. Since the electron number is constant the annulus keeps getting thinner and, what is important, the bandwidth, approximately given (in Rydberg units) by $\frac{1}{r_s^4\bar{\alpha}^2}$ vanishes. This situation is depicted in the right panel of Figure 2. It is quite clear that this phenomenon is quite robust so that, while the description of the corresponding phase transition obtained via mean field theory should be considered as a rough approximation, correlation effects can only enhance the instability.

Although for any density the instability will occur for sufficiently large spin-orbit coupling, on the other hand if $\bar{\alpha}$ is kept constant, the Fermi liquid picture is recovered in the limit of high densities. We conclude by commenting that having established an in-plane ferromagnetic instability does not establish the HF phase diagram of the system. This can only be determined through a thoughtful numerical analysis.

References

[1] G. F. Giuliani and S. Chesi, "Highlights in the Quan- tum Theory of Condensed Matter", F. Beltram (editor), Edizioni della Normale, Pisa, 2005, p. 269.

[2] R. E. Peierls, "Quantum Theory of Solids", Oxford University Press, Oxford, 1955.

[3] B. Braunecker, P. Simon, and D. Loss, Phys. Rev. B **80** (2009), 165119.

[4] B. Braunecker, G. I. Japaridze, J. Klinovaja and D. Loss, Phys. Rev. B **82** (2010), 045127.

[5] C. P. Scheller, *et al.*, Phys. Rev. Lett. **112** (2014), 066801.

[6] H. Min, G. Borghi, M. Polini, and A. H. MacDonald, Phys. Rev. B **77** (2008), 041407R.

[7] S. Chesi, Ph.D. thesis, Purdue University, 2007.

[8] J. Jung, M. Polini and A. H. MacDonald, arXiv:1111.1765 (2011).

[9] E. Berg, M. S. Rudner and S. A. Kivelson, Phys. Rev. B **85** (2012), 035116.

[10] T. A. Sedrakyan, A. Kamenev and L. I. Glazman, Phys. Rev. A **86** (2012), 063639.

[11] F. Bloch, Z. Physik **57** (1929), 545.

[12] J. Ruhman and E. Berg, Phys. Rev. B **90** (2014), 235119.

[13] Y. Bahri and A. C. Potter, arXiv:1408.6826 (2014).

[14] J. B. Miller, D. M. Zumbühl, C. M. Marcus, Y. B. Lyanda-Geller, D. Goldhaber-Gordon, K. Campman and A. C. Gossard, Phys. Rev. Lett. **90** (2003), 076807.

[15] L. P. Rokhinson, V. Larkina, Y. B. Lyanda-Geller, L. N. Pfeiffer and K. W. West, Phys. Rev. Lett. **93** (2004), 146601.
[16] L. Meier, G. Salis, I. Shorubalko, E. Gini, S. Schön and K. Ensslin, Nat. Phys. **3** (2007), 650.
[17] G. F. Giuliani and G. Vignale, "Quantum Theory of the Electron Liquid", Cambridge University Press, Cambridge, 2005.
[18] Some early work on the quasiparticle properties of this system can be found, in G.-H. Chen and M. E. Raikh, Phys. Rev. B **60** (1999), 4826; for a more recent analysis see D. S. Saraga and D. Loss, Phys. Rev. B **72** (2005), 195319.
[19] S. Chesi and G. F. Giuliani, to be published.
[20] R. Winkler, E. Tutuc, S. J. Papadakis, S. Melinte M. Shayegan, D. Wasserman and S. A. Lyon, Phys. Rev. B **72** (2005), 195321.
[21] S. Chesi and G. F. Giuliani, Phys. Rev. B **75** (2007), 155305; in this paper one will also find the correct form taken by the spin-orbit interaction in a number of experimentally relevant modern quantum well devices.
[22] S. Chesi, G. Simion and G. F. Giuliani, cond-mat/0702060 (2007).
[23] G. Simion and G. F. Giuliani, to be published.
[24] Y. A. Bychkov and E. I. Rashba, JETP Lett. **39** (1984), 78; J. Phys. C **17** (1984), 6039.
[25] An equivalent Hamiltonian is obtained if the spin-orbit coupling is of the Dresselhaus type. See G. Dresselhaus, Phys. Rev. **100** (1955), 580.
[26] Notice that had we assumed different amplitudes for the distortions of $\kappa_\pm$ the trial state would have a finite current. It is easy to show that, not unexpectedly, the kinetic and spin-orbit energy cost is indeed minimized for identical magnitudes, *i.e.* for zero current.
[27] Relaxing this assumption does not improve the quality of the trial state in the regime studied. It is clear that for intermediate values of the parameters it will prove necessary to allow the spin orientations to relax.

Josephson phase diffusion in small Josephson junctions: a strongly nonlinear regime

Mikhail V. Fistul

Abstract. I present a theoretical study of current-voltage characteristics (I-V curves) of small Josephson junctions. In the limit of a small Josephson coupling energy $E_J \ll k_B T$ the thermal fluctuations result in a stochastic dependence of the Josephson phase φ on time, *i.e. the Josephson phase diffusion*. These thermal fluctuations destroy the superconducting state, and the low-voltage resistive state is characterized by a nonlinear I-V curve. Such I-V curve is determined by the *resonant interaction* of ac Josephson current with the Josephson phase oscillations excited in the junction. The main frequency of ac Josephson current is $\omega = eV/\hbar$, where V is the voltage drop on the junction. In the phase diffusion regime the Josephson phase oscillations show a broad spectrum of frequencies. The *average I-V* curve is determined by the time-dependent correlations of the Josephson phase. By making use of the method of averaging elaborated in Ref. [1] for Josephson junctions with randomly distributed Abrikosov vortices I will be able to obtain two regimes: a linear regime as the amplitudes of excited phase oscillations are small, and a *strongly nonlinear regime* as both the amplitudes of excited Josephson phase oscillations and the strength of resonant interaction are large. The latter regime can be realized in the case of low dissipation. The crossover between these regimes is analyzed.

1. Introduction

A great attention is devoted to an experimental and theoretical study of small Josephson junctions [2]. In these systems one can observe such interesting physical phenomena as superconductor-insulator phase transition [3], Coulomb blockade of Cooper pairs [4,5], incoherent and coherent Josephson phase-slips [6–8], Josephson phase diffusion [9,10], just to name a few. The physical origin of all these phenomena is the presence of thermal and/or quantum fluctuations that greatly influence the dc and ac Josephson effect. In this paper we consider moderately small Josephson junctions as the charging energy E_c is smaller than the Josephson coupling energy, E_J. For such Josephson junctions one can safety neglect the quantum fluctuations of Josephson phase. However, as E_J is small, *i.e.* $E_J \ll k_B T$, the *Josephson phase diffusion regime* induced by thermal fluctuations, occurs. In the regime of a strong dissipation the Josephson phase diffusion regime has been studied in detail experimentally and theoretically [9–13]. Most pronounce features of the Josephson phase diffusion are the absence of the zero-voltage superconducting state, nonlinear current voltage characteristics (I-V curves) occurring in a low voltage region, and a strong suppression of the maximum current value.

In the presence of Josephson phase diffusion dc I-V curves can be qualitatively explained as follows. As the dc voltage V is applied the ac Josephson current with the main frequency $\omega = 2eV/\hbar$ is flowing in the junction. Such a Josephson current excites the *Josephson phase oscillations* which, in turn, resonate with the alternating part of the Josephson current leading to the finite dc current I. The thermal fluctuations result in a broad spectrum of Josephson phase oscillations and determine the strength of resonant interaction.

It is also well known for many years that in the Josephson phase diffusion regime the dc I-V curves depend crucially on the Josephson phase damping. Such a damping is determined mostly by various dissipative effects and, in particular, the quasi-particles resistance. In the limit of a large dissipation (damping) the amplitudes of excited Josephson phase oscillations are small, and therefore, using the perturbation analysis the dc I-V curve has been carried out quantitatively [2,10,12].

$$I = \frac{I_c}{\alpha} \frac{V V_p}{V^2 + (\delta V_p)^2}, \tag{1.1}$$

where we introduce the characteristic voltage $V_p = \hbar\omega_p/2e$, the plasma frequency ω_p, and the dimensionless parameter α describing the dissipative effects. In the Josephson phase diffusion regime the thermal fluctuations induce a stochastic part of the Josephson phase $\psi(t)$, and an *average* dc I-V curve is determined by the specific time-dependent correlation function of $\psi(t)$, *i.e.* $\rho(t) = <\cos(\psi(t) - \psi(0))>$. As the damping is large the $\rho(t)$ shows a diffusive form: $\rho(t) = \exp(-\delta t)$. The typical I-V curve of a small Josephson junction in the Josephson phase diffusion regime is presented in Figure 1.

Notice here, that a crucial condition allowing one to obtain Equation (1.1) is a large value of the damping parameter, $\alpha \gg 1$. Thus, a next question naturally arises: how vary the I-V curves in the limit of a small damping? In such a case the Josephson phase displays oscillations with a large amplitude, and the perturbation analysis can not be applied. Instead of the perturbation approach I will use the method of averaging elaborated in Refs. [1,14]. Although this method has been used, previously, in order to analyze the current resonances in long Josephson junctions with randomly distributed Abrikosov vortices, *i.e.* coordinate-dependent inhomogeneities, it is possible to adjust such a method to the Josephson junction with thermal fluctuations, *i.e.* time-dependent inhomogeneities.

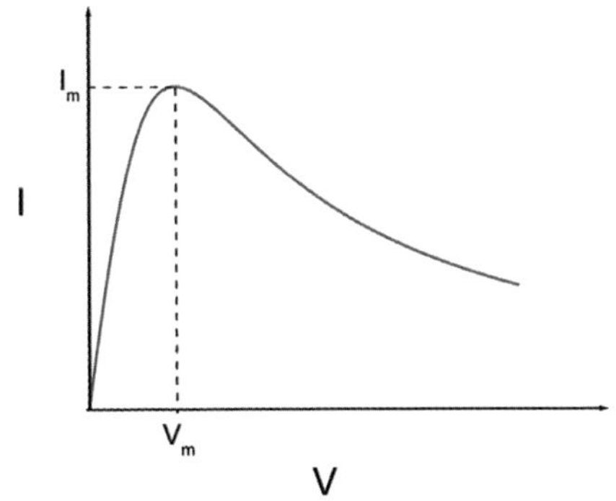

Figure 1. The typical I-V curve of a small Josephson junction in the Josephson phase regime. The voltage drop V_m and the current I_m corresponding to the maximum of I-V curve are shown.

The paper is organized as follows. In Section II the dynamics of the Josephson phase in the low voltage resistive state and in the presence of thermal fluctuations will be analyzed. In Section III we calculate the time-dependent correlation functions of the Josephson phase $\psi(t)$ determining the electrodynamic properties of small Josephson junction. In Section IV, by making use of the averaging method elaborated in Ref. [1] we obtain the dc I-V curves of small Josephson junctions in the Josephson phase diffusion regime. The Section V provides discussion and conclusions.

ACKNOWLEDGEMENTS. I acknowledge a partial financial support of the Ministry of Education and Science of the Russian Federation.

2. The dynamics of the Josephson phase in the resistive state: the Josephson phase diffusion regime

In order to quantitatively analyze the I-V curve of a small Josephson junction in the Josephson phase diffusion regime we write the dynamic equation for the Josephson phase $\varphi(t)$

$$\ddot{\varphi}(t) + \alpha\dot{\varphi}(t) + \sin\varphi(t) = j + \xi(t). \tag{2.1}$$

Here, j is the dc current, and $\xi(t)$ is a random function of time t describing thermal fluctuations (the Langevin force). The dimensionless units were used, *i.e.*, the time is normalized to ω_p^{-1}, the dc bias $j = I/I_c$ is normalized to the critical current value I_c. The solution of this equation corresponding to the resistive state is written as

$$\varphi(t) = vt + \psi(t) + \varphi_1(t) \ , \tag{2.2}$$

where the dc voltage drop V is normalized to V_p as $v = V/V_p$, and the random function $\psi(t)$ determines the Josephson phase diffusion. As the Josephson phase oscillations term $\varphi_1(t)$ is small, the perturbation approach can be used, and the Equation (1.1) is recovered. In a generic nonlinear case $\varphi_1(t)$ is written as

$$\varphi_1(t) = A(t)e^{ivt} + B(t)e^{-ivt}, \quad B(t) = A^*(t)\,. \tag{2.3}$$

Thus, $\varphi_1(t)$ shows rapid oscillations of frequency v and a smooth time-dependence describing by the function $A(t)$. Substituting (2.3) in (2.1) and carrying out the averaging over the rapid oscillations of frequency v we obtain

$$\begin{aligned} |A|^2 = \frac{1}{4}\int\int & dt_1 dt_2 G(t-t_1)G^*(t-t_2)\{J_0[|A|(t_1)]J_0[|A|(t_2)] \\ & + J_2[|A|(t_1)]J_2[|A|(t_2)]\}\cos[\psi(t_1)-\psi(t_2)], \end{aligned} \tag{2.4}$$

where $J_n(x)$ are the Bessel functions, and the kernel $G(x)$ is the Green function of the following homogeneous equation

$$\ddot{G}(t) + (2iv + \alpha)\dot{G}(t) - (v^2 - i\alpha v)G(t) = 0. \tag{2.5}$$

Similarly we calculate the dc current j flowing in the system

$$\begin{aligned} j &= \overline{\sin[A(t)e^{ivt} + B(t)e^{-ivt} + vt + \psi(t)]} \\ &= 2\int_0^\infty dt\, ImG(t)\frac{J_1[|A|(t)]J_0[|A|(t)]}{|A|(t)}\cos[\psi(t)-\psi(0)] \end{aligned} \tag{2.6}$$

Thus, one can see that all electrodynamic properties of small Josephson junctions in the phase diffusion regime are determined by the specific correlation function, *i.e.* $\rho(t) = <\cos(\psi(t) - \psi(0))>$.

3. Time-dependent correlation function of the Josephson phase

In order to obtain the time-dependent correlation function of the Josephson phase we write $\psi(t)$ as

$$\psi(t) = \int dx R(t-x)\xi(x), \tag{3.1}$$

where the kernel $R(t)$ is the Green function of the following homogeneous equation

$$\ddot{R}(t) + \alpha\dot{R}(t) = 0. \tag{3.2}$$

By making use of the method proposed and elaborated in [12, 15] we obtain the correlation function $\rho(t)$ in the following form

$$\rho(t) = \exp\left\{-\int \frac{d\tau}{\tau_0} \int d\xi F(\xi) \left\{1 - e^{i\tau_0\xi[R(t-\tau)-R(-\tau)]}\right\}\right\} , \quad (3.3)$$

where $F(\xi)$ and τ_0 are the distribution function and the correlation time of the current noise, accordingly. Since the $R(t)$ is presented as $R(t) = \frac{(1-e^{-\alpha t})}{\alpha}\theta(t)$ we obtain in the limit of large dissipation

$$\rho(t) = \exp(-\delta|t|) , \delta \ll \alpha , \quad (3.4)$$

where the parameter $\delta = \frac{\tau_0}{2\alpha^2}\int d\xi\xi^2 F(\xi)$ determines the decay of the Josephson phases correlation function. In the opposite regime of $\delta \gg \alpha$ we obtain

$$\rho(t) = \exp(-\delta\alpha^2|t|^3/3) , \delta \gg \alpha . \quad (3.5)$$

4. The I-V curves of small Josephson junctions: phase-diffusion regime

First, we notice that the function $|A(t)|$ smoothly depends on time in respect to both the kernel $G(t)$ and the correlation function $\rho(t)$. Moreover, the kernel $G(t)$ has a simple form: $G(t) = \frac{1}{\alpha}e^{ivt}$. By making use of this assumption and taking into account an explicit expression for $G(t)$ we rewrite the Equations (2.4) and (2.6) as

$$|A|^2 = \frac{J_0^2[|A|] + J_2^2[|A|]}{4\alpha^2} \int_{-\infty}^{\infty} dt e^{ivt} \cos[\psi(t) - \psi(0)] \quad (4.1)$$

and

$$j = \frac{2J_1[|A|]J_0[|A|]}{\alpha|A|} \int_0^{\infty} dt \sin(vt) \cos[\psi(t) - \psi(0)] \quad (4.2)$$

Since we are interested in averaged quantity only, the dc I-V curve can be expressed through the voltage dependent correlation time $\tau(v) = \tau_1(v) + i\tau_2(v)$, where

$$\tau_1(v) = \left\langle \int_0^{\infty} dt \cos(vt) \cos[\psi(t) - \psi(0)] \right\rangle \quad (4.3)$$

and

$$\tau_2(v) = \left\langle \int_0^{\infty} dt \sin(vt) \cos[\psi(t) - \psi(0)] \right\rangle \quad (4.4)$$

By making use of Equations (3.4) and (3.5) we obtain that τ_1 approaches to the finite value for small values of voltage v as

$$\tau_1(0) = \begin{cases} \frac{1}{\delta} & \text{if } \delta \ll \alpha \\ \Gamma(1/3)(9\delta\alpha^2)^{-1/3} & \text{if } \delta \gg \alpha \end{cases} \tag{4.5}$$

In the opposite limit of large values of v the τ_1 decreases as $1/v^2$ for overdamped junctions ($\delta \ll \alpha$) and it becomes exponentially small for underdamped junctions ($\delta \gg \alpha$). The correlation time τ_2 linearly increases for small values of v, and decreases as $1/|v|$ for large values of voltage v.

Next we analyze the Equations (4.1) and (4.2) determining the current-voltage characteristics of a small Josephson junction. In the limit of a small value of $\tau_1(v)$ or more precisely $\tau_1 \ll \alpha^2$ the amplitude of Josephson phase oscillations $|A|$ is small, and expanding the Bessel functions over a small argument A we obtain

$$j(v) = \frac{\tau_2(v)}{\alpha}, \quad \tau_1(v) \ll \alpha^2. \tag{4.6}$$

In this regime the current I increases linearly in the region of small voltages, and in the limit $\delta \ll \alpha$ we recover Equation (1.1). In the opposite regime, $\tau_1(v) \gg \alpha^2$, the amplitude of Josephson phase oscillations becomes large but the current I is still strongly suppressed by oscillations of Bessel functions. In this strongly nonlinear regime the averaged value of I is expressed through the parameters τ_1 and τ_2 as

$$j(v) = \tau_2(v)\left(\frac{\alpha}{\tau_1^2}\right)^{1/3} \exp\left[-\frac{1}{4}\left(\frac{\tau_1}{4\alpha^2}\right)^{2/3}\right], \quad \tau_1 \gg \alpha^2. \tag{4.7}$$

Thus, one can see that in the low dissipative junctions ($\alpha << 1$) the linear resistance is strongly (exponentially) suppressed.

5. Discussion and Conclusions

A theoretical study of the low-voltage resistive state of small Josephson junctions has been developed. In such junctions as $E_J \ll k_B T$ the thermal current fluctuations induce a stochastic time dependence of the Josephson phase. These fluctuations of the Josephson phase destroy the superconducting state and a specific resistive state occurs. The I-V curve in such so-called Josephson phase diffusion regime crucially depends on the dimensionless dissipation parameter α. As this parameter is large, *i.e.* $\alpha \gg 1$, the Josephson phase dynamics is strongly overdamped and the amplitudes of self-excited Josephson phase oscillations are small, therefore one can apply perturbation analysis, and the I-V curve is described by Equation (1.1) or Equation (4.6). Moreover, in the case of a large dissipation $V_m = \delta V_p$ and $I_m = I_c/(\alpha\delta)$ (see Figure 1). Notice here, that introduced parameters δ and α can be expressed through the physical characteristics of a junction as $\delta = (2\pi k_B T)R_n/(V_p \Phi_0)$ and $\alpha = V_p/(I_c R_n)$, where R_n is the quasi-particle resistance, I_c is the nominal critical current, and Φ_0 is the magnetic flux quantum [10].

In an opposite regime of a strongly underdamped junction, *i.e.* $\alpha \ll 1$, the amplitudes of self-excited Josephson oscillations are large, and the perturbation analysis can not be applied. Instead of that I used *the method of averaging* elaborated previously in Ref. [1,14]. In this a strongly nonlinear regime the I-V curve is described by Equations (4.7) and (4.5). In this underdamped regime the both values $V_m = V_p\delta(\alpha/\delta)^{2/3}$ and $I_m \simeq I_c \exp[-4^{-5/3}(\delta\alpha^8)^{-2/9})]$ are strongly suppressed in respect to the overdamped case.

Finally, we notice that the crossover between these two regimes is determined by the ratio of the parameters δ and α. Since this ratio depends on the critical current value I_c, such a crossover can be observed experimentally in a single setup just by application and variation of an external magnetic field.

References

[1] M. V. Fistul and G. F. Giuliani, Phys. Rev. B **56** (1997), 788.

[2] M. Tinkham, "Introduction to Superconductivity", McGraw-Hill, New York, 1996, 2nd ed.

[3] E. Chow, P. Delsing and D. B. Haviland, Phys. Rev. Lett. **81** (1998), 204.

[4] D. V. Averin, A. B. Zorin and K. K. Likharev, Sov. Phys. JETP **61** (1985), 407; D. V. Averin and K. K. Likharev, J. Low Temp. Phys. **62** (1986), 345.

[5] K. A. Matveev, M. Gisselfält, L. I. Glazman, M. Jonson and R. I. Shekhter, Phys. Rev. Lett. **70** (1993), 2940.

[6] V. Ambegaokar and B. I. Halperin, Phys. Rev. Lett. **22** (1969), 1364.

[7] J. E. Mooij and C. J. P. M. Harmans, N. J. Phys. **7** (2005), 219.

[8] O. V. Astafiev, L. B. Ioffe, S. Kafanov, Yu. A. Pashkin, K. Yu. Arutyunov, D. Shahar, O. Cohen and J. S. Tsai, Nature **484** (2012), 355.

[9] J. M. MARTINIS and R. L. KAUTZ, Phys. Rev. Lett. **63** (1989), 1507; R. L. KAUTZ and J. M. MARTINIS, Phys. Rev. B **42** (1990), 9903.
[10] Y. KOVAL, M. V. FISTUL and A. V. USTINOV, Phys. Rev. Lett. **93** (2004), 087004.
[11] D. VION, M. GÖTZ, P. JOYEZ, D. ESTEVE and M. H. DEVORET, Phys. Rev. Lett. **77** (1996), 3435.
[12] G.-L. INGOLD and YU. V. NAZAROV, in "Single Charge Tunneling", H. Grabert and M. H. Devoret (eds.), NATO ASI, Ser. B, Vol. 294, (Plenum, New York, 1991.
[13] H. GRABERT, G.-L. INGOLD and B. PAUL, Europhys. Lett. **44** (1998), 360.
[14] M. P. LISITSKIY and M. V. FISTUL, Phys. Rev. B **81** (2010), 184505.
[15] M. V. FISTUL, Sov. Phys. JETP **69** (1989), 209.

Spin-dependent magnetic focusing

Yuli Lyanda-Geller, L. P. Rokhinson and Stefano Chesi

Abstract. Gabriele Giuliani was fascinated by spin-dependent phenomena. Here we review experiments on spin separation in cyclotron motion, semiclassical theory of effects of spin-orbit interactions on cyclotron resonance, and theory of spin filtering by a quantum point contact in two-dimensional hole systems.

1. Introduction

It is a great honor to contribute to Gabriele Giuliani's memorial volume, and it has been a remarkable experience to work with Gabriele. In the past 15 years, Condensed Matter physicists became greatly interested in spin-dependent phenomena, creating a direction of research named 'spintronics' [1]. Gabriele was genuinely interested in this trend, involved his students in research in this field, and was instrumental in attracting several faculty members with interest in spin-dependent phenomena to the Purdue University Physics Department.

Among several interesting new phenomena discovered over the last decade, there is spin-dependent magnetic focusing [2–4]. Classical electron focusing was first observed in metals [5,6]. Coherent electron focusing is most remarkably pronounced in semiconductor nanostructures, where it became a signature phenomenon for quantum ballistic transport [7]. When two quantum point contacts in a two-dimensional electron gas are separated by multiples of the cyclotron diameter, injection from one point contact results in an additional potential developed across the detector point contact. It has been long appreciated that signature quantum effects, such as the Aharonov-Bohm effect, have remarkable spin counterparts due to spin-orbit interactions [8,9]. In [2], it has been discovered that the effect of magnetic focusing can be used as spin filter.

This work was supported by the U.S. Department of Energy, Office of Basic Energy Sciences, Division of Materials Sciences and Engineering under Award DE-SC0010544 (Y.L-G), and by the National Science Foundation under Grant No. 1307247-DMR (L.P.R.).

The origin of such filtering can be traced to spin-orbit interactions introducing a dependence of the cyclotron radius on the spin of the charge carriers.

Gabriele Giuliani recognized that several features observed in magnetic focusing experiments in two-dimensional hole gases are unaccounted for in a semiclassical theory of magnetic focusing. More specifically, Gabriele was intrigued by the reappearance of a filtered spin component at high in-plane magnetic fields. That led to the paper where a theory of spin-dependent transmission through quantum point contacts in the two-dimensional hole gas (2DHG) has been developed [10]. In the present paper, which is our tribute to Gabriele, we review the results of experiments on spin-dependent focusing, discuss the semiclassical theory of spin-dependent focusing, and the spin filtering by quantum point contacts in the presence of spin-orbit interactions.

2. Experiment

To demonstrate spatial separation of spins experimentally we fabricated several 2DHG devices in the magnetic focusing geometry, see the inset in Figure 1. The structure is formed using atomic force microscopy local anodic oxidation technique (AFM LAO) [11–13]. Oxide lines separate the 2DHG underneath by forming $\sim$ 200 mV potential barriers. A specially designed heterostructure is grown by MBE on [113]A GaAs. Despite very close proximity to the surface (350Å), the 2DHG has an exceptionally high mobility $0.4 \cdot 10^6$ V·s/cm^2 and relatively low hole density $n = 1.38 \cdot 10^{11}$ cm^{-2}. The device consists of two QPCs oriented along the [33$\bar{2}$] crystallographic direction, separated by a central gate; the lithographically defined distance between QPCs is $L = 0.8$ μm. Potential in the point contacts is controlled separately by two gates G_{inj} and G_{det}, or by the central gate G_C. In our experiments the central gate was kept at -0.3 V and $\sim$ 0.2 V were applied to the gates G_{inj} and G_{det}. Asymmetric biasing of point contacts provides sharper confining potential and reduces the distance between the two potential minima by $\Delta L \sim 0.07$ μm.

Magnetic focusing manifests itself as equidistant peaks in the magnetoresistance $R(B_\perp)$ for only one direction of $B_\perp$. R is measured by applying a small current through the injector QPC while monitoring voltage across the detector QPC. At $B_\perp < 0$, cyclotron motion forces the carriers away from the detector. Then, only the 2DHG contributes to R, which has almost no $B_\perp$-dependence at low fields and shows Shubnikov–de Haas oscillations at $|B_\perp| > 0.3$ T. For $B_\perp > 0$, several peaks due to magnetic focusing are observed. The peak separation $\Delta B \approx 0.18$ T is consistent with the distance between the injector and detector QPCs. The data is

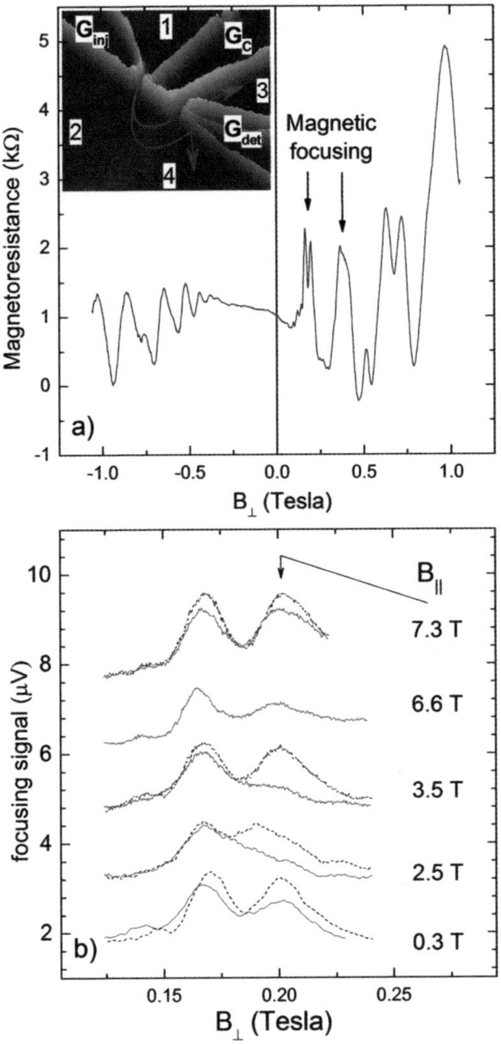

Figure 1. a) Magnetoresistance and layout of focusing devices. The voltage across the detector (contacts 3 and 4) is measured as a function of magnetic field perpendicular to the surface of the sample ($B_\perp$). The lithographical separation between point contacts is 0.8 μm. A current of 1 nA is flowing through the injector (contacts 1 and 2). The positions of the magnetic focusing peaks are marked with arrows. Inset: AFM micrograph of a sample (5μm$\times 5\mu$m). Light lines are the oxide which separates different regions of the 2D hole gas. The semicircles show schematically the trajectories for two spin orientations. b) Focusing signal for the first focusing peak in a tilted magnetic field, plotted versus $B_\perp$. The values of the corresponding $B_\parallel$, for $B_\perp = 0.2$ T, are marked on the right. Curves are offset for clarity. The dashed black (solid red) curves correspond to $G_{inj} = 2e^2/h$ ($< e^2/h$).

symmetric upon exchange of the injector and detector and simultaneous reversal of the magnetic field direction.

When the conductance of both QPCs is tuned to be $2e^2/h$, the first focusing peak splits into two peaks. When the in-plane component is $B_{\|} = 0$ the peaks in the doublet have approximately the same height. If the conductance of the injector QPC is $G_{inj} < 2e^2/h$ the rightmost peak is slightly suppressed, which has been interpreted as due to spontaneous polarization [14].

We use the spin filtering by QPCs in the presence of in-plane magnetic field $B_{\|}$ to probe the spin states which correspond to the first focusing peak doublet. Applying $B_{\|}$ along $[33\bar{2}]$ affects the energies of the spin subbands without affecting the cyclotron motion. As the Zeeman splitting of the spin subbands in a 2D gas increases, preferential transmission of the largest-k_F spin subband is expected for electrons, corresponding to suppression of the *left* peak. Instead, in a hole gas we observe suppression of the *right* peak up to $B_{\|} \approx 2.5$ T, see Figure 1 b. For $B_{\|} > 2.5$ T the right peak reappears and at $B_{\|} = 7.3$ T becomes as prominent as the left one.

3. Theory

There has been a considerable interest to understand hole spectra in low dimensional systems over the past decade, also in connection with research in the field of quantum computing. Of special interest are heterostructures grown along the [001] direction, in which the hole spectra are remarkably different from electron spectra. In this case, several authors concluded that intrinsic Dresselhaus and Rashba spin-orbit interactions are cubic in the wavevector [17], and that the in-plane g-factor describing the Zeeman splitting of holes with an in-plane magnetic field is quadratic in electron momentum and depends on its orientation. As it turns out, however, earlier work [19] pointed out that for this crystallographic orientation of the 2DHG, the Dresselhaus term gives rise to contributions linear in momentum. Furthermore, approaches based on low-order perturbation theory are generally oversimplified because, as was discussed in [20], do not take properly into account the non-perturbative effect of a mutual transformation of heavy and light holes upon reflection from the walls of the quantum well [21–23]. This effect results in the presence of two standing hole waves in the wavefunctions of hole states, corresponding to heavy and light holes moving along the growth direction, as opposed to electron case with only one standing wave. Taking mutual transformation of heavy and light holes into account considerably alters the in-plane effective mass of holes, and the coupling constants of the intrinsic spin-orbit interaction as well. Numerical simulations taking into account a finite number of levels of spatial quantization in the [001] growth direction might also lead to inaccurate results, if proper care is not taken upon truncation of the Hilbert space, because all levels of size quantization result in contributions to the in-plane effective mass and cubic spin-orbit splitting characterized by the same physical scale (*i.e.*, all contributions have in principle the same order of magnitude).

Our knowledge of properties of holes in quantum wells grown along the [113] crystallographic orientation is even less extensive. Existing analytical results [15, 16] were obtained in the so-called axial approximation [17], which may take into account effects of the mutual transformation of heavy and light holes upon reflection from the walls of the quantum well only partially. Numerical work was performed which should give more accurate results [10]. Although a progressively larger number of spatial quantization levels were included, until the numerical spectra did not change significantly, a more careful analysis of truncation errors seems necessary in the light of the non-perturbative nature of the effects described above [20–23]. Nevertheless, conclusions about certain properties can be drawn on symmetry grounds from the properties of [001]-grown structures. In particular, quantization along the [113] crystallographic direction mixes in-plane and out-of-plane properties of the [001] structures, which results in a contribution to the in-plane g-factor independent of wavevector. This contribution is non-zero only because the bulk spectrum of holes is anisotropic. Both cubic- and linear-in-momentum Rashba and Dresselhaus spin-orbit interactions are present in the [113] configuration, and the linear in momentum Rashba spin-orbit term is related to the anisotropy of the bulk hole spectra. Although the precise magnitude and angular dependence of these interactions is not known, we will describe how simple models explain the experimental data on focusing in [113]-oriented hole quantum wells.

4. Semiclassical theory of focusing

It has long been appreciated that intrinsic spin-orbit (SO) interactions can be interpreted as an effective momentum-dependent magnetic field that influences the spin of charge carriers [24]. More recently, it has been recognized [8,9,25–27] that SO interactions can be also viewed as an effective orbital magnetic field with an opposite sign for different spin orientations. In order to explain the effect of spin filtering in magnetic focusing qualitatively, it is reasonable to assume that charge carriers in GaAs quantum well are characterized by an isotropic kinetic energy and the Dresselhaus intrinsic spin-orbit interaction linear in the hole momentum.

Indeed, for the lowest hole states in a 2DHG in both [113] and [001] configurations, a linear Dresselhaus term is present. The simplified hole Hamiltonian can be written as $H = \frac{1}{2m}(p_x + \beta\sigma_x)^2 + \frac{1}{2m}(p_y - \beta\sigma_y)^2$, where m is the effective mass, $\vec{p}$ is the electron momentum, σ_i are the Pauli matrices ($i = x, y$), and β is the SO parameter. For simplicity it is also reasonable to neglect anisotropy of the effective mass as this assumption does not change the qualitative picture. In the semiclassical description, appropriate for the range of magnetic fields $B_\perp$ used for the focusing experiments, the motion is described by simple equations:

$$\frac{d\vec{p}}{dt} = e\vec{v} \times \vec{B} \qquad \vec{v} = \frac{d\vec{r}}{dt} = \frac{\partial \epsilon_\pm(\vec{p})}{\partial \vec{p}}$$
$$\epsilon_\pm = \frac{1}{2m}(p \pm \beta)^2 + \frac{\beta^2}{2m}, \tag{4.1}$$

where $\vec{r}$, $\vec{v}$ and $\epsilon_\pm$ are the charge carrier coordinate, velocity and energy for the two spin projections. This description implies that the carrier wavelength is smaller than the cyclotron radius, and that jumps between orbits with different spin projections are absent, *i.e.*, $\epsilon_f \gg \beta p/m \gg \hbar\omega_c$. Equation (4.1) show that the charge carrier with energy $\epsilon_\pm = \epsilon_f$ is characterized by a spin-dependent trajectory with momentum $\vec{p}_\pm$, coordinate $\vec{r}_\pm$, and cyclotron frequency $\omega_c^\pm$. The solution to these equations is

$$p_\pm^{(x)} + ip_\pm^{(y)} = p_\pm \exp(-i\omega_c^\pm t)$$
$$r_\pm^{(x)} + ir_\pm^{(y)} = \frac{i\sqrt{2m\epsilon_f}}{m\omega_c^\pm} \exp(-i\omega_c^\pm t)$$
$$\omega_c^\pm = \frac{eB_\perp}{m}(1 \pm \beta/p_\pm). \tag{4.2}$$

Thus, the cyclotron motion is characterized by a spin-dependent field $B_\pm = B_\perp(1 \pm \beta/p_\pm) = B_\perp \pm B_{so}$, where B_{so} is the SO effective field characterizing the cyclotron motion. Using a semiclassical limit of the quantum description [28], one obtains identical results.

In the focusing configuration, QPCs are used as monochromatic point sources. Holes, injected in the direction perpendicular to the 2DHG boundary, can reach the detector directly or after specular reflections from the boundary. As follows from Eqs. (4.2), for each of the two spin projections there is a characteristic magnetic field such that the point contact separation is twice the cyclotron radius for a given spin, $L = 2R_\pm^c = 2p_f/eB_\pm$, $p_f = \sqrt{2m\epsilon_f}$. The first focusing peak occurs at

$$B_\perp^\pm = \frac{2(p_f \mp \beta)}{eL}. \tag{4.3}$$

The magnitude of β can be calculated directly from the peak splitting $\beta = (B_\perp^+ - B_\perp^-)eL/4 = 7 \cdot 10^{-9}$ eV·s/m. A larger value of $\beta \approx 25 \cdot 10^{-9}$ eV·s/m was extracted from the splitting of the cyclotron resonance at 3 times higher hole concentration [29]. We note that Equation (4.3) is more general than the Eqs. (4.2). The coefficient β essentially describes the separation in momentum space of the two parts of the Fermi surface which correspond to $\epsilon^\pm = \epsilon_f$, and includes contributions of various spin-orbit terms in the 2DHG.

The difference of the spin-dependent focusing field $B_\perp^\pm$ is proportional to β and does not depend on the cyclotron frequency $\omega_c = eB_\perp/m$. At the same time, the difference of spin-dependent cyclotron frequencies in Equation (4.2) is proportional to both ω_c and β. Thus, the effective magnetic field B_{so} is itself proportional to $B_\perp$. This effect differs from the spin-dependent shift of the Aharonov-Bohm oscillations in the conductance of rings, where the additional spin-orbit flux and the Aharonov-Bohm flux are independent of each other [9]. If the Zeeman effect is taken into account, both $\omega_c^\pm$ and $R_c^\pm$ acquire an additional dependence on $B_\perp$, as well as on the in-plane component $B_{||}$.

5. Focusing peaks in in-plane magnetic field

The behavior of the focusing peaks in Figure 1b requires to consider simultaneously the charge carriers motion in the 2DHG and their transmission through quantum point contacts. The observed results cannot be explained by considering only an intrinsic spin-orbit coupling of the 2D hole system linear-in-momentum. Furthermore, both 2D Dresselhaus and Rashba SO terms which are cubic-in-momentum necessarily generate additional linear-in-momentum contributions within the quantum point contact, similar to the generation of both cubic and linear terms in the 2D electron Hamiltonian from the bulk cubic Dresselhaus terms. To illustrate the physics of filtering by point contacts we consider, for example, an Hamiltonian with the cubic Rashba term of the form $\frac{i\gamma}{2}(\hat{p}_-^3\hat{\sigma}_+ - \hat{p}_+^3\hat{\sigma}_-)$. Here, $\hat{p}_\pm = \hat{p}_x \pm i\hat{p}_y$ and $\hat{\sigma}_\pm = \hat{\sigma}_x \pm i\hat{\sigma}_y$. Such cubic spin-orbit interaction is responsible for a peculiar dispersion of the lowest two one-dimensional (1D) subbands. For a channel with lateral extent W, aligned with the x-axis, we can substitute $\langle p_y^2\rangle \sim (\hbar\pi/W)^2$ and $\langle p_y\rangle \sim 0$ in the 2D Hamiltonian, which gives

$$\hat{H}_{1D} = \frac{\hat{p_x}^2}{2m} + \gamma\left(\frac{3\hbar^2\pi^2}{W^2}\hat{p}_x - \hat{p}_x^3\right)\hat{\sigma}_y + \frac{\hbar^2\pi^2}{2mW^2}. \tag{5.1}$$

Due to the lateral confinement, a linear spin-orbit term appears in Equation (5.1), which is dominant at small momenta and coexist with a cubic contribution with opposite sign. Therefore, spin subbands in such a case cross not only at $k_x = 0$, but also at finite wave vectors $k_x = \pm\frac{\sqrt{3}\pi}{W}$. In [10], the spin splitting due to Rashba term in quantum point contacts was computed numerically taking into account up to 10 levels of size quantization in the quantum well for various applied electric fields, as shown in Figure 2. The 1D bands clearly display the main feature: the presence of a crossing point at finite wave vector.

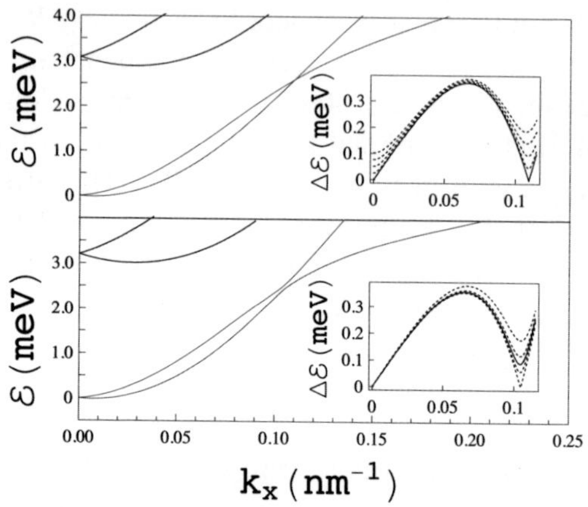

Figure 2. Energy subbands of 1D channels obtained from a 15 nm quantum well grown in the [113] direction. An electric field $\mathcal{E}_z = 1$ V/μm along [113] is present. The lateral confinement has width $W = 40$ nm. Upper panel: wire along [33$\bar{2}$]. The inset shows the energy splitting of the two lowest subbands at several values of $B_\parallel$. The solid curve is for $B_\parallel = 0$ and the dashed curves for $B_\parallel = 0.5, 1, \ldots 2.5$ T. Lower panel: wire along [1$\bar{1}$0]. The inset shows the energy splitting with a lateral electric field. The solid curve is for $\mathcal{E}_y = 0$ and the dashed curves for $\mathcal{E}_y = \pm 0.05, \pm 0.015$ V/μm (the splitting is reduced for negative values of $\mathcal{E}_y$).

As illustrated by the inset of Figure 2 (first panel), the degeneracies at $k_x = 0$ and finite k_x are removed when $B_\parallel \neq 0$. Within the effective Hamiltonian (5.1), an external magnetic field is taken into account by adding a Zeeman term $g^* \mu_B B_\parallel \hat{\sigma}_x/2$, where g^* is the effective g-factor [30] and μ_B the Bohr magneton. The total effective magnetic field, which includes spin-orbit interactions, depends on values of W and k_x as follows

$$\vec{B}_{eff}(W, k_x) = B_\parallel \hat{x} + \frac{2\gamma\hbar^3}{g^*\mu_B}\left(\frac{3\pi^2}{W^2}k_x - k_x^3\right)\hat{y}, \qquad (5.2)$$

where $\hat{x}$, $\hat{y}$ are unit vectors along the coordinate axes. The eigenstates of Equation (5.1), $\psi_W(k_x, \pm) = e^{ik_x x}|k_x, \pm\rangle_W$, have spinor functions

$|k_x, \pm\rangle_W$ parallel/antiparallel to $\vec{B}_{eff}$ and energies

$$\epsilon_\pm(W, k_x) = \frac{\hbar^2 k_x^2}{2m} \mp \frac{1}{2} g^* \mu_B |\vec{B}_{eff}(W, k_x)| \,. \tag{5.3}$$

At $k_x = 0$ and $k_x = \pm\sqrt{3}\pi/W$ the spin splitting is $g^*\mu_B B_\parallel$, *i.e.*, it is only due to the external magnetic field.

In a realistic QPC the width $W(x)$ of the lateral confinement changes along the channel. As in [31], a sufficiently smooth variation of the width is assumed, such that holes adiabatically follow the lowest *orbital* subband. Introducing in Equation (5.1) a x-dependent width $W(x) = W_0 e^{x^2/2\Delta x^2}$, where Δx is a typical length scale of the QPC and W_0 its minimum width, one obtains the following effective Hamiltonian

$$\begin{aligned} \hat{H}_{QPC} = \; & \frac{\hat{p}_x{}^2}{2m} + V(\hat{x}) + \frac{g^*\mu_B}{2} B_\parallel \hat{\sigma}_x \\ & + \gamma \left[3m\{V(\hat{x}), \hat{p}_x\} - \hat{p}_x^3 \right] \hat{\sigma}_y, \end{aligned} \tag{5.4}$$

with $\{a, b\} = ab + ba$ [32]. The potential barrier has the following form:

$$V(x) = \frac{\hbar^2\pi^2}{2mW(x)^2} = \frac{\hbar^2\pi^2}{2mW_0^2} e^{-x^2/\Delta x^2}. \tag{5.5}$$

The main qualitative conclusions are independent of the detailed form of the potential, but Equation (5.5) allows to solve explicitly the 1D transmission problem and obtain a spin-resolved conductance in the Landauer-Büttiker formalism. The scattering eigenstates are obtained with incident wavefunctions $\psi_{W=\infty}(k_\mu, \mu)$ at $x \ll -\Delta x$, where $\mu = \pm$ denotes the spin subband and $k_\pm$ are determined by the Fermi energy ϵ_f, at which the holes are injected in the QPC. For $x \gg \Delta x$, such QPC wavefunctions have the asymptotic form $\sum_{\nu=\pm} t_{\mu,\nu}\psi_\infty(k_\nu, \nu)$, where $t_{\mu,\nu}$ are transmission amplitudes. The spin-resolved conductances are simply given by $G_\pm = \frac{e^2}{h}\sum_{\mu=\pm}\frac{v_\pm}{v_\mu}|t_{\mu,\pm}|^2$ [33], where the Fermi velocities are $v_\pm = \frac{\partial\epsilon_\pm(\infty,k_\pm)}{\partial\hbar k_x}$, from Equation (5.3). The total conductance is $G = G_+ + G_-$. Typical results at several values of $B_\parallel$ are shown in Figure 3. As usual, by opening the QPC, a current starts to flow above a minimum value of W_0 and, with a finite magnetic field, $G_+ \neq G_-$. At zero magnetic field, there is structureless unpolarized conductance ($G_+ = G_-$). At larger magnetic fields, $G_- > G_+$, *i.e.*, holes in the *higher* spin subband have larger transmission at the first plateau. The sign is opposite to the case of linear Rashba spin-orbit coupling (see [34]) and in agreement with the experimental results of Figure 1. For a magnetic field $B_\parallel \approx 7$ T (see the third panel of Figure 3) $G_+ \simeq G_-$ and the transmission becomes unpolarized, as observed in the data of Figure 1. Finally, at even larger values of $B_\parallel > 7$ T, $G_+ \simeq e^2/h$, $G_- \simeq 0$ (fourth panel of Figure 3). For such sufficiently large magnetic field the role of the spin-orbit coupling becomes negligible and the spin direction (parallel/antiparallel to the external magnetic field) of the holes is conserved. The injected holes remain in the original (+ or −) branch and the current at the first plateau is polarized in the + band, which has lower energy. Deviations from this behavior are due to non-adiabatic transmission in the spin subband. In order to gain a qualitative understanding, we consider the semiclassical picture of the hole motion in quantum point contact.

When a hole wave-packet is at position x, it is subject to a magnetic field $\vec{B}_{eff}$ determined by $W(x)$ and $k_x(x)$ as in Equation (5.2). For holes injected at ϵ_f, the momentum is determined by energy conservation. Treating the spin-orbit coupling as a small perturbation compared to the kinetic energy, one has $k_x(x) \simeq \sqrt{k_f^2 - \pi^2/W(x)^2}$, where $k_f = \sqrt{2m\epsilon_f}/\hbar$ is the Fermi wave-vector in the absence of spin-orbit coupling. Therefore, the injected hole experiences a varying magnetic field in its semiclassical motion along x, due to the change of both k_x and $W(x)$. For adiabatic transmission of the spin subbands, the spin follows the direction of the magnetic field, but this is not possible in general if $B_\parallel$ is sufficiently small. In particular, for $B_\parallel = 0$ Equation (5.1) implies that $\hat{\sigma}_y$ is conserved. Therefore, the initial spin orientation along y is not affected by the motion of the hole. On the other hand, $\vec{B}_{eff}$ of Equation (5.2) changes direction when $k_x = \sqrt{3}\pi/W$. After this point, a hole in the + branch continues its motion in the − branch and vice-versa.

At finite in-plane magnetic field the degeneracy of the spectrum is removed but the holes do not follow adiabatically the spin branch, unless the Landau-Zener condition $\frac{dB_y/dt}{B_\parallel} \ll \omega_B$ is satisfied, where $\hbar\omega_B = g^*\mu_B B_\parallel$. The change ΔB_y in the spin-orbit field is obtained from Equation (5.2): $|B_y|$ is equal to $2\gamma\hbar^3 k_f^3/g^*\mu_B$ far from the QPC and vanishes at the degeneracy point. This change occurs on the length scale Δx of the QPC, and the estimate of the time interval is $\Delta t \simeq \Delta x/v$, where v is a typical velocity of the hole. This gives

$$B_\parallel \gg \sqrt{\frac{\hbar\Delta B_y}{g^*\mu_B\Delta t}} \simeq \frac{\hbar^2\sqrt{2\gamma k_f^3 v/\Delta x}}{g^*\mu_B}. \tag{5.6}$$

The estimate of v at the degeneracy point $k_x = \sqrt{3}\pi/W$ is obtained from $\sqrt{3}\pi/W \simeq \sqrt{k_f^2 - \pi^2/W^2}$, which gives $k_x = \frac{\sqrt{3}}{2}k_f$. Therefore, v is

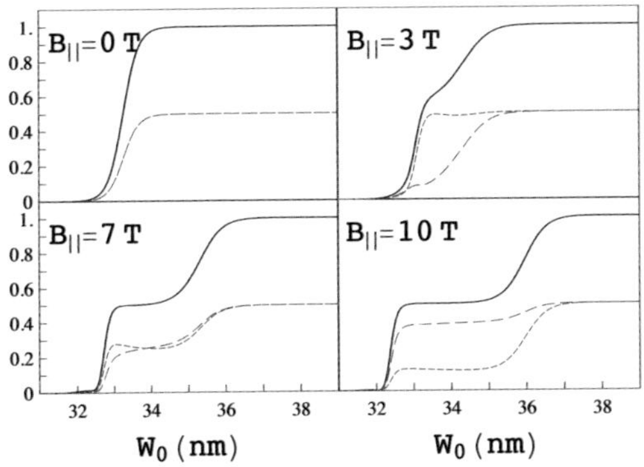

Figure 3. Total conductance G (black solid curves) and spin-resolved conductances G_+ (blue, long-dashed) and G_- (red, short-dashed), plotted in units of $2e^2/h$ as functions of the minimum width W_0 of the QPC [see Equation (5.5)]. In these simulations, $m = 0.14m_0$ [16], where m_0 is the bare electron mass, $g^* = 0.8$ [30], $\gamma\hbar^3 = 0.45$ eV nm ^{3}, $\Delta x = 0.3\mu$m, and $\epsilon_f = 2.3$ meV.

large at the degeneracy point ($v \simeq v_f$, where $v_f = \hbar k_f/m$ is the Fermi velocity), and to follow adiabatically the spin branches requires a large external field. The crossover occurs for

$$B^* \simeq \frac{(\hbar k_f)^2\sqrt{2\gamma\hbar/(m\Delta x)}}{g^*\mu_B}. \tag{5.7}$$

Below B^*, holes injected in the $+$ band cross non-adiabatically to the $-$ spin branch when $k_x \simeq \sqrt{3}\pi/W$. Therefore, holes injected in the lower subband have higher energy at $x \simeq 0$ and are preferentially reflected, as seen in the second panel of Figure 3 (with $B_\parallel = 3$ T). The reflection is not perfect, due to non-adiabaticity at $k_x \simeq 0$: at this second quasi-degenerate point the $-$ holes can cross back to the $+$ branch, and be transmitted.

This discussion shows that, in a model where the cubic Rashba term of 2D holes givs rise to both linear- and cubic-in-momentum terms in the QPC, the degeneracy of the hole spectrum at $k_x = \sqrt{3}\pi/W$ is crucial to obtain the anomalous transmission of Figures 1 and 3. We expect that when all cubic and linear terms are taken into account, arising from both Rashba and Dresselhaus SO interactions, the result will be qualitatively the same.

6. Conclusion

The cyclotron motion makes it possible to spatially separate spin currents in materials with sufficiently strong intrinsic spin-orbit interactions.

We have understood the physical mechanisms which give rise to spin-dependent magnetic focusing and the anomalous spin filtering by quantum point contacts. Professor Gabriele Giuliani made important contributions to the theory and our current understanding of spin-dependent magnetic focusing, as well as in the broader field of spintronics and spin transport.

References

[1] D. D. Awschalom, D. Loss and N. Samarth (eds.), Semiconductor Spintronics and Quantum Computation, Springer, New York, 2002.

[2] L. P. Rokhinson, V. Larkina, Y. B. Lyanda-Geller, L. N. Pfeiffer and K. W. West, Phys. Rev. Lett. **93** (2004), 146601.

[3] B.I. Halperin, Commentary, Journal Club for Condensed Matter Physics, July 2004.

[4] *Directing Matter and Energy: Five Challenges for Science and Imagination*. A report from Department of Energy Basic Energy Sciences Advisory Committee (2007), p. 5.

[5] Y. V. Sharvin, Zh. Eksp. Teor. Fiz. **48** (1965), 984 [Sov. Phys. JETP **21** (1965), 655].

[6] V. S. Tsoi, JETP Letters **22** (1975), 197.

[7] H. VanHouten, C. W. J. Beenakker, J. G. Williamson, M. E. I. Broekaart, P. H. M. VanLoosdrecht, B. J. VanWees, J. E. Mooij, C. T. Foxon and J. J. Harris, Phys. Rev. B **39** (1989), 8556.

[8] Y. Aharonov and A. Casher, Phys. Rev. Lett. **53** (1984), 319.

[9] A. G. Aronov and Y. B. Lyanda-Geller, Phys. Rev. Lett.**70** (1993), 343.

[10] S. Chesi, G. Giuliani, L. P. Rokhinson. L. Pfeiffer and K. West, Phys. Rev. Lett. **106** (2011), 236601.

[11] E. S. Snow and P. M. Campbell, Appl. Phys. Lett. **64** (1994), 1932.

[12] R. Held, T. Heinzel, A. P. Studerus, K. Ensslin and M. Holland, Appl. Phys. Lett. **71** (1997), 2689.

[13] L. P. Rokhinson, D. C. Tsui, L. N. Pfeiffer and K. W. West, Superlattices Microstruct. **32** (2002), 99.

[14] L. P. Rokhinson, L. N. Pfeiffer and K. W. West, Phys. Rev. Lett. **96** (2006), 156602.

[15] S. Chesi, unpublished

[16] T. Minagawa, PhD Thesis, Purdue University, 2010,

[17] R. WINKLER, "Spin-Orbit Coupling Effects in Two-Dimensional Electron and Hole Systems", Springer, Berlin, 2003.

[18] R. WINKLER, Phys. Rev. B **62** (2000), 4245; R. WINKLER, H. NOH, E. TUTUC and M. SHAYEGAN, Phys. Rev. B **65** (2002), 155303; S. CHESI and G. F. GIULIANI, Phys. Rev. B **75** (2007), 155305.

[19] E. I. RASHBA and E. YA. SHERMAN, Phys. Lett. A **129** (1988), 175.

[20] Y. LYANDA-GELLER, ArXiv:1210.7825

[21] S. S. NEDOREZOV, Sov. Phys. Solid State **12** (1971), 1814 [Fizika Tverdogo Tela, 12, 2269 (1970)],

[22] I. A. MERKULOV, V. I. PEREL' and M. E.PORTNOI, Zh. Eksper. Teot. Fiz **99** (1991), 1202 [Sov Phys. JETP, **72** 660 (1991)].

[23] M. I. DYAKONOV and A. V. KHAETSKII, Zh. Eksp. Teor Fiz. **82** (1982), 1584 [Sov. Phys. JETP, 55 917 (1982)].

[24] E. I. RASHBA, Sov. Phys. Solid State **2** (1960), 1109.

[25] D. LOSS, P. GOLDBART and A. V. BALATSKY, Phys. Rev. Lett. **65** (1990), 1655.

[26] S. V. IORDANSKII, YU. B. LYANDA-GELLER and G. E. PIKUS, JETP Letters **60** (1994), 206.

[27] I. L. ALEINER and V. I. FALKO, Phys. Rev. Lett. **87** (2001), 256801.

[28] Y. A. BYCHKOV and E. I. RASHBA, JETP Lett. **39** (1984), 78.

[29] H. L. STORMER, Z. SCHLESINGER, A. CHANG, D. C. TSUI, A. C. GOSSARD and W. WIEGMANN, Phys. Rev. Lett. **51** (1983), 126.

[30] R. DANNEAU *et al.*, Phys. Rev. Lett. **97** (2006), 026403; S. P. KODUVAYUR *et al.*, Phys. Rev. Lett. **100** (2008), 126401.

[31] L. I. GLAZMAN, G. B. LESOVIK, D. E. KHMEL'NITSKII and R. I. SHEKHTER, JETP Lett. **48** (1988), 238.

[32] The anticommutator is introduced to obtain a hermitian Hamiltonian. This has negligible effect in the limit of a smooth contact, when $\partial V/\partial x$ is small.

[33] $G_{\pm}$ is defined for unpolarized incident holes and spin-resolved detection.

[34] A. REYNOSO, G. USAJ and C. A. BALSEIRO, Phys. Rev. B **75** (2007), 085321.

Charge density wave surface phase slips and non-contact nanofriction

Franco Pellegrini, Giuseppe E. Santoro and Erio Tosatti

Abstract. Bulk electrical dissipation caused by charge-density-wave (CDW) depinning and sliding is a classic subject. We present a novel local, nanoscale mechanism describing the occurrence of mechanical dissipation peaks in the dynamics of an atomic force microscope tip oscillating above the surface of a CDW material. Local surface 2π slips of the CDW phase are predicted to take place giving rise to mechanical hysteresis and large dissipation at discrete tip surface distances. The results of our static and dynamic numerical simulations are believed to be relevant to recent experiments on $NbSe_2$; other candidate systems in which similar effects should be observable are also discussed.

1. How it all began: Gabriele Giuliani and CDW in the 70'

Erio Tosatti – When I first met Gabriele — here "I" is ET — it was 1975, when he turned up in my office at the University of Rome, introduced by my senior colleague and former mentor Franco Bassani. Gabriele was then a young undergraduate student of Pisa's Scuola Normale (where I also came from) and was seeking outside advisors for a thesis subject in modern condensed matter theory. My Rome colleague Mario Tosi, who was a professor, accepted to serve as his formal external advisor, and so Gabriele started coming back periodically to Rome, working on the general subject of the electron gas — "*Electron Gas*" even became Gabriele's nickname, as far as I was concerned. The age gap between Gabriele and me was slightly less than ten years; we became friends and spent time together, talking physics, politics, and everything else on our minds. I had recently come back from Cambridge where I had worked with Phil Anderson on a possible surface version of charge-density-waves (CDWs) instabilities of the electron gas invented a decade earlier by Al Overhauser in the US. I was enthusiastic about the subject, and ended up

We acknowledge research support by SNSF, through SINERGIA Project CRSII2 136287/1, by ERC Advanced Research Grant N. 320796 MODPHYSFRICT, by EU-Japan Project 283214 LEMSUPER, and by MIUR, through PRIN-2010LLKJBX_001.

Figure 1. Gabriele Giuliani (center) with Mario Tosi (left) and Franco Bassani (right) on the Appennines during a trip from Rome to Gabriele's hometown, Ascoli Piceno (Autumn 1976). Photograph taken by Erio Tosatti.

getting Gabriele interested too, and so it was that CDWs and Overhauser made their first entry in Gabriele's life.

In 1977 when, after completion of his degree in Pisa, I asked Gabriele to consider my newly founded group in Trieste, to which I had managed to attract colleagues of the caliber of young Michele Parrinello and of mature Mario Tosi. He liked it, and decided to join. In Trieste, we began working on one-dimensional CDWs and similar systems [1–3]. He worked hard and got far ahead of my rudimentary command of many body theory and related suggestions. I recall for example one calculation where he was supposed to reproduce, with some supposedly clever approximation I had cooked up, one exactly known result. I abused him abundantly because he was consistently failing to get the exact result by a factor 2 — only to discover the hard way that he was right and the culprit was my beloved approximation. In spite of all that, when one summer I met Al Overhauser in the US, he approached me quite enthusiastically about our otherwise universally ignored CDW paper [2], and declared that he would be delighted to welcome Gabriele to visit his group, possibly for a PhD curriculum (a title that did not exist in Italy at the time). And this is how Gabriele ended up at Purdue, in 1979, first as a postdoc with Al Overhauser, and then, after a further postdoc at Brown University, with a tenure. Our more than brotherly relationship continued uninterrupted to his very last few days, with visits, contacts, and many many phone and Skype calls which he would never forget to make in connection with all kinds of occasions or even without. Besides Gabriele's friendship, his other main present has been his former student Giuseppe Santoro, who came to Trieste at his suggestion to become a close friend and collaborator to this day.

Giuseppe Santoro – It was 1987 — a few years after he returned to Purdue from Brown — that I joined Gabriele at Purdue, the "I" now being GES. I had met him already in a couple of occasions in Pisa, were I was a student of the Scuola Normale Superiore that Gabriele regularly visited during his trips to Italy, to visit his family in the beloved hometown of Ascoli Piceno (*Alé Ascoli* was, as I soon discovered, Gabriele's first screen upon log-in on any computer on earth). By that time — we were in the Quantum Hall Effect era — the focus of Gabriele research was the two-dimensional electron gas, and this was the initial interest of my first papers at Purdue. But I still cannot recall a single office day, during my stay at Purdue from 1987 to 1991, in which Gabriele was not paying a visit to Al Overhauser: and in Overhauser's office, a most sure topic of discussion was charge (and spin) density waves.

It is fair to say that, since the early days, Gabriele and CDW (first) and two-dimensional electronic systems (later) remained entangled together. It is thus fitting that this article, devoted to him and his legacy, should be on the CDWs of a peculiar two-dimensional layered system, $NbSe_2$, on which we have recently come across.

2. Introduction

Charge-density-waves (CDWs) are static modulations of small amplitude and generally incommensurate periodicities which occur in the electron density distribution and in the lattice positions of a variety of materials [4]. They may derive either by an exchange-driven instability of a metallic Fermi surface [5], or by a lattice dynamical instability leading to a static periodic lattice distortion (PLD) which may equivalently be driven by electrons near Fermi [6,7] or finally just by anharmonicity [8]. A CDW superstructure, characterized by amplitude ρ_0 and phase ϕ relative to the underlying crystal lattice can be made to slide with transport of mass and charge and with energy dissipation under external perturbations and fields [4].

Phase slips in bulk CDWs/PLDs are involved in a variety of phenomena, including noise generation [9], switching [10], current conversion at contacts [11], noise [12,13] and more. While these phenomena are now classic knowledge, there is to date no parallel work addressing the possibility to mechanically provoke CDW phase slips at a chosen local point, see pictorial illustration in Figure 2. In this work we describe a two-dimensional model showing how a localized CDW/PLD phase slip may be provoked by external action of an atomic force microscope (AFM) tip at an arbitrarily chosen point outside a surface.

Figure 2. Pictorial view of an AFM tip provoking phase slips over a surface CDW modulation.

The study of the microscopic mechanisms leading to energy dissipation and friction has very important theoretical and practical implications. In recent years, experiments have started to single out the effects of microscopic probes in contact or near contact with different surfaces, and much theoretical effort has been devoted to the understanding of such experiments [14]. In particular, the minimally invasive non-contact experiments offer a chance to investigate delicate surface properties and promise to bring new insight on localized effects and their interaction with the bulk. The development of ultra-sensitive tools such as the "pendulum" AFM [15, 16] offers a chance to investigate more delicate and intimate substrate properties. Near a CDW material the tip oscillations may actuate, through van der Waals or electrostatic coupling, an electronic and atomic movement in the surface right under the tip, amounting in this case to coupling to the CDW order parameter. Owing to the periodic nature of the CDW state, the coupled tip-CDW system has multiple solutions, characterized by a different winding number (a topological property) which differ by a local phase slip, and correspond to different energy branches. At the precise tip-surface distance where two branches cross, the system will jump from one to the other injecting a local 2π phase slip, and the corresponding hysteresis cycle will reflect directly as a mechanical dissipation, persisting even at low tip oscillation frequencies.

Recently, a non-contact atomic force microscopy (AFM) experiment [17] on a $NbSe_2$ sample has shown dissipation peaks appearing at specific heights from the surface and extending up to 2 nm far from it. These peaks were obtained with tips oscillating both parallel and perpendicular to the surface, and in a range of temperatures compatible with the surface charge density wave (CDW) phase of the sample. In this paper, a model is proposed explaining in detail the mechanism responsible for these peaks: the tip oscillations induce a charge perturbation in the surface right under the tip, but, due to the nature of the CDW order parameter, multiple stable charge configurations exist characterized by different "topological" properties. When the tip oscillates at distances corresponding to the crossover of this different manifolds, the system is not allowed to follow the energy minimum configuration, even at the low experimental frequencies of oscillation, and this gives rise to a hysteresis loop for the tip, leading to an increase in the dissipation.

3. The Ginzburg-Landau model

In the following, we will use the term CDW to indicate a periodic modulation of the charge density ρ, irrespective of the process behind its generation. This modulation is described, in the unperturbed system and for the simplest form of CDW, as $\Delta\rho(\mathbf{r}) = \rho_0 \cos(\mathbf{Q}\cdot\mathbf{r}+\phi_0)$, where ρ_0 is the intensity, $\lambda = 2\pi Q^{-1}$ the characteristic wavelength, and ϕ_0 an initially constant phase, fixed by some far away agent. Perturbations to CDWs have been studied extensively [4, 18–20], but most studies are concerned either with uniform perturbations (e.g., the dynamics of a CDW under an external electric field) or point-like perturbations (*e.g.*, the static pinning of the CDW by defects), and often consider one-dimensional models, appropriate for quasi-one-dimensional materials, where the coherence length in the perpendicular directions is smaller than the atomic distance. Here we wish to study, instead, the effect of a localized perturbation represented by a weakly interacting and slowly oscillating nano or mesoscopic sized probe hovering above the surface, acting on a length scale σ similar to the CDW wavelength, $\sigma \sim 2\pi Q^{-1}$, and on a material where the coherence length is macroscopic in more than one dimension.

Starting from the standard Fukuyama-Lee-Rice model [18, 19] for CDW, the charge modulation is described as a classical elastic medium, through a Ginzburg-Landau (GL) theory. A complex space-dependent (and later, time-dependent) order parameter $\psi(\mathbf{r}) = A(\mathbf{r})e^{i\phi(\mathbf{r})}$ will take into account both the amplitude degree of freedom A, as well as the phase ϕ, in terms of which the charge density modulation is expressed as $\Delta\rho(\mathbf{r}) = A(\mathbf{r})\cos(\mathbf{Q}\cdot\mathbf{r}+\phi(\mathbf{r})) = \mathrm{Re}\left[\psi(\mathbf{r})e^{i\mathbf{Q}\cdot\mathbf{r}}\right]$. The unperturbed system has $A(\mathbf{r}) = \rho_0$ and $\phi(\mathbf{r}) = \phi_0$, both constant. The GL free-energy functional, in absence of any external perturbation, will read:

$$\mathcal{F}_0[\psi(\mathbf{r})] = \int d\mathbf{r}\, \left[-2f_0\,|\psi(\mathbf{r})|^2 + f_0\,|\psi(\mathbf{r})|^4 + \kappa\,|\nabla\psi(\mathbf{r})|^2\right] , \quad (3.1)$$

where f_0 sets the energy scale and κ accounts for the elastic energy cost. If we now consider the effect of an external perturbation — in our case, the AFM tip, generically described as a potential $V(\mathbf{r})$ coupling to the charge density modulation $\Delta\rho(\mathbf{r})$ —, we will need to add to the GL free-energy functional an extra term of the form:

$$\mathcal{F}_V[\psi(\mathbf{r})] = \int d\mathbf{r}\, V(\mathbf{r})\, \Delta\rho(\mathbf{r}) = \int d\mathbf{r}\, V(\mathbf{r})\, \mathrm{Re}\left[\psi(\mathbf{r})e^{i\mathbf{Q}\cdot\mathbf{r}}\right] . \quad (3.2)$$

The literature [20,21] has dealt extensively with the case in which $V(\mathbf{r})$ represents the potential due to impurities present in the sample, where it is appropriate to take $V(\mathbf{r}) = \sum_i \delta(\mathbf{r} - \mathbf{r}_i)$, since the typical scale in which the impurity potential acts is much smaller than $2\pi Q^{-1}$. In that case, phase-only oscillations — with an essentially constant amplitude $A(\mathbf{r})$ — are often enough to study the ground state of the system, resulting from the balance of elastic and potential energy, and described by a phase-only GL functional of the form:

$$\mathcal{F}_\phi[\phi(\mathbf{r})] = \int d\mathbf{r}\, \left[\kappa\, |\nabla\phi(\mathbf{r})|^2 + V(\mathbf{r})\rho_0 \cos(\mathbf{Q}\cdot\mathbf{r} + \phi(\mathbf{r}))\right] . \quad (3.3)$$

Extremely localized impurity perturbations of this sort, however, only impose a likewise point-like constraint on the phase of the order parameter, and cannot lead to a phase slip, in the absence of an external driver [20]. To model an AFM tip, on the contrary, we should consider the case where $V(\mathbf{r})$ has a given specific shape with a finite width σ of the order of the wavelength $2\pi Q^{-1}$, and minimize the total GL free energy $\mathcal{F} = \mathcal{F}_0 + \mathcal{F}_V$, including the amplitude degree of freedom $A(\mathbf{r})$. The fact that a phase-only functional $\mathcal{F}_\phi$ is inadequate in describing this specific effect can be argued as follows. If we consider a purely one-dimensional model $\mathcal{F}_\phi[\phi(x)]$, we would end-up with a linear behavior of $\phi(x)$ — the solution of the Laplace equation in one-dimension — in the regions where the potential is zero. Since we expect a decay of ϕ towards some constant ϕ_0 far from the perturbation, this is a clearly unphysical result. But moving to a two-dimensional phase-only functional does not improve the situation very much. Indeed, due to the nature of the phase, which is defined modulo 2π, given some boundary conditions the solution is not univocally defined unless the total variation of ϕ along the sample is also specified. Assuming the phase to have the unperturbed value ϕ_0 far from the perturbation, we can define the integer *winding number* N of a solution as the integral

$$N = \frac{1}{2\pi} \int \nabla\phi(x) dx \,, \quad (3.4)$$

taken along the CDW direction $\mathbf{Q}$ (with $N = 0$ typically representing the unperturbed case). Since any change in the winding number along the $\mathbf{Q}$ direction would extend to the whole sample, and unnaturally raise the energy of such a solution, to recover a physical result we definitely need to take into account the amplitude degree of freedom, which will allow for the presence of dislocations and local changes in the winding number.

For these reasons, we consider the full GL problem $\mathcal{F} = \mathcal{F}_0 + \mathcal{F}_V$ in two dimensions, with the complete complex order parameter $\psi(\mathbf{r}) = A(\mathbf{r})e^{i\phi(\mathbf{r})}$, and in a subspace with a definite winding number. The final result is expected to be similar to what previously considered in the wider context of phase slips [20] and more specifically in the case of localized phase slip centers [22,23]. Namely, the local strain induced by the perturbation on the phase will reduce the order parameter amplitude, to the point where a local phase slip event becomes possible. In more than one dimension, the boundary between areas with different winding number will be marked by *vortices* of the phase.

From this preliminary analysis, the mechanism responsible for the dissipation peaks can be understood: as the tip approaches the surface, it encounters points where the energies of solutions with different winding number undergo a crossover. At these points the transition between manifolds is not straightforward, due to the mechanism required to create the vortices; therefore the tip oscillations lead to jumps between different manifolds, resulting in hysteresis for the tip, and ultimately dissipation.

4. Equilibrium and time-dependent GL simulations

To asses the validity of the proposed mechanism, we have performed numerical simulations of the tip-surface interaction with a full GL free-energy $\mathcal{F}[\psi(\mathbf{r})] = \mathcal{F}_0 + \mathcal{F}_V$. For simplicity, a two-dimensional GL functional is considered, since this takes into account the relevant elastic effects while keeping the simulation simple enough: indeed, the experimental substrate $NbSe_2$ [17] has a quasi-two-dimensional structure, so that volume effects are expected to be not crucial. Differently from the experimental system [24], we will model the CDW as being characterized by a single wavevector $\mathbf{Q}$, leading to a simpler order parameter and a clearer effect. To represent the effect of the tip, the shape of a van der Waals potential C/r^6 is integrated over a conical tip at distance d from the surface. We have found that the result of such a calculation can be reasonably approximated in the main area under the tip by a Lorentzian curve:

$$V(\mathbf{r}; d) = \frac{V_0(d)}{\mathbf{r}^2 + \sigma^2(d)}, \tag{4.1}$$

where $\mathbf{r}$ is the distance in the plane from the point right below the tip and the parameters are found to scale like $V_0(d) = \overline{V}/d$ and $\sigma(d) = \overline{\sigma}d^2$. Knowing the shape of the perturbation, the total free energy $\mathcal{F} = \mathcal{F}_0 +$

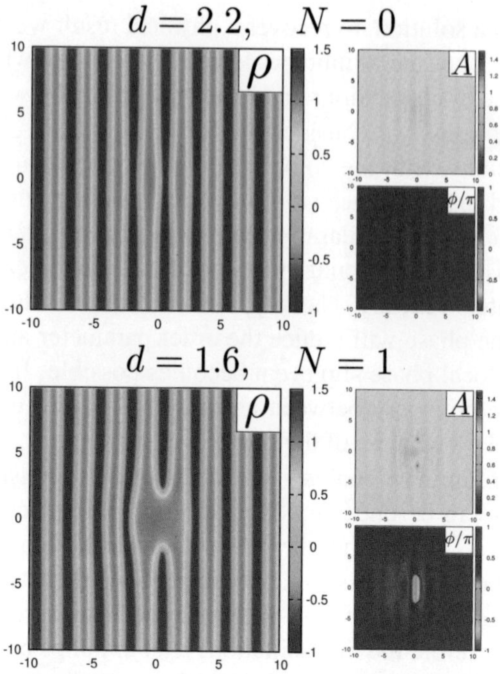

Figure 3. Charge density $\Delta\rho$, order parameter amplitude A and potraits of the phase ϕ for minimal free energy solutions with different winding-number N and tip-surface distance d (in nm). Results from simulations on a 201×201 grid with parameters (see text) $f_0 = 2$ eV/nm, $\kappa = 0.2$ eV, $Q = 2.5$ nm, $\overline{V} = -9.4$ eV·nm, $\overline{\sigma} = 1, 2$ nm^{-1} and boundary conditions $\psi_0 = i$ (right and left sides).

$\mathcal{F}_V$ is minimized numerically on a square grid of points with spacing much smaller than the characteristic wavelength of the CDW, imposing a constant boundary condition ψ_0 on the sides perpendicular to $\mathbf{Q}$, while setting periodic boundary conditions in the other direction to allow for possible phase jumps. The minimization is carried out through a standard conjugated gradients algorithm [25]. The parameters we have employed are order of magnitude estimates of the real parameters, reproducing the relevant experimental effects on $NbSe_2$ [17] in a qualitative fashion.

Figure 3 shows the charge density modulation $\Delta\rho$, together with a plot of the amplitude $A(\mathbf{r})$ and a phase portrait of $\phi(\mathbf{r})$, corresponding to GL minima with different winding-number N, for a non-contact (attractive) tip at different distances d. The winding-number is calculated along the line passing through the point right below the tip (center of the simulation cell) according to Eq. (3.4), with $N = 0$ being the unperturbed case. As predicted, we see upon decreasing d through the first and successive critical distances d_{01}, d_{12}, etc. the appearance of vortex-antivortex pairs

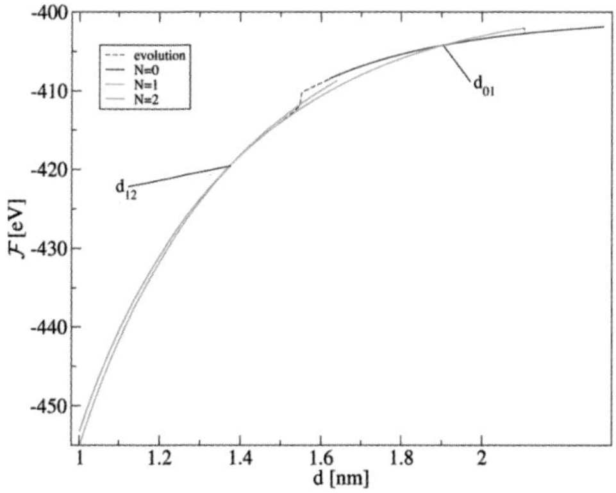

Figure 4. Minimal free energy $\mathcal{F}$ as a function of tip distance d for subspaces with different winding number N (full lines) and evolution of $\mathcal{F}$ during a tip oscillation (dashed line) with $d_0 = 1.8$ nm, $\bar{d} = 0.4$ nm, $\omega = 6 \cdot 10^4$ Hz (other parameters are the same as Figure 3).

for every unit increase of the winding number. These vortices are characterized by a zero of the amplitude $A(\mathbf{r})$ and a total change of the phase by 2π on a path around them, as they separate the phase-slippage center from the unaffected area far from the tip.

Since the solution with a given winding number N represents a local minimum, it is possible to use the minimization algorithm, for example by starting from a reasonable configuration, to find solutions in a certain N-subspace, even when that is not the global minimum for that given case. This allows us to extend the calculation of the local free energy minima in a given N subspace well beyond their crossing points, generating a family of free energy curves of definite N as a function of the distance d. Figure 4 (full lines) is an example, showing two successive crossing points. We expect each crossing to give rise to a first order transition, and thus to a hysteretic peak in the experimental dissipation trace. Of course, a more complex CDW configuration or different parameters could give rise to more and different peaks.

To justify more firmly the validity of the proposed dissipation mechanism, we need to look into the dynamics of the CDW, upon varying the tip-surface distance according to the law $d(t) = d_0 + \bar{d}\cos(\omega t)$, to guarantee that the evolution through a crossing point does not lead to immediate relaxation between different N-manifolds. To do this, the time evolution of the system was simulated, following the time-dependent Ginzburg-Landau equation [23]

$$-\Gamma \frac{\partial \psi}{\partial t} = \frac{\delta \mathcal{F}}{\delta \psi^*} . \tag{4.2}$$

This equation can be interpreted as an overdamped relaxation of the order parameter towards the equilibrium position, with a relaxation rate Γ^{-1}. Integrating this equation (through a standard Runge-Kutta algorithm [25]), the instantaneous force $F = -\nabla_d \mathcal{F}$ as a function of the distance can be computed for a tip performing a full oscillation perpendicular to the surface according to the law $d(t) = d_0 + \bar{d}\cos(\omega t)$. Figure 5 shows the force evolution during such oscillations at different frequencies. As we can see, the tip suffers a hysteresis even at low frequencies, since the decay from one manifold to the other happens far from the crossing point. The area of the loops represents directly the dissipated energy per cycle W, as reported in the inset.

5. Discussion and conclusions

We have shown that local surface CDW phase slips and vortex pairs can be introduced by the external potential of an approaching tip. In the context of macroscopic CDW conduction noise [4, 12, 13], the creation and movement of vortices has been invoked earlier in connection with phase slips near the CDW boundaries. In a broader context, our system can be placed in between these macroscopic situations and the simple models of defect pinning and phase-slip [20] by a localized perturbation.

Experimentally, Langer *et al.* [17] recently reported AFM dissipation peaks appearing at discrete tip-surface distances above the CDW material 2H-$NbSe_2$, qualitatively suggesting, in a 1D toy model, the injection of 2π phase slips. The present results describe at the minimal level a theory that can explain this type of phenomenon, connecting the phase slip to

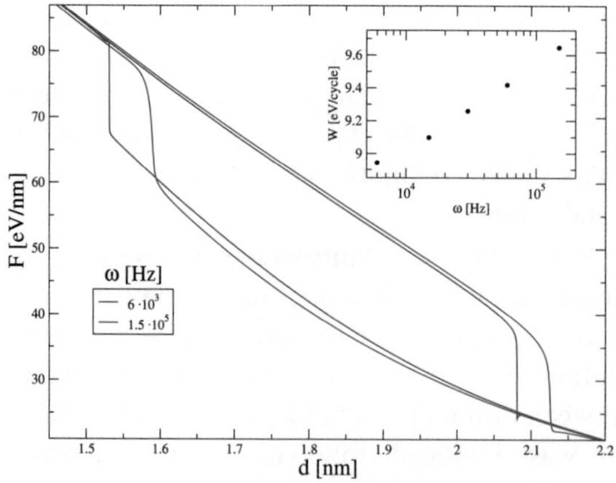

Figure 5. Force as a function of distance for evolutions with $d_0 = 1$ nm, $\bar{d} = 0.4$ nm and different values of ω with $\Gamma = 10^{-7}$ eV·s. Inset: total work W as a function of oscillation frequency ω.

a vortex-pair formation, and providing the time dependent portrait of the injection process.

It would be of considerable future interest to explore further this effect in other systems with different characteristics. In insulating, quasi-one dimensional CDW systems the injected phase slip should also amount to the injection of a quantized, possibly fractional pairs of opposite charges [26]. In a spin density wave system, such as the chromium surface, a non-magnetic tip would still couple to the accompanying CDW [27] where surface phase slips could be injected. In superconductors, the induction of single vortices over Pb thin film islands has been experimentally verified [28] and the feasibility of controlling single vortices through magnetic force microscopy (MFM) tips demonstrated [29]: it would be interesting to probe for dissipation peaks, as we have addressed above, induced by the MFM tip creation of vortex pairs in thin superconducting films.

References

[1] G. GIULIANI, E. TOSATTI and M. P. TOSI, Lettere al Nuovo Cimento **16** (1976), 385.

[2] G. GIULIANI and E. TOSATTI, Il Nuovo Cimento **47B** (1978), 135.

[3] G. GIULIANI, E. TOSATTI and M. P. TOSI, J. Phys. C - Solid State Phys. **12** (1979), 2769.

[4] G. GRÜNER, Rev. Mod. Phys. **60** (1988), 1129.

[5] A. W. OVERHAUSER, Phys. Rev. **167** (1968), 691.

[6] R. E. PEIERLS, "Quantum Theory of Solids, Oxford University Press, 1955.

[7] E. J. WOLL and W. KOHN, Phys. Rev. **126** (1962), 1693.

[8] F. WEBER, S. ROSENKRANZ, J.-P. CASTELLAN, R. OSBORN, R. HOTT, R. HEID, K.-P. BOHNEN, T. EGAMI, A. H. SAID and D. REZNIK, Phys. Rev. Lett. **107** (2011), 107403.

[9] S. N. COPPERSMITH, Phys. Rev. Lett. **65** (1990), 1044.

[10] M. INUI, R. P. HALL, S. DONIACH and A. ZETTL, Phys. Rev. B **38** (1988), 13047.

[11] M. P. MAHER, T. L. ADELMAN, S. RAMAKRISHNA, J. P. MCCARTEN, D. A. DICARLO and R. E. THORNE, Phys. Rev. Lett. **68** (1992), 3084.

[12] N. P. ONG, G. VERMA and K. MAKI, Phys. Rev. Lett. **52** (1984), 663.

[13] G. GRÜNER, A. ZAWADOWSKI and P. M. CHAIKIN, Phys. Rev. Lett. **46** (1981), 511.

[14] A. Vanossi, N. Manini, M. Urbakh, S. Zapperi and E. Tosatti, Rev. Mod. Phys. **85** (2013), 529.
[15] B. C. Stipe, H. J. Mamin, T. D. Stowe, T. W. Kenny and D. Rugar, Phys. Rev. Lett. **87** (2001), 096801.
[16] U. Gysin, S. Rast, M. Kisiel, C. Werle and E. Meyer, Rev. Sci. Instrum. **82** (2011), 023705.
[17] M. Langer, M. Kisiel, R. Pawlak, F. Pellegrini, G. E. Santoro, R. Buzio, A. Gerbi, G. Balakrishnan, A. Baratoff, E. Tosatti *et al.*, Nature Mat. **13** (2014), 173.
[18] H. Fukuyama and P. A. Lee, Phys. Rev. B **17** (1978), 535.
[19] P. A. Lee and T. M. Rice, Phys. Rev. B **19** (1979), 3970.
[20] J. R. Tucker, Phys. Rev. B **40** (1989), 5447.
[21] I. Tüttő and A. Zawadowski, Phys. Rev. B **32** (1985), 2449.
[22] K. Maki, Phys. Lett. A **202** (1995), 313.
[23] L. P. Gor'kov, Zh. Eksp. Teor. Fiz. **86** (1984), 1818.
[24] W. L. McMillan, Phys. Rev. B **12** (1975), 1187.
[25] W. H. Press, S. A. Teukolsky, W. T. Vetterling and B. P. Flannery, "Numerical Recipes: The Art of Scientific Computing" (3rd ed.), Cambridge University Press, 2007.
[26] This feature is not expected in $NbSe_2$, which is metallic and in reality an anharmonicity-driven PLD [8]. Better candidates for these effects would be $NbSe_3$ or TaS_3.
[27] H. C. Kim, J. M. Logan, O. G. Shpyrko, P. B. Littlewood and E. D. Isaacs, Phys. Rev. B **88** (2013), 140101(R).
[28] T. Nishio, S. Lin, T. An, T. Eguchi and Y. Hasegawa, Nanotechnology **21** (2010), 465704.
[29] O. M. Auslaender, L. Luan, E. W. J. Straver, J. E. Hoffman, N. C. Koshnick, E. Zeldov, D. A. Bonn, R. Liang, W. N. Hardy and K. A. Moler, Nature Phys. **5** (2009), 35.

The quasiparticle lifetime in a doped graphene sheet

Marco Polini and Giovanni Vignale

Abstract. We present a calculation of the quasiparticle decay rate due to electron-electron interactions in a doped graphene sheet. In particular, we emphasize subtle differences between the perturbative calculation of this quantity in a doped graphene sheet and the corresponding one in ordinary parabolic-band two-dimensional (2D) electron liquids. In the random phase approximation, dynamical over-screening near the light cone yields a universal quasiparticle lifetime, which is independent of the dielectric environment surrounding the 2D massless Dirac fermion fluid.

1. Introduction

Gabriele Giuliani loved the Landau theory of normal Fermi liquids [1–4]. The notion that a system of strongly interacting particles could behave like an ideal gas of plain non-interacting particles, was to him a source of endless fascination. This was largely a reflection of his "down-to-earth" approach to theoretical physics. Gabriele disliked all forms of mystification and particularly the widespread one of couching trivial or wrong ideas in high-sounding theoretical language. Fermi liquid theory, with its deceptive simplicity, was precisely the opposite of mystification: it was the sophisticated plainness he was striving for. At the heart of Fermi liquid theory lies the concept of "quasiparticle" – a quasi-exact eigenstate of a single excited particle that decays very slowly in time. How slowly? The critical requirement is that the decay rate of the state remain much smaller than its energy in the limit that the latter tends to zero. If this condition is satisfied, then an "adiabatic switching-on" process becomes viable, whereby, starting from an infinitely long-lived excited eigenstate of the non-interacting system, and slowly turning on the interaction ("slowly" meaning at a rate that is much longer than the excitation frequency – yet faster than the decay rate), one generates the long-lived eigenstate of the interacting system.

A standard argument for estimating the decay rate (also known as inverse lifetime) of a quasiparticle goes as follows. *Assuming that long-lived quasiparticles exist* with a small energy ξ in the vicinity of the

Fermi surface it is evident that they can only decay by scattering into other available (*i.e.*, empty) quasiparticle states. This is because Pauli's exclusion principle pre-empts scattering of a fermion into an occupied state (we ignore spin for simplicity). The number of available states is thus proportional to ξ (at zero temperature) or to T, if $\xi \ll k_{\rm B}T$. Further, conservation of momentum and energy require that the decay be accompanied by the production of a quasi-electron-quasi-hole pair, whose energy is also of the order of ξ or $k_{\rm B}T$, whichever is larger. The density of such pairs is proportional to ξ or $k_{\rm B}T$. Taking the two factors together, we conclude that the quasiparticle decay rate is proportional to ξ^2 or $(k_{\rm B}T)^2$, which is indeed much smaller than the excitation frequency, ξ or $k_{\rm B}T$, in the limit that the latter tends to zero.

Notice that this somewhat circular argument is valid (when it is valid) regardless of the strength of the electron-electron interaction. And indeed, for three-dimensional Fermi systems the naive argument gives the right answer, even when the interactions are very strong (as in ^{3}He and in heavy fermion compounds) and the renormalizations of the effective (*i.e.*, quasiparticle) mass are correspondingly large. The situation is completely different in one spatial dimension, where the same argument fails to predict the collectivization of the electron and the formation of the Luttinger liquid state (the situation is very well described in Giamarchi's book [5]).

What about the two-dimensional (2D) electron liquid? In the early 1980s, when Gabriele was just beginning his career, two-dimensional electron gases (2DEGs) in GaAs-based heterostructures and Si inversion layers were among the most fashionable systems studied by condensed matter physicists. The twin discoveries of the localizing effect of impurities in two spatial dimensions [6] (scaling theory of localization) and, more subtly, of the quantum Hall effect [7], which critically depended on the former, appeared to undermine the Fermi liquid picture of the 2DEG. The very existence of the metallic state of the 2DEG was in doubt [6]. With his "no-nonsense" attitude Gabriele followed those developments closely, but never bought into the most adventurous ideas. To those who denied the existence of the metallic state of electrons in 2D GaAs he was likely to suggest the following thought experiment: "*OK, let us stick this end of the sample into the power socket, while you hold the other end...*" But at the same time he would not accept uncritically the conventional wisdom about the Fermi liquid state in two spatial dimensions. And it was so that, during his postdoc with John Quinn at Brown University, he began to investigate the key question of the quasiparticle lifetime in the 2DEG. Working within the Fermi liquid picture, he was able to establish [8] that the decay rate of a quasiparticle in the 2DEG does not scale as ξ^2 or $(k_B T)^2$ as the naive argument would suggest, but rather as $-\xi^2 \ln(\xi)$ or $-(k_B T)^2 \ln(k_B T)$, depending on whether $k_B T \ll \xi$ or $k_B T \gg \xi$, respectively [9]:

$$\frac{1}{\tau_k} = \begin{cases} -\dfrac{\varepsilon_F}{\hbar}\dfrac{1}{4\pi}\left(\dfrac{\xi_k}{\varepsilon_F}\right)^2 \ln\left(\dfrac{|\xi_k|}{\varepsilon_F}\right), & \text{for } k_B T \ll |\xi_k| \\ -\dfrac{\varepsilon_F}{\hbar}\dfrac{1}{2\pi}\left(\dfrac{k_B T}{\varepsilon_F}\right)^2 \ln\left(\dfrac{k_B T}{\varepsilon_F}\right), & \text{for } k_B T \gg |\xi_k|. \end{cases} \tag{1.1}$$

Here $\xi_k = \hbar^2 k^2/(2m) - \varepsilon_F$ is the parabolic-band energy measured from the Fermi energy ε_F, m being the electron's (band) mass and $\hbar \boldsymbol{k}$ the 2D momentum. The unexpected logarithmic enhancement of the decay rate is due to a subtle feature of the 2D phase space available for the scattering of quasi-particles near the Fermi surface—a feature that is not captured by the naive argument. Another surprising feature of the Giuliani-Quinn formula for the decay rate is that the coefficient of the leading terms $-\xi^2 \ln(\xi)$ or $-T^2 \ln(T)$ is *independent* of the electron-electron coupling constant or, as Giuliani and Quinn aptly put it, of the magnitude of the electron charge. This counterintuitive feature arises from the fact that, in the Giuliani-Quinn theory, the dominant contribution to the decay rate arises from scattering processes with small momentum transfer q: these are the processes for which the Coulomb interaction between two quasiparticles is most strongly screened [4] by the electronic medium that surrounds them, leading to an effective interaction that depends only on the non-interacting density of states.

Equation (1.1) provides the justification for applying Fermi liquid theory to the 2DEG, at least when disorder is not too strong. The logarithmic enhancement of the decay rate does not create any serious danger to the stability of quasiparticles, probably less than Gabriele's thought experiment to its hypothetical subjects. Over the years, the paper [8] in which Equation (1.1) was first reported grew to be the standard reference on the subject. Adjustments had to be made [10] over the years to include the contributions of $2k_F$ scattering, vertex corrections, exchange effects, etc..., but none of these refinements changed the basic picture established in the original paper. Furthermore, numerous experiments since then have established the validity of the Fermi liquid concept in the 2DEG [11], the quasiparticle lifetime has been probed in detail [12], and the existence of the metallic state has been demonstrated [13].

Fast-forward 40^+ years to 2004, the year in which, for the first time, few-layer graphene sheets were electrically contacted and the field effect was demonstrated [14]. In its pristine state, graphene, *i.e.* a single layer of Carbon atoms arranged in a honeycomb structure, is a semimetal [15]. Its conical conduction and valence bands have dispersions $\sim \pm\hbar v_{\rm F}(k - k_{\rm D})$ in the vicinity of the Dirac point $k_{\rm D}$, where they touch. Due to the vanishing density of states at the Fermi level one would expect a complete break-down of the Fermi liquid paradigm. And, indeed, many-body calculations [16] suggest that the so-called massless Dirac fermion (MDF) quasiparticles exhibit singular features, such as a logarithmically diverging velocity [17] and linear-in-energy decay rates [18], which are hardly compatible with the Landau Fermi liquid paradigm. Nevertheless, these singularities are found to be relevant only for extremely low carrier densities and, when a sizeable Fermi surface is created (by doping, or, more conveniently, by electrostatic gating), the conventional Fermi liquid description seems to take hold again, even in suspended sheets, where the strength of electron-electron interactions is the largest. To be convinced that this is truly the case, one must calculate carefully the decay rate for quasiparticles near the Fermi surface. One might suppose that the presence of the Fermi surface erases any difference between the ordinary Schrödinger electrons of a 2DEG and the MDFs of graphene: after all the parabolic dispersion of Schrödinger electrons is approximately linear in the vicinity of the Fermi surface. However, the lesson of the Giuliani-Quinn paper is that such a-priori arguments must be taken with a good dose of skepticism, because subtle differences in the structure of the phase space can lead to quantitative differences in the decay rate. And indeed, a careful calculation, presented in the next few sections, exposes several differences between the calculation of the quasiparticle lifetime in graphene and in the 2DEG—differences that arise from the suppression of backscattering (characteristic of MDFs) as well as from the large enhancement of screening in MDF systems at frequencies near the light cone $\omega = \pm v_{\rm F} q$. The final upshot of the calculation, however, is that the Giuliani-Quinn picture remains valid, with the added feature that collinear scattering processes with small momentum transfer are now more important than ever, and completely dominate the behavior of the quasiparticle lifetime, while $2k_{\rm F}$ processes (initially neglected by Giuliani and Quinn) are happily suppressed.

The Fermi liquid properties and Coulomb decay rates of quasiparticles in graphene sheets have been studied by many authors. We have provided a (certainly incomplete) list of pertinent works in Refs. [19–32]. In this Article we present a pedagogical description of the calculation leading to an explicit formula for the Coulomb decay rate (*i.e.* inverse lifetime) of a weakly-excited plane-wave state in a doped graphene sheet. Our main results, Equation (2.47) and Equation (2.49), have been derived earlier by other authors (see, *e.g.*, Ref. [27]): the emphasis of this work is on the intermediate steps of the calculation.

ACKNOWLEDGEMENTS. We gratefully acknowledge Leonid Levitov, Kostya Novoselov, Alessandro Principi, Justin Song, and Andrea Tomadin for useful discussions. This Article is dedicated to the memory of our friend Gabriele F. Giuliani, physicist, soccer player, car racer and provocateur, who passed away on November 22, 2012, after a heroic battle with a very aggressive form of cancer that lasted 12 years.

2. Coulomb-enabled two-body decay rates in a doped graphene sheet

In this Section we present a theory of the decay rate $1/\tau_{\boldsymbol{k},\lambda}$ of a plane-wave state with momentum $\hbar\boldsymbol{k}$ and band index λ in a doped graphene sheet, at a temperature T. We will consider decay rates solely due to two-body Coulomb collisions.

For future purposes, we introduce the so-called graphene's fine-structure constant [33] $\alpha_{\rm ee}$,

$$\alpha_{\rm ee} = \frac{e^2}{\epsilon \hbar v_{\rm F}} \,. \tag{2.1}$$

Here, the dielectric constant ϵ is the average of the dielectric constants ϵ_1 and ϵ_2 of the media above and below the graphene flake, *i.e.* $\epsilon \equiv (\epsilon_1 + \epsilon_2)/2$. The dimensionless parameter $\alpha_{\rm ee}$ determines the strength of electron-electron interactions with respect to the kinetic energy.

We start by considering the so-called "G_0W-RPA" approximation for the imaginary part of the self-energy $\Sigma_\lambda(\boldsymbol{k}, \omega)$ in a doped graphene sheet [20] (from now on we set $\hbar = 1$, unless otherwise stated):

$$\begin{aligned} &\Im m[\Sigma_\lambda(\boldsymbol{k}, \omega)] \\ &= \int \frac{d^2\boldsymbol{q}}{(2\pi)^2} \sum_{\lambda'} \Im m\left[\frac{v_q}{\varepsilon(q, \omega - \xi_{\boldsymbol{k}-\boldsymbol{q},\lambda'}, T)}\right] \mathcal{F}_{\lambda\lambda'}(\theta_{\boldsymbol{k},\boldsymbol{k}-\boldsymbol{q}}) \\ &\times [n_{\rm B}(\omega - \xi_{\boldsymbol{k}-\boldsymbol{q},\lambda'}) + n_{\rm F}(-\xi_{\boldsymbol{k}-\boldsymbol{q},\lambda'})] \,, \end{aligned} \tag{2.2}$$

where λ, λ' are band indices ($\lambda = +$ denotes conduction-band states, $\lambda = -$ denotes valence-band states), $\theta_{\boldsymbol{k},\boldsymbol{k}-\boldsymbol{q}}$ is the angle between $\boldsymbol{k}$ and $\boldsymbol{k} - \boldsymbol{q}$,

$$\mathcal{F}_{\lambda\lambda'}(\varphi) \equiv \frac{1 + \lambda\lambda' \cos(\varphi)}{2} \tag{2.3}$$

is the usual chirality factor, and

$$\xi_{\boldsymbol{k},\lambda} \equiv \varepsilon_{\boldsymbol{k},\lambda} - \mu = \lambda v_{\rm F} k - \mu \tag{2.4}$$

are Dirac-band single-particle energies measured from the chemical potential μ.

In Equation (2.2) $n_{\mathrm{B/F}}(x) \equiv 1/[\exp(\beta x) \mp 1]$ are the usual Bose (Fermi) statistical factors with $\beta = (k_{\mathrm{B}}T)^{-1}$ and

$$\varepsilon(q,\omega,T) \equiv 1 - v_q \chi^{(0)}(q,\omega,T) \tag{2.5}$$

is the finite-temperature dynamical screening function in the random phase approximation (RPA) [4]. Here $v_q = 2\pi e^2/(\epsilon q)$ is the 2D Fourier transform of the Coulomb interaction and $\chi^{(0)}(q,\omega,T)$ the non-interacting finite-temperature density-density response function of a 2D gas of MDFs [34]. It contains both intra- and inter-band contributions.

Note that

$$\Im m\left[\frac{1}{\varepsilon(q,\omega,T)}\right] = v_q \frac{\Im m[\chi^{(0)}(q,\omega,T)]}{|\varepsilon(q,\omega,T)|^2}\,. \tag{2.6}$$

Using the previous identity in Equation (2.2) we find the following expression for the decay rate due to two-body Coulomb collisions:

$$\begin{aligned}\frac{1}{\tau_{\boldsymbol{k},\lambda}} &\equiv -2\,\Im m[\Sigma_\lambda(\boldsymbol{k},\xi_{\boldsymbol{k},\lambda})]\\ &= -2\sum_{\lambda'}\int \frac{d^2\boldsymbol{q}}{(2\pi)^2} v_q^2 \frac{\Im m[\chi^{(0)}(q,\xi_{\boldsymbol{k},\lambda}-\xi_{\boldsymbol{k}-\boldsymbol{q},\lambda'},T)]}{|\varepsilon(q,\xi_{\boldsymbol{k},\lambda}-\xi_{\boldsymbol{k}-\boldsymbol{q},\lambda'},T)|^2}\mathcal{F}_{\lambda\lambda'}(\theta_{\boldsymbol{k},\boldsymbol{k}-\boldsymbol{q}})\\ &\quad\times[n_{\mathrm{B}}(\xi_{\boldsymbol{k},\lambda}-\xi_{\boldsymbol{k}-\boldsymbol{q},\lambda'}) + n_{\mathrm{F}}(-\xi_{\boldsymbol{k}-\boldsymbol{q},\lambda'})]\,.\end{aligned} \tag{2.7}$$

We now use the exact identity

$$\begin{aligned}&n_{\mathrm{B}}(\xi_{\boldsymbol{k},\lambda}-\xi_{\boldsymbol{k}-\boldsymbol{q},\lambda'}) + n_{\mathrm{F}}(-\xi_{\boldsymbol{k}-\boldsymbol{q},\lambda'})\\ &= \frac{1-n_{\mathrm{F}}(\xi_{\boldsymbol{k}-\boldsymbol{q},\lambda'})}{1-\exp[-\beta(\xi_{\boldsymbol{k},\lambda}-\xi_{\boldsymbol{k}-\boldsymbol{q},\lambda'})]} - \frac{n_{\mathrm{F}}(\xi_{\boldsymbol{k}-\boldsymbol{q},\lambda'})}{1-\exp[\beta(\xi_{\boldsymbol{k},\lambda}-\xi_{\boldsymbol{k}-\boldsymbol{q},\lambda'})]}\end{aligned} \tag{2.8}$$

and introduce the following auxiliary delta function on the energy transfer ω:

$$1 = \int_{-\infty}^{\infty} d\omega\, \delta(\xi_{\boldsymbol{k},\lambda}-\xi_{\boldsymbol{k}-\boldsymbol{q},\lambda'}-\omega)\,. \tag{2.9}$$

We can therefore rewrite Equation (2.7) as follows

$$\begin{aligned}\frac{1}{\tau_{\boldsymbol{k},\lambda}} &= -\frac{2}{(2\pi)^2}\sum_{\lambda'}\int_{-\infty}^{+\infty} d\omega\, \frac{1-n_{\mathrm{F}}(\xi_{\boldsymbol{k},\lambda}-\omega)}{1-\exp(-\beta\omega)}\\ &\quad\times\int_0^{+\infty} dq\, q\left|\frac{v_q}{\varepsilon(q,\omega,T)}\right|^2 \Im m[\chi^{(0)}(q,\omega,T)]A_{\lambda\lambda'}(k,q,\omega)\\ &\quad+\frac{2}{(2\pi)^2}\sum_{\lambda'}\int_{-\infty}^{+\infty} d\omega\, \frac{n_{\mathrm{F}}(\xi_{\boldsymbol{k},\lambda}-\omega)}{1-\exp(\beta\omega)}\\ &\quad\times\int_0^{+\infty} dq\, q\left|\frac{v_q}{\varepsilon(q,\omega,T)}\right|^2 \Im m[\chi^{(0)}(q,\omega,T)]A_{\lambda\lambda'}(k,q,\omega)\,.\end{aligned} \tag{2.10}$$

Note that the second term in the previous equation can be obtained from the first term by performing the replacements $1-n_{\rm F}(\xi_{\boldsymbol{k},\lambda}-\omega) \to n_{\rm F}(\xi_{\boldsymbol{k},\lambda}-\omega)$, $1-\exp(-\beta\omega) \to 1-\exp(\beta\omega)$, and changing the overall sign. For this reason, it is customary [4] to define the first term in Equation (2.10) as the quasiparticle decay rate, the second term as the quasihole decay rate and the sum of the two as the decay rate of the plane-wave state $\boldsymbol{k}$, λ:

$$\frac{1}{\tau_{\boldsymbol{k},\lambda}} \equiv \frac{1}{\tau^{(\rm e)}_{\boldsymbol{k},\lambda}} + \frac{1}{\tau^{(\rm h)}_{\boldsymbol{k},\lambda}} \,. \tag{2.11}$$

In Equation (2.10) we have introduced the following angular integral

$$A_{\lambda\lambda'}(k,q,\omega) \equiv \int_0^{2\pi} d\theta\, \delta(\xi_{\boldsymbol{k},\lambda} - \xi_{\boldsymbol{k}-\boldsymbol{q},\lambda'} - \omega) \times \mathcal{F}_{\lambda\lambda'}(\theta_{\boldsymbol{k},\boldsymbol{k}-\boldsymbol{q}}) \,, \tag{2.12}$$

where θ is the angle between $\boldsymbol{q}$ and $\boldsymbol{k}$, which can be oriented along the $\hat{\boldsymbol{x}}$ axis without loss of generality, *i.e.* $\boldsymbol{k} = k\hat{\boldsymbol{x}}$. For future purposes it is important to note that

$$\cos(\theta_{\boldsymbol{k},\boldsymbol{k}-\boldsymbol{q}}) = \frac{k - q\cos(\theta)}{\sqrt{k^2+q^2-2kq\cos(\theta)}} \,. \tag{2.13}$$

Since the integrand in Equation (2.12) is a function of $\cos(\theta)$ only, we can write

$$A_{\lambda\lambda'}(k,q,\omega) = 2\int_0^{\pi} d\theta\, \delta(\xi_{\boldsymbol{k},\lambda} - \xi_{\boldsymbol{k}-\boldsymbol{q},\lambda'} - \omega) \times \mathcal{F}_{\lambda\lambda'}(\theta_{\boldsymbol{k},\boldsymbol{k}-\boldsymbol{q}}) \,. \tag{2.14}$$

The function $A_{\lambda\lambda'}(k,q,\omega)$ can be easily evaluated analytically. One first realizes that the delta function in Equation (2.12) gives a non-zero contribution to $A_{\lambda\lambda'}$ if and only if the equality

$$v_{\rm F}\lambda k - v_{\rm F}\lambda'\sqrt{k^2+q^2-2kq\cos(\theta)} = \omega \tag{2.15}$$

is satisfied. This condition does not depend on the chemical potential μ.

2.1. Intra-band contribution

For $\lambda' = +1$ (intra-band scattering) Equation (2.15) reduces to

$$\sqrt{k^2+q^2-2kq\cos(\theta)} = k - \frac{\omega}{v_{\rm F}} \,, \tag{2.16}$$

which requires $k \geq \omega/v_{\rm F}$. When this condition is satisfied,

$$\cos(\theta) = \frac{q^2 - \omega^2/v_{\rm F}^2 + 2k\omega/v_{\rm F}}{2kq}, \tag{2.17}$$

which in turn requires

$$\left|\frac{q^2 - \omega^2/v_{\rm F}^2 + 2k\omega/v_{\rm F}}{2kq}\right| \leq 1. \tag{2.18}$$

Equation (2.17) admits always one solution in the interval $[0, \pi]$. When Equation (2.17) is satisfied,

$$\cos(\theta_{\boldsymbol{k},\boldsymbol{k}-\boldsymbol{q}}) \mapsto 1 - \frac{q^2 - \omega^2/v_{\rm F}^2}{2k(k - \omega/v_{\rm F})}, \tag{2.19}$$

and therefore

$$\begin{aligned} \mathcal{F}_{++}(\cos(\theta_{\boldsymbol{k},\boldsymbol{k}-\boldsymbol{q}})) &= \frac{1 + \cos(\theta_{\boldsymbol{k},\boldsymbol{k}-\boldsymbol{q}})}{2} \\ &\mapsto 1 - \frac{q^2 - \omega^2/v_{\rm F}^2}{4k(k - \omega/v_{\rm F})}. \end{aligned} \tag{2.20}$$

For future purposes, we introduce the following definition:

$$\widetilde{\mathcal{F}}(k, q, \omega) \equiv 1 - \frac{q^2 - \omega^2/v_{\rm F}^2}{4k(k - \omega/v_{\rm F})}. \tag{2.21}$$

Note that $\widetilde{\mathcal{F}}(k, q, \omega) = 1$ for $\omega = \pm v_{\rm F} q$.

For intra-band scattering the result of the angular integration in Equation (2.14) is therefore

$$\begin{aligned} &A_{++}(k, q, \omega) \\ &= 2 \times \frac{2c(k - \omega/v_{\rm F})\widetilde{\mathcal{F}}(k, q, \omega)}{v_{\rm F}\sqrt{(2k+q-\omega/v_{\rm F})(2k-q-\omega/v_{\rm F})(q-\omega/v_{\rm F})(q+\omega/v_{\rm F})}} \\ &\quad \times \Theta(k - \omega/v_{\rm F})\Theta\left(1 - \left|\frac{q^2 - \omega^2/v_{\rm F}^2 + 2k(\omega/v_{\rm F})}{2kq}\right|\right), \end{aligned} \tag{2.22}$$

where the first factor of two is the same as the one appearing in Equation (2.14). In Equation (2.22) $\Theta(x) = 1$ if $x \geq 0$ and 0 otherwise. Furthermore, c is a numerical coefficient:

$$c = \begin{cases} 1/2, & \text{for } q = \omega/v_{\rm F} \text{ and } q = 2k - \omega/v_{\rm F} \\ 1, & \text{elsewhere} \end{cases}. \tag{2.23}$$

Indeed, for $q = \omega/v_{\rm F}$ and $q = 2k - \omega/v_{\rm F}$ we have $\cos(\theta) = +1$— see Equation (2.17)—and therefore the solution, *i.e.* $\theta = 0$, falls on the boundary of the integration domain in Equation (2.14) (and therefore the integral of the delta function gives an extra factor 1/2).

A careful analysis of Equation (2.22) allows us to conclude that

$$A_{++} = \frac{4c(k - \omega/v_{\rm F})\widetilde{\mathcal{F}}(k, q, \omega)}{v_{\rm F}\sqrt{[(2k - \omega/v_{\rm F})^2 - q^2](q^2 - \omega^2/v_{\rm F}^2)}} \tag{2.24}$$

for

$$\frac{|\omega|}{v_{\rm F}} \leq q \leq 2k - \frac{\omega}{v_{\rm F}}, \tag{2.25}$$

and zero elsewhere. Note that for $|\omega|/v_{\rm F} \leq q \leq 2k - \omega/v_{\rm F}$ the argument of the square root in Equation (2.24) is positive.

2.2. Inter-band contribution

For $\lambda' = -1$ (inter-band scattering) Equation (2.15) requires $0 \leq k \leq \omega/v_{\rm F}$ since it must be

$$\sqrt{k^2 + q^2 - 2kq\cos(\theta)} = \frac{\omega}{v_{\rm F}} - k\,. \tag{2.26}$$

We therefore find

$$\cos(\theta) = \frac{q^2 - \omega^2/v_{\rm F}^2 + 2k\omega/v_{\rm F}}{2kq} \tag{2.27}$$

as in Equation (2.17), and

$$\cos(\theta_{\boldsymbol{k},\boldsymbol{k}-\boldsymbol{q}}) \mapsto -1 + \frac{q^2 - \omega^2/v_{\rm F}^2}{2k(k - \omega/v_{\rm F})}\,. \tag{2.28}$$

Note the difference between the previous result and Equation (2.19). We therefore find that

$$\mathcal{F}_{+-}(\cos(\theta_{\boldsymbol{k},\boldsymbol{k}-\boldsymbol{q}})) = \frac{1 - \cos(\theta_{\boldsymbol{k},\boldsymbol{k}-\boldsymbol{q}})}{2} \mapsto \widetilde{F}(k, q, \omega)\,. \tag{2.29}$$

We emphasize that, when Equations (2.16) and (2.26) are satisfied, $\mathcal{F}_{+-}(\cos(\theta_{\boldsymbol{k},\boldsymbol{k}-\boldsymbol{q}})) = \mathcal{F}_{++}(\cos(\theta_{\boldsymbol{k},\boldsymbol{k}-\boldsymbol{q}}))$.

In summary, we find

$$\begin{aligned} A_{+-}(k, q, \omega) &= \frac{4c(\omega/v_{\rm F} - k)\widetilde{\mathcal{F}}(k, q, \omega)}{v_{\rm F}\sqrt{(2k + q - \omega/v_{\rm F})(2k - q - \omega/v_{\rm F})(q - \omega/v_{\rm F})(q + \omega/v_{\rm F})}} \\ &\times \Theta(\omega/v_{\rm F} - k)\Theta\left(1 - \left|\frac{q^2 - \omega^2/v_{\rm F}^2 + 2k(\omega/v_{\rm F})}{2kq}\right|\right)\,. \end{aligned} \tag{2.30}$$

A careful analysis of Equation (2.30) allows us to conclude that

$$A_{+-} = \frac{4c(\omega/v_{\rm F} - k)\widetilde{\mathcal{F}}(k, q, \omega)}{v_{\rm F}\sqrt{[(2k - \omega/v_{\rm F})^2 - q^2](q^2 - \omega^2/v_{\rm F}^2)}} \,. \tag{2.31}$$

for

$$\left|2k - \frac{\omega}{v_{\rm F}}\right| \leq q \leq \frac{\omega}{v_{\rm F}} \,, \tag{2.32}$$

and zero elsewhere.

2.3. Intra-band scattering and the collinear scattering singularity

As we will see below in Sect. 2.4, for weakly-excited states only intra-band terms contribute to the decay rate of a plane-wave state. We can therefore focus only on Equations (2.24)-(2.25).

We clearly see from Equation (2.24) that the denominator of $A_{++}(q, k, \omega)$ vanishes like $\sqrt{q^2 - \omega^2/v_{\rm F}^2}$ for $\omega \to \pm v_{\rm F} q$. Now, the imaginary part of the non-interacting density-density response function, $\Im m[\chi^{(0)}(q, \omega, T)]$, in Equation (2.10) diverges [34] like $1/\sqrt{q^2 - \omega^2/v_{\rm F}^2}$. The combination of these two facts produces an overall factor $1/(q^2 - \omega^2/v_{\rm F}^2)$ in Equation (2.10). Because of this factor, the standard static screening approximation, [4,8] which consists in replacing $\varepsilon(q, \omega, T)$ by $\varepsilon(q, 0, T)$ in Equation (2.10), is seen to fail miserably in doped graphene, yielding a logarithmically-divergent intra-band scattering rate [35, 36]. The divergence arises from the regions of phase space in which $\omega = \pm v_{\rm F} q$. This condition characterizes scattering events in which all the involved electronic momenta are parallel to each other. The "collinear scattering" singularity has been known for a long time in systems with linear-in-momentum energy bands (see, for example, Ref. [37]) and has been extensively discussed in the recent graphene-related literature (see, for example, Ref. [26] and references therein to earlier work). This divergence can be handled in a variety of ways: one can, for example, introduce a cut-off in the integration over q or use dynamical screening, as the G_0W-RPA theory we have adopted since the very beginning seems to suggest. Dynamical RPA screening, indeed, naturally cures the collinear scattering singularity because $\varepsilon(q, \omega, T)$ in Equation (2.10) diverges precisely as $1/\sqrt{q^2 - \omega^2/v_{\rm F}^2}$ upon approaching $q = |\omega|/v_{\rm F}$.

2.4. Asymptotic behavior of the decay rate for weakly-excited states

With this body of knowledge at our disposal, we can now evaluate Equation (2.10) analytically. For the sake of definiteness, we consider an n-

doped graphene sheet with electron density n, Fermi energy $\varepsilon_{\rm F} = \hbar v_{\rm F} k_{\rm F}$, and Fermi momentum $k_{\rm F} = \sqrt{4\pi n/N_{\rm f}}$. Here $N_{\rm f} = 4$ is the number of fermion flavors.

Let us start by considering the first term in Equation (2.10), *i.e.* the quasiparticle decay rate. We consider a weakly-excited state composed of a quasiparticle with $k \gtrsim k_{\rm F}$ and $k_{\rm B}T \ll \varepsilon_{\rm F}$. The thermal factor

$$\frac{1 - n_{\rm F}(\xi_{\boldsymbol{k},+} - \omega)}{1 - \exp(-\beta\omega)} \tag{2.33}$$

imposes some natural bounds on the integration domain with respect to the energy transfer ω. For $\omega < 0$, the natural lower bound of integration is $k_{\rm B}T$. For $\omega > 0$ the upper bound of integration is of the order of $\xi_{\boldsymbol{k},+} = v_{\rm F}(k - k_{\rm F}) \ll \varepsilon_{\rm F}$. For $k \to k_{\rm F}$, moreover, we can approximate Equations (2.24)-(2.25) as

$$\begin{aligned} A_{++}(k, q, \omega) &\simeq \frac{2c(1 - q^2/4k_{\rm F}^2)}{v_{\rm F}\sqrt{(1 - q^2/4k_{\rm F}^2)(q^2 - \omega^2/v_{\rm F}^2)}} \\ &= \frac{2c\sqrt{1 - q^2/4k_{\rm F}^2}}{v_{\rm F}\sqrt{q^2 - \omega^2/v_{\rm F}^2}} \, . \end{aligned} \tag{2.34}$$

¿From now on, we set $c = 1$ in Equation (2.34) since $c \neq 1$ on a set of zero measure with respect to the 2D integral in Equation (2.10).

In the same limits, the inter-band contribution to the quasiparticle decay rate vanishes since, on the one hand, the thermal factor $1 - n_{\rm F}(\xi_{\boldsymbol{k},+} - \omega)$ imposes $\omega < \xi_{\boldsymbol{k},+}$ for $T \to 0$, while, on the other hand, $A_{+-}(q, k, \omega)$ is non zero if and only if $\omega \geq v_{\rm F}k$ (at any temperature—see Sect. 2.2). Finally, we emphasize that *only* the spectral density of intra-band electron-hole pairs contributes to $1/\tau^{(\rm e)}_{\boldsymbol{k},+}$, since $A_{++}(k, q, \omega) \neq 0$ if and only if $q > |\omega|/v_{\rm F}$.

For small values of the energy transfer ω, $|\omega|/v_{\rm F} \leq q \leq 2k_{\rm F} - \omega/v_{\rm F}$, and $k_{\rm B}T \ll \varepsilon_{\rm F}$ the imaginary part of the non-interacting density-density response function can be approximated as following:

$$\begin{aligned} \Im m[\chi^{(0)}(q, \omega, T)] &\simeq -N(0)\frac{\omega}{v_{\rm F}q}\sqrt{1 - \frac{q^2}{4k_{\rm F}^2}} \\ &\quad \times \frac{q}{\sqrt{q^2 - \omega^2/v_{\rm F}^2}} \, , \end{aligned} \tag{2.35}$$

where

$$N(0) = \frac{N_{\rm f} k_{\rm F}}{2\pi v_{\rm F}} \tag{2.36}$$

is the density-of-states at the Fermi energy. Equation (2.35) is a contribution of purely intra-band origin. Note that: i) $\Im m[\chi^{(0)}(q,\omega,T)]$ is proportional (and not inversely proportional, as in the ordinary 2DEG [4]) to the factor $\sqrt{1-q^2/4k_{\rm F}^2}$: this fact beautifully reflects the impossibility of MDFs to be backscattered; ii) we have retained the frequency dependence of the factor on the second line of the previous equation: this is crucial to regularize the collinear scattering singularity for $q \to |\omega|/v_{\rm F}$ in Equation (2.34).

In the same range of values of ω, q, and T we have

$$\Re e[\chi^{(0)}(q,\omega,T)] = -N(0)\,. \tag{2.37}$$

We therefore conclude that, in the relevant range of values of ω, q, and T, the RPA dielectric function can be well approximated by

$$\begin{aligned}\varepsilon(q,\omega,T) \;&\simeq\; 1+\frac{2\pi e^2 N(0)}{\epsilon q} - i\,\frac{2\pi e^2 N(0)}{\epsilon q}\\ &\times \frac{\omega}{v_{\rm F} q}\sqrt{1-\frac{q^2}{4k_{\rm F}^2}}\,\frac{q}{\sqrt{q^2-\omega^2/v_{\rm F}^2}}\,.\end{aligned} \tag{2.38}$$

It is useful at this stage to introduce the Thomas-Fermi screening wave vector:

$$q_{\rm TF} \equiv \frac{2\pi e^2 N(0)}{\epsilon} = N_{\rm f}\alpha_{\rm ee}k_{\rm F}\,. \tag{2.39}$$

We therefore find

$$\begin{aligned}\frac{1}{\tau^{(\rm e)}_{\boldsymbol{k},+}} &\simeq \frac{4N(0)}{(2\pi)^2 v_{\rm F}^2}\int_{-\infty}^{+\infty} d\omega\,\omega\frac{1-n_{\rm F}(\xi_{\boldsymbol{k},+}-\omega)}{1-\exp(-\beta\omega)}\\ &\times\int_{|\omega|/v_{\rm F}}^{2k_{\rm F}-\omega/v_{\rm F}} dq\,q\frac{v_q^2}{\left(1+\dfrac{q_{\rm TF}}{q}\right)^2+\dfrac{q_{\rm TF}^2}{q^2}\dfrac{\omega^2}{v_{\rm F}^2}\dfrac{1-q^2/4k_{\rm F}^2}{q^2-\omega^2/v_{\rm F}^2}}\,\frac{1-q^2/4k_{\rm F}^2}{q^2-\omega^2/v_{\rm F}^2}\\ &= 4N(0)\alpha_{\rm ee}^2\int_{-\infty}^{+\infty} d\omega\,\omega\frac{1-n_{\rm F}(\xi_{\boldsymbol{k},+}-\omega)}{1-\exp(-\beta\omega)}\\ &\times\int_{|\omega|/v_{\rm F}}^{2k_{\rm F}-\omega/v_{\rm F}} dq\,\frac{1}{q}\frac{1}{\left(1+\dfrac{q_{\rm TF}}{q}\right)^2+\dfrac{q_{\rm TF}^2}{q^2}\dfrac{\omega^2}{v_{\rm F}^2}\dfrac{1-q^2/4k_{\rm F}^2}{q^2-\omega^2/v_{\rm F}^2}}\,\frac{1-q^2/4k_{\rm F}^2}{q^2-\omega^2/v_{\rm F}^2}\,.\end{aligned} \tag{2.40}$$

The integral over q in the previous equation is easily seen to diverge logarithmically for $\omega \to 0$. Indeed, we can estimate the integral over q

as follows:

$$\int_{|\omega|/v_F}^{2k_F-\omega/v_F} dq \, \frac{1}{q} \frac{1}{\left(1+\frac{q_{TF}}{q}\right)^2 + \frac{q_{TF}^2}{q^2}\frac{\omega^2}{v_F^2}\frac{1-q^2/4k_F^2}{q^2-\omega^2/v_F^2}} \frac{1-q^2/4k_F^2}{q^2-\omega^2/v_F^2} \simeq \frac{1}{q_{TF}^2}\ln\left(\frac{\Lambda}{|\omega|}\right), \tag{2.41}$$

where Λ is an arbitrary ultraviolet cut-off whose value does not affect the results to *leading* order in the low-energy and low-temperature limits.

To obtain Equation (2.41) we have neglected the first term in the denominator, which is much smaller than the second term since the latter diverges as $(q^2 - \omega^2/v_F^2)^{-1}$ when q approaches the lower bound of integration. In other words, "dynamical overscreening", which occurs near the light cone $\omega = \pm v_F q$ of a MDF system, completely dominates over the conventional static screening $(1 + q_{TF}/q)^2$. From Equation (2.41) it is clear that the logarithmic divergence for $|\omega| \to 0$ originates from the region of small momenta. In the ordinary 2DEG, a similar divergence [4] picks a finite contribution also from the region $q \sim 2k_F$. Chirality of MDFs strongly suppresses this contribution in the case of a doped graphene sheet.

In summary, we find

$$\frac{1}{\tau_{k,+}^{(e)}} \to \frac{4N(0)}{N_f^2 k_F^2} \int_{-\infty}^{+\infty} d\omega \, \omega \ln\left(\frac{\Lambda}{|\omega|}\right) \frac{n_F(\omega - \xi_{k,+})}{1-\exp(-\beta\omega)}, \tag{2.42}$$

where we have used that $n_F(x) + n_F(-x) = 1$. Note that the final result (2.42) does not depend on the fine-structure constant α_{ee}. As emphasized in the Introduction, this feature arises from the fact that the dominant contribution to the quasiparticle decay rate arises from scattering processes with small momentum transfer q. For these processes the Coulomb interaction between two quasiparticles is strongly screened by the electronic medium, leading to an effective interaction that depends only on the non-interacting density of states $N(0)$.

The above Equation (2.42) is valid regardless of the relative magnitude of temperature and quasiparticle energy ξ_k, provided they are both much smaller than the Fermi energy. We now specialize to the case in which one of these two quantities is much larger than the other. Following Ref. [4], we first consider the zero-temperature limit in which $\beta\xi_{k,+} \gg 1$. In this case the main contribution to the previous integral comes from the region $\omega \sim \xi_{k,+}$. Since the logarithm is a slowly-varying function of its argument we find

$$\frac{1}{\tau^{(\rm e)}_{\boldsymbol{k},+}} \simeq \frac{4N(0)}{N_{\rm f}^2 k_{\rm F}^2} \ln\left(\frac{\Lambda}{\xi_{\boldsymbol{k},+}}\right) \int_{-\infty}^{+\infty} d\omega\, \omega \frac{n_{\rm F}(\omega - \xi_{\boldsymbol{k},+})}{1-\exp(-\beta\omega)} . \tag{2.43}$$

We now use the "beautiful integral" (Ref. [4], p. 497)

$$\int_{-\infty}^{+\infty} dy \frac{x-y}{(1+e^{-y})(1-e^{y-x})} = \frac{x^2+\pi^2}{2(1+e^{-x})} , \tag{2.44}$$

with $x = \beta\xi_{\boldsymbol{k},+}$ and $y = \beta(\xi_{\boldsymbol{k},+} - \omega)$. We find

$$\int_{-\infty}^{+\infty} d\omega\, \omega \frac{n_{\rm F}(\omega - \xi_{\boldsymbol{k},+})}{1-\exp(-\beta\omega)} = \frac{1}{\beta^2} \frac{\beta^2\xi^2_{\boldsymbol{k},+} + \pi^2}{2(1+e^{-\beta\xi_{\boldsymbol{k},+}})} . \tag{2.45}$$

In the regime $\beta\xi_{\boldsymbol{k},+} \gg 1$ we therefore have

$$\frac{1}{\tau^{(\rm e)}_{\boldsymbol{k},+}} \simeq \frac{\varepsilon_{\rm F}}{\hbar} \frac{1}{\pi N_{\rm f}} \left(\frac{\xi_{\boldsymbol{k},+}}{\varepsilon_{\rm F}}\right)^2 \ln\left(\frac{\Lambda}{\xi_{\boldsymbol{k},+}}\right) , \tag{2.46}$$

where we have restored $\hbar$. A careful inspection of Equation (2.10) shows that, in the limit $\beta|\xi_{\boldsymbol{k},+}| \gg 1$, the decay rate of a plane-wave state is given by is

$$\frac{1}{\tau_{\boldsymbol{k},+}} \simeq \frac{\varepsilon_{\rm F}}{\hbar} \frac{1}{\pi N_{\rm f}} \left(\frac{\xi_{\boldsymbol{k},+}}{\varepsilon_{\rm F}}\right)^2 \ln\left(\frac{\Lambda}{|\xi_{\boldsymbol{k},+}|}\right) , \tag{2.47}$$

i.e. it is equal to that in the right-hand side of Equation (2.46), provided that we replace $\xi_{\boldsymbol{k},+} \to |\xi_{\boldsymbol{k},+}|$ in this equation. To leading order in the zero-temperature limit the functional dependence on $\xi_{\boldsymbol{k},+}$, *i.e.* $-\xi^2_{\boldsymbol{k},+} \ln(|\xi_{\boldsymbol{k},+}|)$, coincides with that of an ordinary 2DEG [8].

We now turn to analyze the other relevant regime, *i.e.* $\beta|\xi_{\boldsymbol{k},+}| \ll 1$. In this case the main contribution to the integral comes from a region of the order of $k_{\rm B}T$ centered around the origin. Once again, the logarithm can be taken out of the integral giving a factor $\ln(\Lambda/k_{\rm B}T)$.

In the regime $\beta|\xi_{\boldsymbol{k},+}| \ll 1$ we therefore find

$$\frac{1}{\tau^{(\rm e)}_{\boldsymbol{k},+}} \simeq \frac{\varepsilon_{\rm F}}{\hbar} \frac{\pi}{2N_{\rm f}} \left(\frac{k_{\rm B}T}{\varepsilon_{\rm F}}\right)^2 \ln\left(\frac{\Lambda}{k_{\rm B}T}\right) , \tag{2.48}$$

where we have used an expansion of Equation (2.45) for $\beta|\xi_{\boldsymbol{k},+}| \to 0$ and we have restored $\hbar$. Inspecting Equation (2.10) for $\xi_{\boldsymbol{k},+} = 0$ we conclude that

$$\lim_{\xi_{\boldsymbol{k},+}\to 0} \frac{1}{\tau_{\boldsymbol{k},+}} = \frac{2}{\tau^{(\rm e)}_{\boldsymbol{k},+}} \simeq \frac{\varepsilon_{\rm F}}{\hbar} \frac{\pi}{N_{\rm f}} \left(\frac{k_{\rm B}T}{\varepsilon_{\rm F}}\right)^2 \ln\left(\frac{\Lambda}{k_{\rm B}T}\right) . \tag{2.49}$$

Once again, the functional dependence on temperature of Equation (2.48), *i.e.* $-T^2 \ln(T)$, coincides with that obtained by Giuliani and Quinn [8] for an ordinary 2DEG.

3. Summary and conclusions

In summary, we have presented a pedagogical derivation of the quasiparticle lifetime in a doped graphene sheet. Three main differences with respect to the classic Giuliani-Quinn calculation [8] for an ordinary two-dimensional electron gas have been identified: i) a simple Fermi golden rule approach with statically screened Coulomb interactions is not viable in graphene as it yields logarithmically-divergent intra-band scattering rates due to the collinear scattering singularity; ii) the leading-order contribution to the quasiparticle decay rate in the low-energy and low-temperature limits is completely controlled by scattering events with small momentum transfer: $2k_{\rm F}$ contributions are suppressed by the chiral nature of massless Dirac carriers in graphene; iii) because of ii), the leading order contribution to the quasiparticle decay rate is completely independent on the strength on the background dielectric constant ϵ: the result is therefore *universal* in that it does not depend on the substrate on which graphene is placed.

Finally, we emphasize how the recently developed ability to align the crystals of two graphene sheets [38] paves the way for two-dimensional-to-two-dimensional tunneling experiments [12] in which inter-layer tunneling does not spoil momentum conservation. These experiments may allow a direct measurement of the temperature and doping dependence of the quasiparticle lifetime in high-quality graphene sheets.

References

[1] P. NOZIÉRES, "Theory of Interacting Fermi Systems", W. A. Benjamin, Inc., New York, 1964.

[2] D. PINES and P. NOZIÉRES, "The Theory of Quantum Liquids", W. A. Benjamin, Inc., New York, 1966.

[3] R. SHANKAR, Rev. Mod. Phys. **66** (1994), 129.

[4] G. F. GIULIANI and G. VIGNALE, "Quantum Theory of the Electron Liquid", Cambridge University Press, Cambridge, 2005.

[5] T. GIAMARCHI, "Quantum Physics in One Dimension", Clarendon Press, Oxford, 2004.

[6] E. ABRAHAMS, P. W. ANDERSON, D. C. LICCIARDELLO and T. V. RAMAKRISHNAN, Phys. Rev. Lett. **42** (1979), 673.

[7] K. VON KLITZING, G. DORDA and M. PEPPER, Phys. Rev. Lett. **45** (1980), 494.

[8] G. F. GIULIANI and J. J. QUINN, Phys. Rev. B **26** (1982), 4421.
[9] For other earlier works on the quasiparticle lifetime of a 2DEG see, for example, A. V. CHAPLIK, Sov. Phys. JETP **33** (1971), 997; C. HODGES, H. SMITH and J. W. WILKINS, **4** (1971), 302; H. FUKUYAMA and E. ABRAHAMS, **27** (1983), 5976.
[10] See, for example, T. JUNGWIRTH and A. H. MACDONALD, Phys. Rev. B **53** (1996), 7403; L. ZHENG and S. DAS SARMA, Phys. Rev. B **53** (1996), 9964; M. REIZER and J. W. WILKINS, Phys. Rev. B **55** (1997), R7363; Z. QIAN and G. VIGNALE, Phys. Rev. B **71** (2005), 075112.
[11] See, for example, V. M. PUDALOV, M. E. GERSHENSON, H. KOJIMA, N. BUTCH, E. M. DIZHUR, G. BRUNTHALER, A. PRINZ and G. BAUER, Phys. Rev. Lett. **88** (2002), 196404; J. ZHU, H. L. STORMER, L. N. PFEIFFER, K. W. BALDWIN and K. W. WEST, Phys. Rev. Lett. **90** (2003), 056805; K. VAKILI, Y. P. SHKOLNIKOV, E. TUTUC, E. P. DE POORTERE and M. SHAYEGAN, Phys. Rev. Lett. **92** (2004), 226401; Y.-W. TAN, J. ZHU, H. L. STORMER, L. N. PFEIFFER, K. W. BALDWIN and K. W. WEST, Phys. Rev. Lett. **94** (2005), 016405 and references therein.
[12] S. Q. MURPHY, J. P. EISENSTEIN, L. N. PFEIFFER and K. W. WEST, Phys. Rev. B **52** (1995), 14825.
[13] S. V. KRAVCHENKO, G. V. KRAVCHENKO, J. E. FURNEAUX, V. M. PUDALOV and M. D'IORIO, Phys. Rev. B **50** (1994), 8039; S. V. KRAVCHENKO, W. E. MASON, G. E. BOWKER, J. E. FURNEAUX, V. M. PUDALOV and M. D'IORIO, Phys. Rev. B **51** (1995), 7038.
[14] K. S. NOVOSELOV, A. K. GEIM, S. V. MOROZOV, D. JIANG, Y. ZHANG, S. V. DUBONOS, I. V. GRIGORIEVA and A. A. FIRSOV, Science **306** (2004), 666.
[15] A. K. GEIM and K. S. NOVOSELOV, Nature Mater. **6**, 183 (2007), 183; A. H. CASTRO NETO, F. GUINEA, N. M. R. PERES, K. S. NOVOSELOV and A. K. GEIM, Rev. Mod. Phys. **81** (2009), 109; S. DAS SARMA, S. ADAM, E. H. HWANG and E. ROSSI, Rev. Mod. Phys. **83** (2011), 407; F. BONACCORSO, A. LOMBARDO, T. HASANA, Z. SUNA, L. COLOMBO and A. C. FERRARI, Mater. Today **15** (2012), 564; M. I. KATSNELSON, "Graphene: Carbon in Two Dimensions", Cambridge University Press, Cambridge, 2012.
[16] J. GONZÁLEZ, F. GUINEA and M. A. H. VOZMEDIANO, Phys. Rev. Lett. **77** (1996), 3589 and Phys. Rev. B **59** (1999), R2474.
[17] D. C. ELIAS, R. V. GORBACHEV, A. S. MAYOROV, S. V. MOROZOV, A. A. ZHUKOV, P. BLAKE, L. A. PONOMARENKO, I. V.

GRIGORIEVA, K. S. NOVOSELOV, F. GUINEA and A. K. GEIM, Nature Phys. **7** (2011), 701.
[18] D. A. SIEGEL, C.-H. PARK, C. HWANG, J. DESLIPPE, A. V. FEDOROV, S. G. LOUIE and A. LANZARA, Proc. Natl. Acad. Sci. (USA) **108** (2011), 11365.
[19] Y. BARLAS, T. PEREG-BARNEA, M. POLINI, R. ASGARI and A. H. MACDONALD, Phys. Rev. Lett. **98** (2007), 236601.
[20] M. POLINI, R. ASGARI, Y. BARLAS, T. PEREG-BARNEA and A. H. MACDONALD, Solid State Commun. **143** (2007), 58.
[21] M. POLINI, R. ASGARI, G. BORGHI, Y. BARLAS, T. PEREG-BARNEA and A. H. MACDONALD, Phys. Rev. B **77** (2008), 081411(R).
[22] E. H. HWANG, B.Y.-K. HU and S. DAS SARMA, Phys. Rev. B **76** (2007), 115434.
[23] S. DAS SARMA, E. H. HWANG and W.-K. TSE, Phys. Rev. B **75** (2007), 121406.
[24] E. H. HWANG and S. DAS SARMA, Phys. Rev. B **77** (2008), 081412(R).
[25] M. SCHÜTT, P. M. OSTROVSKY, I. V. GORNYI and A. D. MIRLIN, Phys. Rev. B **83** (2011), 155441.
[26] A. TOMADIN, D. BRIDA, G. CERULLO, A. C. FERRARI and M. POLINI, Phys. Rev. B **88** (2013), 035430.
[27] Q. LI and S. DAS SARMA, Phys. Rev. B **87** (2013), 085406.
[28] J. C. W. SONG, K. J. TIELROOIJ, F. H. L. KOPPENS and L. S. LEVITOV, Phys. Rev. B **87** (2013), 155429.
[29] D. M. BASKO, Phys. Rev. B **87** (2013), 165437.
[30] U. BRISKOT, I. A. DMITRIEV and A. D. MIRLIN, Phys. Rev. B **89** (2014), 075414.
[31] D. BRIDA, A. TOMADIN, C. MANZONI, Y. J. KIM, A. LOMBARDO, S. MILANA, R. R. NAIR, K. S. NOVOSELOV, A. C. FERRARI, G. CERULLO RM AND M. POLINI, Nature Commun. **4** (2013), 1987.
[32] K. J. TIELROOIJ, J. C. W. SONG, S. A. JENSEN, A. CENTENO, A. PESQUERA, A. ZURUTUZA ELORZA, M. BONN, L. S. LEVITOV and F. H. L. KOPPENS, Nature Phys. **9** (2013), 248.
[33] V. N. KOTOV, B. UCHOA, V. M. PEREIRA, F. GUINEA and A. H. CASTRO NETO, Rev. Mod. Phys. **84** (2012), 1067.
[34] M. R. RAMEZANALI, M. M. VAZIFEH, R. ASGARI, M. POLINI and A. H. MACDONALD, J. Phys. A **42** (2009), 214015.
[35] A. B. KASHUBA, Phys. Rev. B **78** (2008), 085415.
[36] L. FRITZ, J. SCHMALIAN, M. MÜLLER and S. SACHDEV, Phys. Rev. B **78** (2008), 085416.

[37] S. Sachdev, Phys. Rev. B **57** (1998), 7157.
[38] A. Mishchenko, J. S. Tu, Y. Cao, R. V. Gorbachev, J. R. Wallbank, M. T. Greenaway, V. E. Morozov, S. V. Morozov, M. J. Zhu, S. L. Wong, F. Withers, C. R. Woods, Y.-J. Kim, K. Watanabe, T. Taniguchi, E. E. Vdovin, O. Makarovsky, T. M. Fromhold, V. I. Falko, A. K. Geim, L. Eaves and K. S. Novoselov, Nature Nanotech **9** (2014), 808.

Chirality, charge and spin-density wave instabilities of a two-dimensional electron gas in the presence of Rashba spin-orbit coupling

George E. Simion and Gabriele F. Giuliani

Abstract. We show that a result equivalent to Overhauser's famous Hartree-Fock instability theorem can be established for the case of a two-dimensional electron gas in the presence of Rashba spin-obit coupling. In this case it is the spatially homogeneous paramagnetic chiral ground state that is shown to be differentially unstable with respect to a certain class of distortions of the spin-density-wave and charge-density-wave type. The result holds for all densities. Basic properties of these inhomogeneous states are analyzed.

1. Introduction

Recent interest in the properties of the quasi-two dimensional electron and hole devices in the presence of structural (Rashba-Bychkov) [1,2] or intrinsic (Dresselhaus) [3] spin orbit has brought to the fore the problem of the interacting chiral electron liquid. It is therefore important to revisit several of the fundamental notions of many-body theory for this intriguing system. The purpose of present paper is to begin a theoretical exploration of the relevance and special properties of a class of spatially non homogeneous spontaneously broken symmetry states of the electron liquid in the presence of Rashba spin-orbit coupling in two dimensions. Specifically we will focus our attention on spin density and charge density wave type states henceforth referred to for simplicity as SDW and CDW. SDW and CDW states, originally conceived by A. W. Overhauser, [4,5] are generally stabilized by the electron-electron interaction and are characterized by spatial oscillations of the the spin density, the charge density, or both. In the absence of spin-orbit coupling, one can begin to describe SDW and CDW states by simply considering the elec-

GS work was partially supported by a grant from the Purdue Research Foundation.

tron number density for both spin projections:

$$n_{\uparrow} = \frac{n}{2} + A\cos\left(\vec{Q}\cdot\vec{R} + \frac{\phi}{2}\right), \tag{1.1}$$

$$n_{\downarrow} = \frac{n}{2} + A\cos\left(\vec{Q}\cdot\vec{R} - \frac{\phi}{2}\right). \tag{1.2}$$

In these expressions the wave vector $\vec{Q}$ spans the Fermi surface, *i.e.* does satisfy the condition $|\vec{Q}| \simeq 2k_F$. A CDW corresponds to $\phi = 0$, while a SDW state obtains for $\phi = \pi$. Mixed state are also possible: one such state is beautifully realized in chromium [6–8]. As we will show, in the presence of linear Rashba spin orbit the corresponding distorted states are characterized by a more complex spatial dependence of the number density, the spin density and, where appropriate, a chiral density. As a first step towards establishing the fundamental properties of SDW- and CDW-like states in the presence of spin-orbit interaction we present here a generalization of the famous Overhauser's Hartree-Fock (HF) instability theorem. The latter represents an important exact result in many-body theory for it establishes that, within HF, the homogeneous paramagnetic plane wave state does not represents a minimum of the energy for an otherwise uniform electron gas for it can be always variationally bettered by a suitably constructed distorted SDW or CDW [9].

The paper is structured as follows: In Section 2 we discuss the relevant aspects of the theory of a two dimensional non-interacting electron gas in the presence of Rashba spin-obit coupling. Section 3 briefly discusses useful notions of the electron-electron interaction within Hartree-Fock approximation. Section 4 is dedicated to the actual proof of the theorem and contains the main results. Finally the last Section contains the conclusions while a number of useful mathematical relations are derived in the two Appendices.

ACKNOWLEDGEMENTS. The authors would like to acknowledge useful discussions with A. W. Overhauser.

2. Two dimensional electron gas in the presence of Rashba spin-orbit

In the presence of linear Rashba spin-orbit coupling, the one-particle hamiltonian can be written as follows:

$$\hat{H}_0 = \frac{\vec{p}^2}{2m} + \alpha\vec{p}\cdot(\vec{\sigma}\times\hat{z}), \tag{2.1}$$

where $\hat{z}$ is the unit direction along the z-axis, the motion taking place in the x, y plane.

The non interacting problem can be readily diagonalized to obtain the energy spectrum and the eigenfunctions:

$$E_{k,\mu} = \frac{\hbar^2 k^2}{2m} - \alpha\mu k \; , \tag{2.2}$$

and

$$\psi_{\vec{k},+} = \frac{1}{\sqrt{2}L} e^{i\vec{k}\cdot\vec{r}} \begin{pmatrix} 1 \\ -ie^{i\phi_{\vec{k}}} \end{pmatrix} , \tag{2.3}$$

$$\psi_{\vec{k},-} = \frac{1}{\sqrt{2}L} e^{i\vec{k}\cdot\vec{r}} \begin{pmatrix} -ie^{-i\phi_{\vec{k}}} \\ 1 \end{pmatrix} , \tag{2.4}$$

where $\mu = \pm$ labels a state's *chirality* and $\phi_{\vec{k}}$ is the angle spanned by the x-axis and the two dimensional wave vector $\vec{k}$. A schematic of the lower energy sector of the spectrum is plotted in Figure 1.

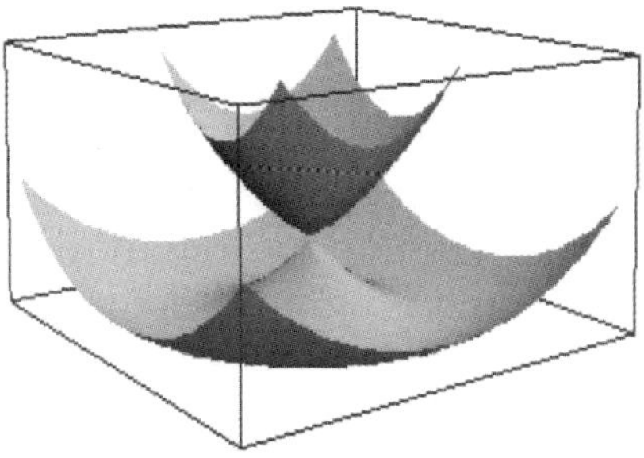

Figure 1. Non interacting energy spectrum in the presence of linear Rashba spin-orbit coupling.

By making use of the states of Equations (2.3) and (2.4) as a basis for a second quantization representation, the familiar fully interacting electron gas hamiltonian reads:

$$\hat{H} = \sum_{\vec{k},\mu} E_{\vec{k},\mu} \hat{b}_{\vec{k},\mu} \hat{b}^{\dagger}_{\vec{k},\mu} + \frac{1}{2L^2} \sum_{\substack{\vec{q},\vec{k}_1,\vec{k}_2 \\ \mu_1,\mu_2 \\ \mu_3,\mu_4}} v_q \Phi^{\vec{k}_1,\vec{k}_2,\vec{q}}_{\mu_1\mu_2\mu_3\mu_4} \hat{b}^{\dagger}_{\vec{k}_1+\vec{q},\mu_1} \hat{b}^{\dagger}_{\vec{k}_2-\vec{q},\mu_2} \hat{b}_{\vec{k}_2,\mu_3} \hat{b}_{\vec{k}_1,\mu_4}, \tag{2.5}$$

where the phase factor $\Phi^{\vec{k}_1,\vec{k}_2,\vec{q}}_{\mu_1\mu_2\mu_3\mu_4}$ is defined as follows:

$$\begin{aligned}\Phi^{\vec{k}_1,\vec{k}_2,\vec{q}}_{\mu_1\mu_2\mu_3\mu_4} &= \frac{1}{4} i^{\frac{\mu_1+\mu_2-\mu_3-\mu_4}{2}} \\ &\times e^{i\left(\frac{1-\mu_1}{2}\phi_{\vec{k}_1+\vec{q}}+\frac{1-\mu_2}{2}\phi_{\vec{k}_2-\vec{q}}-\frac{1-\mu_3}{2}\phi_{\vec{k}_2}-\frac{1-\mu_4}{2}\phi_{\vec{k}_2}\right)} \\ &\times \left[1+\mu_1\mu_4 e^{i\left(\phi_{\vec{k}_1}-\phi_{\vec{k}_1+\vec{q}}\right)}\right]\left[1+\mu_2\mu_3 e^{i\left(\phi_{\vec{k}_2}-\phi_{\vec{k}_2-\vec{q}}\right)}\right].\end{aligned} \tag{2.6}$$

Uncorrelated many body wavefunctions for the system at hand can be represented by Slater determinants constructed by occupying any combinations of the chiral states Equations (2.3) and (2.4).

3. Hartree-Fock theory of a two-dimensional electron liquid in the presence of Rashba spin-orbit coupling

An accurate description of a realistic electronic system requires that electron-electron interaction be taken into account. A first step towards developing such a many-body theory is to investigate the results of a mean-field approach. The main idea behind the mean field procedure is to find an effective Hamiltonian which is quadratic in the electron creation and annihilation operators and can therefore be easily diagonalized. Within the HF theory, the ground state is approximated by a single Slater determinant made out of single particle wavefunctions, which in turn, are determined by imposing the requirement that the expectation value of the Hamiltonian over the Slater determinant be a minimum [10]. Using these wavefunctions as our basis set, a standard Wick decoupling procedure [10] allows us to determine the effective HF potential. It can be easily proved that the non-interacting chiral states are indeed among the solutions of the corresponding HF equations. In this case, the HF potential is diagonal in wave vectors and chiral indices:

$$V^{HF}_{\vec{k}\mu,\vec{k}'\mu'} = -\frac{\delta_{\vec{k},\vec{k}'}\delta_{\mu,\mu'}}{2L^2}\sum_{\vec{\kappa}\nu} v_{\vec{k}-\vec{\kappa}}\left[1+\mu\nu\cos(\phi_{\vec{k}}-\phi_{\vec{\kappa}})\right]. \tag{3.1}$$

The corresponding HF eigenvalues are given by:

$$\epsilon_{\vec{k}\mu} = \frac{\hbar^2k^2}{2m} - \alpha\mu k - \frac{1}{2L^2}\sum_{\vec{\kappa}\nu} v_{\vec{k}-\vec{\kappa}}\left[1+\mu\nu\cos(\phi_{\vec{k}}-\phi_{\vec{\kappa}})\right]. \tag{3.2}$$

An evaluation of the Fermi energy of the two sub-bands leads to an interesting problem. Since one band will, in general, acquire more exchange energy than the other, this may result (in a first iteration) in two different

Fermi levels. In order to equalize them (for elementary stability reasons), electrons from one subband will have to be moved to the other. This is the phenomenon of *repopulation*.

The spatially homogeneous chiral states are just one of the possible Hartree-Fock solutions. A detailed analysis of the possible solutions corresponding to symmetric occupations in momentum space can be done by systematically minimizing the total energy as a function of spin orientation and generalized chirality of the system [11]. More general solutions correspond to non-symmetric occupations of the single particle chiral states. The problem has been studied and the corresponding very interesting phase diagram has been explored [11,12]. As we will presently discuss there also exists an interesting class of spatially non-homogenous solutions to the problem.

4. Proof of the instability theorem

We will proceed by showing that it is always possible to lower the energy of the homogeneous paramagnetic chiral ground state by introducing a suitable real space distortion which is periodic with wave vector $\vec{Q} = 2k_F\hat{x}$. The general approach follows that of Fedders and Martin [13] and is based on an Ansatz which represents a generalization of that given by Giuliani and Vignale for the case of the three dimensional electron gas [10].

Let us consider first the putative HF ground state of our many-body system $|\Phi_S\rangle$. A complete, and, as we shall see convenient, description of this determinantal state can be achieved in terms of the corresponding single-particle density matrix elements here given by:

$$\rho_{\alpha\beta} = \left\langle \Phi_S \left| \hat{a}^\dagger_\alpha \hat{a}_\beta \right| \Phi_S \right\rangle , \tag{4.1}$$

where here α and β label the one-particle states which are used to build the Slater determinant.

Now, within the space of Slater determinants, any slightly modification of the state $|\Phi_S\rangle$ can be described in terms of a corresponding infinitesimal change of the single-particle density matrix elements. Let us indicate such a change by $\delta\rho_{\alpha\beta}$. At this point the next task consists in trying to evaluate the change of the total HF energy in terms of these quantities.

Since $|\Phi_S\rangle$ is a solution of the HF equations, [14] the first order variation in the energy must vanish so that the problem at hand is reduced to determining the sign of the energy change to second order in the $\delta\rho_{\alpha\beta}$'s.

The relevant expression is therefore given by [10]:

$$\Delta^{(2)} E_{HF} = \frac{1}{2} \sum_{(\alpha,\beta)} \delta\rho_{\alpha\beta} \frac{\epsilon_\beta - \epsilon_\alpha}{n_\alpha - n_\beta} \delta\rho_{\beta\alpha} + \frac{1}{2} \sum_{\substack{(\alpha,\delta)\\(\beta,\gamma)}} \left(v_{\alpha\beta\gamma\delta} - v_{\alpha\beta\delta\gamma}\right) \delta\rho_{\alpha\delta}\delta\rho_{\delta\gamma}\ , \tag{4.2}$$

where the notation (α, β) means that only states situated on opposite sides of the Fermi sea are considered in the summation. In this formula, we indicate the Hartree-Fock eigenvalues as ϵ_α and the corresponding occupation numbers as n_α.

The next step in our procedure consists in constructing a Slater determinant for which the HF energy is lower than $\langle\Phi_S|H|\Phi_S\rangle$. In order to do, we follow Overhauser's idea and choose the new one-particle states to be suitable linear combinations of chiral plane waves states situated near opposite points on the Fermi surface, the distortion being limited to a very narrow strip. The width of this strip will play the role of a variational parameter. We then carefully devise an expression for the wave vector dependent amplitude of the coupling between the plane waves and construct the corresponding Slater determinant. In the last step, we calculate the change of the Hartree-Fock energy due to this perturbation to leading order in the distortion amplitude from Equation (4.2).

4.1. Instability for the case of chirality equal one

Because it presents a formally simpler problem, the first case to be treated is that in which only the lower subband is occupied while the upper one is just about to be filled. [15] This situation is depicted in Figure 2.

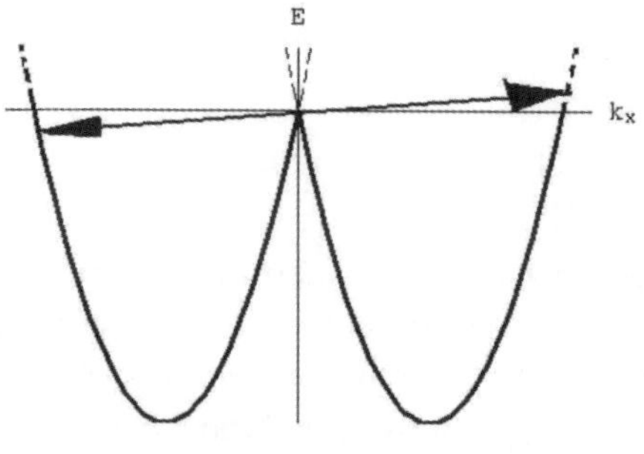

Figure 2. Case of unit chirality when the states of the lower chiral subband are occupied up to the lower threshold of the upper subband. The arrows indicate the states that are coupled by the distortion along the k_x axis.

Let us build the new trial wavefunctions as mixtures of the wavefunctions corresponding to wave vector $\vec{k}$ with those corresponding to wave vector $\vec{k} \pm \vec{Q}$, *i.e.*

$$\Psi_{\vec{k}} \simeq \psi_{\vec{k},+} + A_{\vec{k}+\vec{Q},\vec{k}} \psi_{\vec{k}+\vec{Q},+} + A_{\vec{k}-\vec{Q},\vec{k}} \psi_{\vec{k}-\vec{Q},+} \,. \tag{4.3}$$

Here, as anticipated, $|\vec{Q}| = 2k_F^+$ as shown in Figure 2.

In evaluating the HF energy change, only the states situated on opposite sides of the Fermi sea are relevant. We will consider $\left|\vec{k} \pm \vec{Q}\right| > k_F^+$ and $k < k_F$. Here, the occupation numbers are $n_{\vec{k}} = \theta\left(k_F - k\right)$, while the amplitude satisfies the condition: $A_{\vec{k}\pm\vec{Q},\vec{k}} = A_{\vec{k},\vec{k}\pm\vec{Q}}$. The only non-zero matrix elements of $\delta\hat{\rho}$ have the form:

$$\delta\rho_{\vec{k}+\frac{\vec{Q}}{2},\vec{k}-\frac{\vec{Q}}{2}} = \delta\rho_{\vec{k}-\frac{\vec{Q}}{2},\vec{k}+\frac{\vec{Q}}{2}} = \left(n_{\vec{k}+\frac{\vec{Q}}{2}} + n_{\vec{k}-\frac{\vec{Q}}{2}}\right) A_{\vec{k}} \,. \tag{4.4}$$

These wavefunctions indeed describe SDW/CDW-like states. A simple calculation shows that retaining only the linear order in the amplitude of the distortion, the spin and the charge densities exhibit spatial oscillations with wave vector $2k_F^+$. Specifically:

$$\begin{aligned}
\frac{\delta S_x\left(\vec{r}\right)}{\hbar} &= \sum_{\vec{k}} A_{\vec{k}} \left[\left(\cos\phi_{\vec{k}+\frac{\vec{Q}}{2}} - \cos\phi_{\vec{k}-\frac{\vec{Q}}{2}}\right) \sin\vec{Q}\cdot\vec{r} \right. \\
&\quad \left. + \left(\sin\phi_{\vec{k}+\frac{\vec{Q}}{2}} + \sin\phi_{\vec{k}-\frac{\vec{Q}}{2}}\right) \cos\vec{Q}\cdot\vec{r}\right] , \\
\frac{\delta S_y\left(\vec{r}\right)}{\hbar} &= \sum_{k} A_{\vec{k}} \left[\left(\sin\phi_{\vec{k}+\frac{\vec{Q}}{2}} - \sin\phi_{\vec{k}-\frac{\vec{Q}}{2}}\right) \sin\vec{Q}\cdot\vec{r} \right. \\
&\quad \left. - \left(\cos\phi_{\vec{k}+\frac{\vec{Q}}{2}} + \cos\phi_{\vec{k}-\frac{\vec{Q}}{2}}\right) \cos\vec{Q}\cdot\vec{r}\right] , \\
\frac{\delta S_z\left(\vec{r}\right)}{\hbar} &= \sum_{k} 2A_{\vec{k}} \left[1 - \cos\left(\phi_{\vec{k}+\frac{\vec{Q}}{2}} - \phi_{\vec{k}-\frac{\vec{Q}}{2}}\right)\right] \cos\vec{Q}\cdot\vec{r} \,, \\
\delta n\left(\vec{r}\right) &= \sum_{k} 4A_{\vec{k}} \left[1 + \cos\left(\phi_{\vec{k}+\frac{\vec{Q}}{2}} - \phi_{\vec{k}-\frac{\vec{Q}}{2}}\right)\right] \cos\vec{Q}\cdot\vec{r} \,.
\end{aligned} \tag{4.5}$$

The change in the HF energy is obtained by substituting the expression of the non-zero density matrix elements from Equation (4.4) into Equation (4.2). The resulting expression can be expressed as:

$$\Delta^{(2)} E_{HF} = \Delta_0^{(2)} E_{HF} + \Delta_H^{(2)} E_{HF} + \Delta_X^{(2)} E_{HF} \,, \tag{4.6}$$

where we have defined

$$\Delta_0^{(2)} E_{HF} = \sum_{\vec{k}} \left(n_{\vec{k}+\frac{\vec{Q}}{2}} + n_{\vec{k}-\frac{\vec{Q}}{2}} \right) \frac{\epsilon_{\vec{k}-\frac{\vec{Q}}{2},+} - \epsilon_{\vec{k}+\frac{\vec{Q}}{2},+}}{n_{\vec{k}+\frac{\vec{Q}}{2}} - n_{\vec{k}-\frac{\vec{Q}}{2}}} A_k^2 \tag{4.7}$$

$$\Delta_H^{(2)} E_{HF} = \frac{1}{2} \sum_{\vec{k},\vec{p}} \left(n_{\vec{k}+\frac{\vec{Q}}{2}} + n_{\vec{k}-\frac{\vec{Q}}{2}} \right) \left(n_{\vec{p}+\frac{\vec{Q}}{2}} + n_{\vec{p}-\frac{\vec{Q}}{2}} \right) \times \left(V^{H,+}_{\vec{k},\vec{p},\vec{Q}} + V^{H,+}_{\vec{k},\vec{p},-\vec{Q}} \right) A_{\vec{k}} A_{\vec{p}} \, , \tag{4.8}$$

$$\Delta_X^{(2)} E_{HF} = -\frac{1}{2} \sum_{\vec{k},\vec{p}} \left(n_{\vec{k}+\frac{\vec{Q}}{2}} + n_{\vec{k}-\frac{\vec{Q}}{2}} \right) \left(n_{\vec{p}+\frac{\vec{Q}}{2}} + n_{\vec{p}-\frac{\vec{Q}}{2}} \right) \times \left(V^{X,-}_{\vec{k},\vec{p},\vec{Q}} + V^{X,-}_{\vec{k},\vec{p},-\vec{Q}} \right) A_{\vec{k}} A_{\vec{p}} \, . \tag{4.9}$$

with

$$V^{H,+}_{\vec{k},\vec{p},\vec{Q}} = v_{\vec{k}+\frac{\vec{Q}}{2},+,\vec{p}-\frac{\vec{Q}}{2},+,\vec{p}+\frac{\vec{Q}}{2},+,\vec{k}-\frac{\vec{Q}}{2},+} \tag{4.10}$$

$$V^{X,+}_{\vec{k},\vec{p},\vec{Q}} = v_{\vec{k}+\frac{\vec{Q}}{2},+,\vec{p}-\frac{\vec{Q}}{2},+,\vec{k}-\frac{\vec{Q}}{2},+\vec{p}+\frac{\vec{Q}}{2},+} \tag{4.11}$$

The Hartree and exchange terms in Equations (4.8)-(4.9) contain combinations of the matrix elements of the electron-electron interaction. By employing Equation (2.6), after simple algebraic manipulations, we obtain:

$$\begin{aligned} V^{H,+}_{\vec{k},\vec{p},\vec{Q}} + V^{H,+}_{\vec{k},\vec{p},-\vec{Q}} = \frac{v_Q}{2L^2} \Big[& 1 + \cos\left(\phi_{\vec{k}+\frac{\vec{Q}}{2}} - \phi_{\vec{k}-\frac{\vec{Q}}{2}} \right) \\ & + \cos\left(\phi_{\vec{p}+\frac{\vec{Q}}{2}} - \phi_{\vec{p}-\frac{\vec{Q}}{2}} \right) \\ & + \cos\left(\phi_{\vec{k}+\frac{\vec{Q}}{2}} - \phi_{\vec{k}-\frac{\vec{Q}}{2}} + \phi_{\vec{p}+\frac{\vec{Q}}{2}} - \phi_{\vec{p}-\frac{\vec{Q}}{2}} \right) \Big] \, , \end{aligned} \tag{4.12}$$

and

$$\begin{aligned} V^{X,+}_{\vec{k},\vec{p},\vec{Q}} + V^{X,+}_{\vec{k},\vec{p},-\vec{Q}} = \frac{v_{|\vec{k}-\vec{p}|}}{2L^2} \Big[& 1 + \cos\left(\phi_{\vec{p}+\frac{\vec{Q}}{2}} - \phi_{\vec{k}+\frac{\vec{Q}}{2}} \right) \\ & + \cos\left(\phi_{\vec{k}-\frac{\vec{Q}}{2}} - \phi_{\vec{p}-\frac{\vec{Q}}{2}} \right) \\ & + \cos\left(\phi_{\vec{p}+\frac{\vec{Q}}{2}} - \phi_{\vec{k}+\frac{\vec{Q}}{2}} + \phi_{\vec{k}-\frac{\vec{Q}}{2}} - \phi_{\vec{p}-\frac{\vec{Q}}{2}} \right) \Big] \, . \end{aligned} \tag{4.13}$$

In order to explicitly evaluate the change in the Hartree-Fock energy, we need to assume a specific expression for the distortion amplitudes. As a first condition, we will perturb only a narrow region near the Fermi surface. Following the same pattern of the proof of reference [10], we propose for the present problem the following educated variational guess:

$$A_{\vec{k}} = \begin{cases} \frac{(bk_F^+)^{\frac{3}{2}}}{\ln\frac{2}{b}} \frac{\left|n_{\vec{k}+\frac{\vec{Q}}{2}} - n_{\vec{k}-\frac{\vec{Q}}{2}}\right|}{k^{\frac{3}{2}}} \frac{\sqrt{|\sin\phi_{\vec{k}}|}}{|\cos\phi_{\vec{k}}|}, & bk_F^+ < k < \varsigma bk_F^+ \\ 0, & \text{otherwise}, \end{cases} \tag{4.14}$$

where $b \ll 1$ is our small parameter and the second arbitrary $\varsigma > 1$. This expression is intentionally chosen to display singularities for $k = 0$ and $\phi_k = 0$. These singularities are crucial to the present proof.

We now notice that in the expressions of the interaction matrix elements, there appear factors of the type:

$$\cos\left(\phi_{\vec{p}+\frac{\vec{Q}}{2}} - \phi_{\vec{k}+\frac{\vec{Q}}{2}}\right) = \frac{\left(p\cos\phi_{\vec{p}} + k_F^+\right)\left(k\cos\phi_{\vec{k}} + k_F^+\right) + pk\sin\phi_{\vec{p}}\sin\phi_{\vec{k}}}{\sqrt{p^2 + k_F^{+2} + 2k_F^+ p\cos\phi_{\vec{p}}}\sqrt{k^2 + k_F^{+2} + 2k_F^+ k\cos\phi_{\vec{k}}}}. \tag{4.15}$$

Since we are only interested in the leading order expansion with respect to b, these cosines can be simply taken to be equal to unity, since in the region where the amplitude of the distortion is non-vanishing, both p/k_F^+ and k/k_F^+ are of order b. Accordingly we will assume

$$\cos\left(\phi_{\vec{p}+\frac{\vec{Q}}{2}} - \phi_{\vec{k}+\frac{\vec{Q}}{2}}\right) \simeq 1 + O\left(b^2\right). \tag{4.16}$$

At this point, we recall that $\vec{k} + \frac{\vec{Q}}{2}$ and $\vec{k} - \frac{\vec{Q}}{2}$ must lie on opposite sides of the Fermi sea (i.e $|\vec{k} - \frac{\vec{Q}}{2}| < k_F^+$ and $|\vec{k} + \frac{\vec{Q}}{2}| > k_F^+$), which implies that $\left|\cos\phi_{\vec{k}}\right| > \frac{b}{2}$.

We can now proceed to the evaluation of the three components of $\Delta^{(2)}E_{HF}$ from Equations (4.8)-(4.9) to leading order in b.

For $\Delta_0^{(2)}E_{HF}$ the first step is to calculate $\epsilon_{\vec{k}-\frac{\vec{Q}}{2},+} - \epsilon_{\vec{k}+\frac{\vec{Q}}{2},+}$. Using (3.2), with energies expressed in Ry, $x = \frac{k}{k_F}$ and $x' = \frac{k'}{k_F}$, we have:

$$\begin{aligned} &\epsilon_{\vec{k}-\frac{\vec{Q}}{2}} - \epsilon_{\vec{k}+\frac{\vec{Q}}{2}} \\ &= \frac{4u_1^2}{r_s^2} - \frac{4\tilde{\alpha}u_1}{r_s} - \frac{1}{\pi r_s}\int_0^{2\pi}\int_0^1 \frac{x'\left(1+\cos\varphi'\right)dx'd\phi'}{\sqrt{u_1^2 + x'^2 - 2u_1x'\cos\varphi'}} \\ &\quad - \frac{4u^2}{r_s^2} + \frac{4\tilde{\alpha}u_2}{r_s} + \frac{1}{\pi r_s}\int_0^{2\pi}\int_0^1 \frac{x'\left(1+\cos\varphi'\right)dx'd\phi'}{\sqrt{u_2^2 + x'^2 - 2u_2x'\cos\varphi'}}. \end{aligned} \tag{4.17}$$

with $u_1 = \sqrt{1+x^2-2x\cos\phi_{\vec{k}}}$ and $u_2 = \sqrt{1+x^2+2x\cos\phi_{\vec{k}}}$. The quadratures appearing in this expression can then be manipulated by making use of the results of Appendix A.1. The result is:

$$\epsilon_{\vec{k}-\frac{\vec{Q}}{2}} - \epsilon_{\vec{k}+\frac{\vec{Q}}{2}} = -\frac{16x\cos\phi}{r_s^2}\left(1 - \frac{\tilde{\alpha} r_s}{2} - \frac{r_s}{2\pi}\ln|\xi x\cos\phi|\right), \quad (4.18)$$

where the logarithmic term accounts for the divergence of the derivative of the HF single particle energy near the Fermi level. Here, ξ is a constant approximately equal to 0.51.

By substituting (4.18) and (4.14) in (4.7) we obtain:

$$\Delta_0^{(2)} E_{HF} = \frac{16b^3L^2k_F^2}{\left(\ln\frac{2}{b}\right)^2 r_s^2\pi^2}\int_b^{\varsigma b} dx \int_0^{\arccos\frac{b}{2}} d\varphi \mathcal{F}_0(x,\varphi), \quad (4.19)$$

where

$$\mathcal{F}_0(x,\varphi) = \frac{\sin\varphi}{x\cos\varphi}\left(1 - \frac{\tilde{\alpha} r_s}{2} - \frac{r_s}{2\pi}\ln|\xi x\cos\varphi|\right), \quad (4.20)$$

an expression that, to leading order in b, reduces to:

$$\Delta_0^{(2)} E_{HF} \simeq \frac{48Nb^3\ln\varsigma}{r_s\pi^2}. \quad (4.21)$$

where N is the number of particles.

The Hartree term $\Delta_H^{(2)} E_{HF}$ (containing v_Q) can in turn be evaluated as follows. By making use of the assumed amplitude in (4.14), we write:

$$\Delta_H^{(2)} E_{HF} = \frac{16Nb^3}{\pi^2\left(\ln\frac{2}{b}\right)^2 r_s}\int_b^{b\varsigma} dx \int_b^{b\varsigma} dy \int_0^{\arccos\frac{b}{2}} d\varphi_x \int_0^{\arccos\frac{b}{2}} d\varphi_y \mathcal{F}_H, \quad (4.22)$$

where

$$\mathcal{F}_H(x,y,\varphi_x,\varphi_y) = \frac{1}{\sqrt{xy}}\frac{\sqrt{\sin\varphi_x}}{\cos\varphi_x}\frac{\sqrt{\sin\varphi_y}}{\cos\varphi_y}. \quad (4.23)$$

At this point, using the result (B.5), the leading order in b of this quantity is given by:

$$\Delta_H^{(2)} E_{HF} \simeq \frac{64Nb^4\left(\sqrt{\varsigma}-1\right)^2}{\pi^2 r_s}. \quad (4.24)$$

The last term of (4.6), the exchange energy contribution, is clearly negative and therefore will certainly lower the energy. Its evaluation is formidable, for it involves several complicated and seemingly difficult quadratures. Rather than attempting to actually calculate it, we will establish a lower limit for its magnitude.

We will restrict ourselves to the region in which both angles $\phi_{\vec{k}}$ and $\phi_{\vec{p}}$ are in the first quadrant. Since this excludes some contributions of the same sign, the exchange energy will be underestimated. We therefore have:

$$\left|\Delta_X^{(2)} E_{HF}\right| \geq \frac{2Nb^3}{\pi^2 \left(\ln \frac{2}{b}\right)^2 r_s} \int_b^{b\varsigma} dx \int_b^{b\varsigma} dy \int_0^{\arccos\frac{b}{2}} d\varphi_x \int_0^{\arccos\frac{b}{2}} d\varphi_y \mathcal{F}_X, \tag{4.25}$$

where

$$\begin{aligned} \mathcal{F}_X(x, y, \varphi_x, \varphi_y) = & \frac{1}{\sqrt{xy}} \sqrt{\frac{\left|\sin\varphi_x \sin\varphi_y\right|}{\left|\cos\varphi_x \cos\varphi_y\right|}} \\ & \times \frac{1}{\sqrt{x^2 + y^2 - 2xy\cos\left(\varphi_x - \varphi_y\right)}}. \end{aligned} \tag{4.26}$$

It is simple to see that this expression will turn out to be proportional to $(\ln b)^3$. This is due to the presence of three singularities in the denominator of the integrand. Of these one stems from the divergence of the Coulomb potential, while the other two come from the upper limit of the angular integrations.

Another simplification is provided by the use of the inequality:

$$\frac{1}{\sqrt{x^2+y^2-2xy\cos\left(\varphi_x-\varphi_y\right)}} \geq \frac{1}{\sqrt{x^2+y^2-2xy\sin\varphi_x \sin\varphi_y}}. \tag{4.27}$$

Using the same changes of variable as in Appendix B, *i.e.* $t_x = \tan(\phi_x/2)$ and $t_y = \tan(\phi_y/2)$, we can rewrite this integral as follows:

$$\left|\Delta_X^{(2)} E_{HF}\right| \geq \frac{16Nb^3}{\pi^2 \left(\ln \frac{2}{b}\right)^2 r_s} \int_b^{b\varsigma} dx \int_b^{b\varsigma} dy \int_0^{\sqrt{\frac{2-b}{2+b}}} dt_x \int_0^{\sqrt{\frac{2-b}{2+b}}} dt_y \tilde{\mathcal{F}}_X. \tag{4.28}$$

with

$$\tilde{\mathcal{F}}_X(x,y,t_x,t_y) = \frac{1}{\sqrt{xy}} \frac{1}{\sqrt{x^2+y^2-\frac{8xyt_xt_y}{(1+t_x^2)(1+t_y^2)}}} \times \frac{1}{1-t_x^2}\sqrt{\frac{t_x}{1+t_x^2}}\frac{1}{1-t_y^2}\sqrt{\frac{t_y}{1+t_y^2}}. \tag{4.29}$$

It is clear now that the main contribution to the integral comes from the region around the upper limit of integration for both t_x and t_y. In order to retain the leading order term, a good approximation will be to replace both t's with $1-b/2$ in the denominator of the first square root. In this way, we can separate the angular integrations in (4.25) to obtain:

$$\left|\Delta_X^{(2)} E_{HF}\right| \geq \frac{2Nb^3}{\pi^2 \left(\ln \frac{2}{b}\right)^2 r_s} \int_b^{b\varsigma} dx \int_b^{b\varsigma} dy \int_0^{1-\frac{b}{2}} dt_x \int_0^{1-\frac{b}{2}} dt_y \bar{\mathcal{F}}_X \,. \tag{4.30}$$

where

$$\bar{\mathcal{F}}_X(x,y,t_x,t_y) = \frac{1}{\sqrt{xy}} \frac{1}{\sqrt{x^2+y^2-2xy\left(1-\frac{b^2}{8}\right)^2}} \times \frac{1}{1-t_x^2}\sqrt{\frac{t_x}{1+t_x^2}}\frac{1}{1-t_y^2}\sqrt{\frac{t_y}{1+t_y^2}}. \tag{4.31}$$

The last two integrals are evaluated using (B.5) and other changes of variables ($x = ub, y = vb$) :

$$\left|\Delta_X^{(2)} E_{HF}\right| \geq \frac{8Nb^3}{\pi^2 r_s} \int_1^{\sqrt{\varsigma}} du \int_1^{\sqrt{\varsigma}} dv \frac{1}{\sqrt{\left(u^2-v^2\right)^2+\frac{u^2v^2b^2}{2}}} \,. \tag{4.32}$$

The last quadrature is calculated in (B.8), leading us to a very simple inequality for the exchange contribution:

$$\left|\Delta_X^{(2)} E_{HF}\right| \geq \frac{8Nb^3}{\pi^2 r_s} \ln \varsigma \ln \frac{1}{b} \,. \tag{4.33}$$

This term contains a logarithmic factor $\ln \frac{1}{b}$, which allows the negative change in the exchange contribution to control all the remaining terms. This conclues the proof for this case.

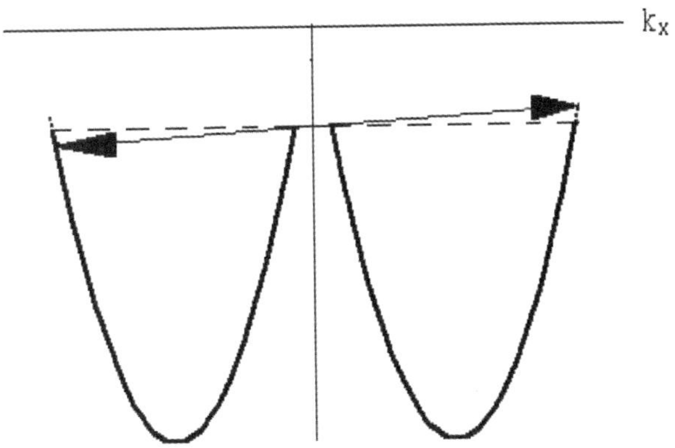

Figure 3. HF instability: symmetry breaking coupling for the case in which only the lower subband is occupied.

The same chain of arguments does apply to the case in which the generalized chirality is greater than one. The coupling that produces this kind of instability is schematically shown in Figure 3. All the formulas we derived in the previous case do still apply. The only difference lies in the lower integration limits of Equation (4.17), but no relevant contribution ensues from this. The matrix elements related to the Hartree and exchange contributions are the same, and, as a consequence, the leading order approximation is identical.

4.2. Instability for the case of chirality less than one

The argument of the previous Section can be applied when the chirality is less than one, *i.e.* when both chiral subbands are occupied. We can try to break the symmetry by coupling states with the same chirality as well as states with different chiralities. When same chirality states are coupled, there is nothing new, as one simply just adds a chirality index to the various quantities. In this case, the wave vectors characterizing the oscillations are given by: $Q_\mu = 2k_F^\mu$ and the trial wavefuntions can be written as:

$$\Psi_{\vec{k}\mu} \simeq \psi_{\vec{k}\mu} + A_{\vec{k}+\vec{Q}_\mu,\vec{k}\mu}\psi_{\vec{k}+\vec{Q}_\mu\mu} + A_{\vec{k}-\vec{Q}_\mu,\vec{k}\mu}\psi_{\vec{k}-\vec{Q}_\mu\mu} \,. \tag{4.34}$$

This type of coupling is depicted in Figure 4.

The corresponding distortion of the components of the spin density and the number density can be again calculated up to the first order in the

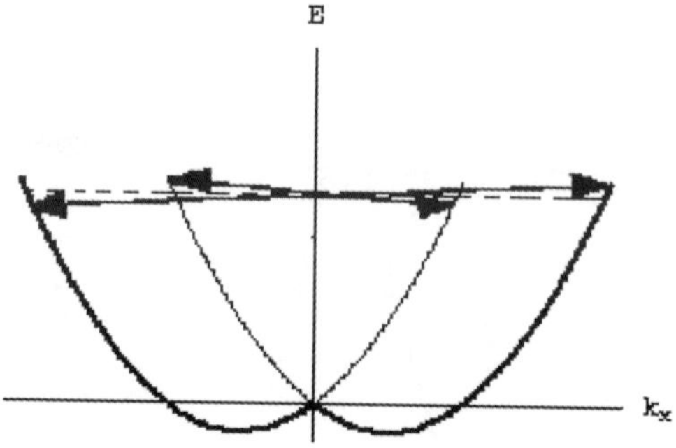

Figure 4. Symmetry breaking coupling of states with the same chirality.

amplitudes:

$$\frac{\delta S_x(\vec r)}{\hbar} = \sum_{k,\mu} A_{\vec k\mu}\left(\cos\phi_{\vec k^+_\mu} - \cos\phi_{\vec k^-_\mu}\right)\sin\vec Q_\mu \vec r + \sum_{k,\mu} A_{\vec k\mu}\left(\sin\phi_{\vec k^+_\mu} + \sin\phi_{\vec k^-_\mu}\right)\cos\vec Q_\mu \vec r, \tag{4.35}$$

$$\frac{\delta S_y(\vec r)}{\hbar} = \sum_{k,\mu} A_{\vec k\mu}\left(\sin\phi_{\vec k^+_\mu} - \sin\phi_{\vec k^-_\mu}\right)\sin\vec Q_\mu \vec r - \sum_{k,\mu} A_{\vec k\mu}\left(\cos\phi_{\vec k^+_\mu} + \cos\phi_{\vec k^-_\mu}\right)\cos\vec Q_\mu \vec r, \tag{4.36}$$

$$\frac{\delta S_z(\vec r)}{2\hbar} = \sum_{k,\mu} A_{\vec k\mu}\left[1 - \cos\left(\phi_{\vec k^+_\mu} - \phi_{\vec k^-_\mu}\right)\right]\cos\vec Q_\mu \vec r, \tag{4.37}$$

$$\frac{\delta n(\vec r)}{4} = \sum_{k,\mu} A_{\vec k\mu}\left[1 - \cos\left(\phi_{\vec k^+_\mu} - \phi_{\vec k^-_\mu}\right)\right]\cos\vec Q_\mu \vec r, \tag{4.38}$$

where $k^\pm_\mu = \vec k \pm \frac{\vec Q_\mu}{2}$. We proceed in this case by choosing an amplitude not unlike the one assumed above:

$$A_{\vec k\mu} = \begin{cases} \frac{(bk_F^\mu)^{\frac{3}{2}}}{\ln\frac{2}{b}} \frac{\left|n_{\vec k+\frac{\vec Q}{2}\mu} - n_{\vec k-\frac{\vec Q}{2}\mu}\right|}{k^{\frac{3}{2}}} \frac{\sqrt{|\sin\phi_{\vec k}|}}{|\cos\phi_{\vec k}|}, & bk_F^\mu < k < \varsigma b k_F^\mu \\ 0, & \text{otherwise}. \end{cases} \tag{4.39}$$

The proof of the corresponding instability theorem proceeds then in exactly the same manner.

As anticipated, there is also not much difference when we try to couple states with opposite chirality (see Figure 5). Although some of the expressions involved in the derivation do change, the main features of the

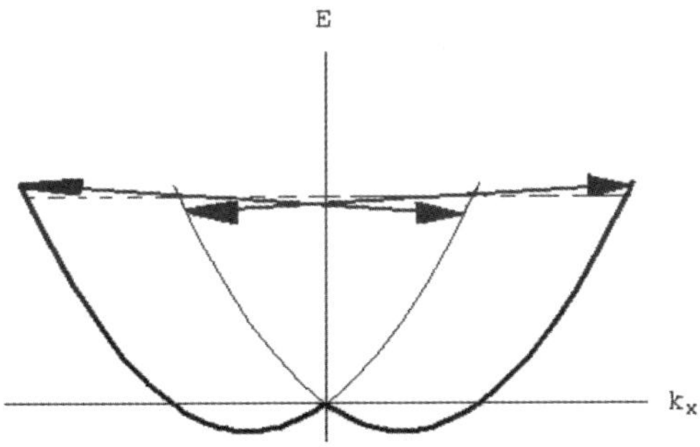

Figure 5. States with opposite chiralities coupled to obtain a HF instability

argument remain unchanged. The coupling vector in this case is given by $Q = k_F^+ + k_F^-$. Here, we try to find a lower energy state by coupling wavefunctions with wave vector $\vec{k}$ with those with wave vector $\vec{k} \pm \vec{Q}$ and opposite chirality. The trial wavefunctions then read:

$$\Psi_{\vec{k}+} \simeq \psi_{\vec{k}+} + A_{\vec{k}+\vec{Q},\vec{k}} \psi_{\vec{k}+\vec{Q},-} + A_{\vec{k}-\vec{Q},\vec{k}} \psi_{\vec{k}-\vec{Q},-} . \tag{4.40}$$

For $A_{\vec{k}\pm\vec{Q},\vec{k}} = A_{\vec{k},\vec{k}\pm\vec{Q}}$ the only non zero variations of the matrix elements of the single-particle density matrix operator acquire the following form:

$$\begin{aligned} \delta\rho_{\vec{k}+\frac{\vec{Q}}{2},\mu,\vec{k}-\frac{\vec{Q}}{2},-\mu} &= \delta\rho_{\vec{k}-\frac{\vec{Q}}{2},\mu,\vec{k}+\frac{\vec{Q}}{2},-\mu} \\ &= \left(n_{\vec{k}+\frac{\vec{Q}}{2},\mu} + n_{\vec{k}-\frac{\vec{Q}}{2},-\mu} \right) A_{\vec{k}} . \end{aligned} \tag{4.41}$$

In this case, the new state is characterized by a similar spin and density modulation:

$$\frac{\delta S_x(\vec{r})}{\hbar} = \sum_k A_{\vec{k}} \left[\cos\left(\vec{Q}\vec{r} - \tilde{\phi}_{\vec{k}} \right) + \cos \vec{Q}\vec{r} \right] , \tag{4.42}$$

$$\frac{\delta S_y(\vec{r})}{\hbar} = \sum_k A_{\vec{k}} \left[\sin\left(-\vec{Q}\vec{r} + \tilde{\phi}_{\vec{k}} \right) + \sin \vec{Q}\vec{r} \right] , \tag{4.43}$$

$$\begin{aligned} \frac{\delta S_z(\vec{r})}{\hbar} &= \sum_k A_{\vec{k}} \left[\sin\left(\vec{Q}\vec{r} - \phi_{\vec{k}-\frac{\vec{Q}}{2}} \right) \right. \\ &\quad \left. + \sin\left(\vec{Q}\vec{r} - \phi_{\vec{k}+\frac{\vec{Q}}{2}} \right) \right] , \end{aligned} \tag{4.44}$$

$$\begin{aligned} \frac{\delta n(\vec{r})}{2} &= \sum_k A_{\vec{k}} \left[\sin\left(\vec{Q}\vec{r} - \phi_{\vec{k}-\frac{\vec{Q}}{2}} \right) \right. \\ &\quad \left. - \sin\left(\vec{Q}\vec{r} - \phi_{\vec{k}+\frac{\vec{Q}}{2}} \right) \right] , \end{aligned} \tag{4.45}$$

where $\tilde{\phi}_{\vec{k}} = \phi_{\vec{k}+\frac{\vec{Q}}{2}} + \phi_{\vec{k}-\frac{\vec{Q}}{2}}$. The corresponding terms in the Hartree-Fock energy change are:

$$\Delta_0^{(2)} E_{HF} = \sum_{\vec{k}\mu} \frac{\epsilon_{\vec{k}-\frac{\vec{Q}}{2}\mu} - \epsilon_{\vec{k}+\frac{\vec{Q}}{2}-\mu}}{n_{\vec{k}+\frac{\vec{Q}}{2}-\mu} - n_{\vec{k}-\frac{\vec{Q}}{2}\mu}} A_k^2 \times \left(n_{\vec{k}+\frac{\vec{Q}}{2},\mu} + n_{\vec{k}-\frac{\vec{Q}}{2},-\mu}\right), \tag{4.46}$$

$$\Delta_H^{(2)} E_{HF} = \frac{1}{2} \sum_{\vec{k}\vec{p}\mu\nu} \left(V^{H,\mu\nu}_{\vec{k},\vec{p},\vec{Q}} + V^{H,\mu\nu}_{\vec{k},\vec{p},-\vec{Q}}\right) A_{\vec{k}\mu} A_{\vec{p}\nu} \times \left(n_{\vec{k}+\frac{\vec{Q}}{2},\mu} + n_{\vec{k}-\frac{\vec{Q}}{2},-\mu}\right)\left(n_{\vec{p}+\frac{\vec{Q}}{2}\nu} + n_{\vec{p}-\frac{\vec{Q}}{2}-\nu}\right), \tag{4.47}$$

$$\Delta_X^{(2)} E_{HF} = -\frac{1}{2} \sum_{\vec{k}\vec{p}\mu\nu} \left(V^{X,\mu\nu}_{\vec{k},\vec{p},\vec{Q}} + V^{X,\mu\nu}_{\vec{k},\vec{p},-\vec{Q}}\right) A_{\vec{k}\mu} A_{\vec{p}\nu} \times \left(n_{\vec{k}+\frac{\vec{Q}}{2},\mu} + n_{\vec{k}-\frac{\vec{Q}}{2},-\mu}\right)\left(n_{\vec{p}+\frac{\vec{Q}}{2}\nu} + n_{\vec{p}-\frac{\vec{Q}}{2}-\nu}\right), \tag{4.48}$$

where

$$V^{H,\mu\nu}_{\vec{k},\vec{p},\vec{Q}} = v_{\vec{k}+\frac{\vec{Q}}{2},\mu,\vec{p}-\frac{\vec{Q}}{2}\nu,\vec{p}+\frac{\vec{Q}}{2},-\nu,\vec{k}-\frac{\vec{Q}}{2},-\mu}$$

and

$$V^{X,\mu\nu}_{\vec{k},\vec{p},\vec{Q}} = v_{\vec{k}+\frac{\vec{Q}}{2}\mu,\vec{p}-\frac{\vec{Q}}{2},-\nu,\vec{k}-\frac{\vec{Q}}{2},-\mu,\vec{p}+\frac{\vec{Q}}{2},\nu}.$$

The relevant interaction matrix elements which appear in the expression of the Hartree term become:

$$\frac{2v_Q}{L^2}\left[\left(\sin\phi_{\vec{p}+\frac{\vec{Q}}{2}} - \sin\phi_{\vec{p}-\frac{\vec{Q}}{2}}\right)\left(\sin\phi_{\vec{k}-\frac{\vec{Q}}{2}} - \phi_{\vec{k}+\frac{\vec{Q}}{2}}\right)\right., \tag{4.49}$$

while those determining the exchange contribution reduce to:

$$\begin{aligned}\frac{v_{|\vec{k}-\vec{p}|}}{L^2}&\left[\cos\left(\phi_{\vec{p}+\frac{\vec{Q}}{2}} - \phi_{\vec{k}+\frac{\vec{Q}}{2}}\right) + \cos\left(\phi_{\vec{p}-\frac{\vec{Q}}{2}} - \phi_{\vec{k}-\frac{\vec{Q}}{2}}\right)\right.\\ &+\cos\left(\phi_{\vec{p}+\frac{\vec{Q}}{2}} - \phi_{\vec{k}-\frac{\vec{Q}}{2}}\right) + \cos\left(\phi_{\vec{p}-\frac{\vec{Q}}{2}} - \phi_{\vec{k}+\frac{\vec{Q}}{2}}\right)\\ &-\cos\left(\phi_{\vec{k}+\frac{\vec{Q}}{2}} - \phi_{\vec{k}-\frac{\vec{Q}}{2}}\right) - \cos\left(\phi_{\vec{p}+\frac{\vec{Q}}{2}} - \phi_{\vec{p}-\frac{\vec{Q}}{2}}\right)\\ &\left.+\cos\left(\phi_{\vec{p}+\frac{\vec{Q}}{2}} - \phi_{\vec{k}+\frac{\vec{Q}}{2}} + \phi_{\vec{k}-\frac{\vec{Q}}{2}} - \phi_{\vec{p}-\frac{\vec{Q}}{2}}\right) + 1\right].\end{aligned} \tag{4.50}$$

As in the previous cases, we assume the following coupling amplitude:

$$A^{\mu}_{\vec{k}} = \begin{cases} \frac{(bk_F^{\mu})^{\frac{3}{2}}}{\ln \frac{2}{b}} \frac{\left| n_{\vec{k}+\frac{\vec{Q}}{2},\mu} - n_{\vec{k}-\frac{\vec{Q}}{2},-\mu} \right|}{k^{\frac{3}{2}}} \frac{\sqrt{|\sin \phi_{\vec{k}}|}}{|\cos \phi_{\vec{k}}|}, & \begin{array}{l} bk_F^+ < k \\ k < \varsigma b k_F^+ \\ |\cos \phi_{\vec{k}}| < \frac{b}{2} \end{array} \\ 0, & \text{otherwise}\,. \end{cases} \tag{4.51}$$

The first term in the Hartree-Fock energy (4.7) has a positive contribution which is proportional to b^3. This originates from the same logarithmic factor associated with the divergence of the derivative of the single-particle energy at the Fermi level. The Hartree term introduces higher order terms in b due to the presence of the sine factors in its matrix elements. Finally, the leading order contribution to the exchange matrix elements is $4v_{|\vec{k}-\vec{p}|}$. By evaluating integrals similar to those in Equation (4.25) we again obtain a negative energy change of order $b^3 \ln(1/b)$.

5. Conclusions

We have been able to formally construct a number of distorted states which, irrespective of the electron density, have a lower energy than the spatially homogeneous paramagnetic chiral HF ground state, thereby affording a rigorous proof of an equivalent of Overhauser's Hartree-Fock instability theorem for a two dimensional electron liquid in the presence of linear Rashba spin-orbit coupling. It is important to notice that, as mentioned in Section 3, to establish the instability we have not allowed for momentum space repopulation: inclusion of this phenomenon would have further lowered the energy of the trial states while greatly increasing the difficulty of the analysis.

It is worth mentioning that the states that have been analyzed in this paper differ in a number of ways form the original spin/charge density waves proposed by Overhauser. Our states are chiral density states, that display both spin and charge modulations. The presence of charge modulations cannot be ignored and the Hartree term has to be evaluated explicitly. The exchange gain is shown to be bigger than kinetic and Hartree terms. It is precisely the electrostatic term that is believed to be fatal to the charge density waves once one adds some measure of correlations. We prove that this is not the case for the chiral waves. The exchange energy is different from the one in the absence of spin-orbit interaction and does not necessarily favor the chiral instability as shown in Reference [11] of the manuscript. Our calculations show that e can induce the chiral instability for all densities of the electron gas.

Our results only represent the first step in understanding non homogeneous states in this interesting many-body system. How the inclusion of correlations will modify the HF scenario is of course a most important question. Advances in this respect can in principle be pursued by following the program outlined in Reference [16]. Such a study will require a much more complicated analysis. An alternative route is make use of perturbative techniques to establish the role of correlation in the high density limit. Part of this program has been carried out in Reference [17].

A. Appendix

A.1. Elliptic integral expansion

For the purpose of our calculations the following expression must be evaluated in the limit of small u:

$$\left.\int_0^{2\pi}\!\!\int_0^1 \frac{x'\,(1+\cos\varphi)\,dx'd\varphi}{\sqrt{z^2+x'^2-2zx'\cos\varphi}}\right|_{z=1+u} - \left.\int_0^{2\pi}\!\!\int_0^1 \frac{x'\,(1+\cos\varphi)\,dx'd\varphi}{\sqrt{z^2+x'^2-2zx'\cos\varphi}}\right|_{z=1-u} . \quad \text{(A.1)}$$

An obvious problem is the presence of singularities in the integrand for $z = 1$.

Let us define:

$$A\,(z) = \int_0^{2\pi}\!\!\int_0^1 \frac{x'\,(1+\cos\varphi)\,dx'd\varphi}{\sqrt{z^2+x'^2-2zx'\cos\varphi}} . \quad \text{(A.2)}$$

Of course, A is not differentiable in $z = 1$. Still we have to evaluate $\frac{\partial A(z)}{\partial z}$ and expand it in an asymptotic series around $z = 1$.

$$\begin{aligned}\frac{\partial}{\partial z}A\,(z) &= \frac{\partial}{\partial z}\left(z\int_0^{2\pi}\!\!\int_0^{1/z} \frac{y\,(1+\cos\varphi)\,dyd\varphi}{\sqrt{1+y^2-2y\cos\varphi}}\right)\\ &= \int_0^{2\pi}\!\!\int_0^{1/z} \frac{y\,(1+\cos\varphi)\,dyd\varphi}{\sqrt{1+y^2-2y\cos\varphi}}\\ &\quad + z\frac{\partial}{\partial z}\left(\int_0^{2\pi}\!\!\int_0^{1/z} \frac{y\,(1+\cos\varphi)\,dyd\varphi}{\sqrt{1+y^2-2y\cos\varphi}}\right) .\end{aligned} \quad \text{(A.3)}$$

The derivative in the second term can be evaluated in the following way:

$$\frac{\partial}{\partial z}\left(\int_0^{2\pi}\int_0^{1/z}\frac{y\,(1+\cos\varphi)\,dy d\varphi}{\sqrt{1+y^2-2y\cos\varphi}}\right)=-\frac{1}{z^2}\int_0^{2\pi}\frac{y\,(1+\cos\varphi)\,d\varphi}{\sqrt{1+y^2-2y\cos\varphi}}\Bigg|_{y=\frac{1}{z}}$$
$$=-\frac{1}{z^2}\int_0^{2\pi}\frac{(1+\cos\varphi)\,d\varphi}{\sqrt{1+z^2-2z\cos\varphi}}\,. \quad \text{(A.4)}$$

We can integrate this expression over its angular part so that the result is expressed as:

$$\int_0^{2\pi}\frac{(1+\cos\varphi)\,d\varphi}{\sqrt{1+z^2-2z\cos\varphi}}=-\frac{2\,|z-1|}{z}E\left(\frac{-4z}{(z-1)^2}\right)$$
$$+\frac{2\,(z+1)^2}{z\;|z-1|}K\left(\frac{-4z}{(z-1)^2}\right), \quad \text{(A.5)}$$

where E and K are the elliptic integrals of first and second kind defined as in [18, 19].

The asymptotic expansion of the elliptic integrals around $z = 1$ leads to:

$$\frac{\partial A\,(z)}{\partial z}=\rho-12\ln 2+4+4\ln|z-1|\,, \quad \text{(A.6)}$$

where $\rho = \int_0^{2\pi}\int_0^{1}\frac{y(1+\cos\varphi)dy d\varphi}{\sqrt{1+y^2-2y\cos\varphi}} \simeq 5.6639$. Integrating over z we finally obtain:

$$A\,(1+u)-A\,(1-u)\simeq 2\,(\rho-12\ln 2+4\ln|u|)\,u$$
$$\simeq 8u\,(\ln|\xi u|)\,, \quad \text{(A.7)}$$

with $\xi \simeq 0.51$.

B. Evaluation of useful quadratures

We begin by calculating here the leading order term in the expansion of the expression

$$I=\int_0^{\arccos\frac{b}{2}}\frac{\sqrt{\sin\varphi}}{\cos\varphi}d\varphi \quad \text{for b} \ll 1\,. \quad \text{(B.1)}$$

Expanding the upper limit and setting $t = \tan\frac{\varphi}{2}$, the integral becomes:

$$\begin{aligned} I &\simeq 2\sqrt{2}\int_0^{1-\frac{b}{2}} \frac{dt}{1-t^2}\sqrt{\frac{t}{1+t^2}} \\ &\simeq \sqrt{2}\int_0^{1-\frac{b}{2}} dt\sqrt{\frac{t}{1+t^2}}\left(\frac{1}{1-t}+\frac{1}{1+t}\right), \end{aligned} \tag{B.2}$$

The divergence in the limit $b \to 0$ stems only from the first term and we therefore proceed to try isolating the singularity:

$$\begin{aligned} I &\simeq \sqrt{2}\int_0^{1-\frac{b}{2}} dt\sqrt{\frac{t}{1+t^2}}\frac{1}{1-t} + \sqrt{2}\int_0^{1} dt\frac{1}{1+t}\sqrt{\frac{t}{1+t^2}} + O(b) \simeq \\ &\simeq -\sqrt{2}\int_0^{1-\frac{b}{2}} dt\sqrt{\frac{t}{1+t^2}}\left(\ln(1-t)\right)' + 0.5256 + O(b)\,. \end{aligned} \tag{B.3}$$

An integration by parts is used in the remaining integral to further isolate the singular contribution:

$$\begin{aligned} I &\simeq -\sqrt{\frac{2t}{1+t^2}}\left(\ln(1-t)\right)\Bigg|_0^{1-\frac{b}{2}} \\ &+ \sqrt{2}\int_0^{1-\frac{b}{2}} dt\frac{d}{dt}\left(\sqrt{\frac{t}{1+t^2}}\right)\ln(1-t) + 0.5256 + O(b)\,. \end{aligned} \tag{B.4}$$

At this point the non singular second term is calculated numerically so that the final result is:

$$I \simeq \ln b + 0.9475\,. \tag{B.5}$$

The following integral is used in the proof of the HF instability theorem:

$$J = \int_1^{\sqrt{5}}\int_1^{\sqrt{5}} \frac{du\,dv}{\sqrt{\left(u^2-v^2\right)^2 + \frac{u^2v^2b^2}{2}}}, \tag{B.6}$$

for $b \ll 1$.

Because the singular behavior originates from the region where $u \simeq v$, in order to find the leading order term, we approximate $u^2 - v^2$ with

$2u(u-v)$. Since the function is symmetric with respect to the interchange of u and v we can use the relation

$$\int_1^{\sqrt{\varsigma}} du \int_1^{\sqrt{\varsigma}} dv\ f(u,v) = 2\int_1^{\sqrt{\varsigma}} du \int_1^{u} dv\ f(u,v).$$

Then:

$$\begin{aligned} J \simeq \int_1^{\sqrt{\varsigma}} \frac{du}{u} \int_1^{u} \frac{dv}{\sqrt{(u-v)^2 + \frac{u^2 b^2}{8}}} \simeq -\int_1^{\sqrt{\varsigma}} \frac{du}{u} \ln\left(\frac{bu}{2\sqrt{2}}\right) \\ + \int_1^{\sqrt{\varsigma}} \frac{du}{u} \ln\left(u - 1 + \sqrt{(u-1)^2 + \frac{u^2 b^2}{8}}\right). \end{aligned} \tag{B.7}$$

The integrand of the first term has a logarithmic singularity in the limit of small b while the second one is instead analytic in this limit. The main contribution to the integral is therefore:

$$J \simeq -\int_1^{\sqrt{\varsigma}} \frac{du}{u} \ln\left(\frac{bu}{2\sqrt{2}}\right) \simeq -\ln\varsigma \ln b\,. \tag{B.8}$$

References

[1] E. I. Rashba, Sov. Phys. Solid State **1** (1959), 368.
[2] E. I. Rashba, Sov. Phys. Solid State **2** (1960), 1224.
[3] G. Dresselhaus, Phys. Rev. **100** (1955), 580.
[4] A. W. Overhauser, Phys. Rev. Lett. **4** (1960), 462.
[5] A. W. Overhauser, Phys. Rev. **128**, (1962), 1437.
[6] A. W. Overhauser and A. Arrott, Phys. Rev. Lett. **4** (1960), 226.
[7] L. M. Corliss, J. M. Hastings and R. J. Weiss, Phys. Rev. Lett. **3** (1959), 211.
[8] C. G. Shull and M. K. Wilkinson, Rev. Mod. Phys. **25** (1953), 100.
[9] Stability of a CDW state can only be achieved if the associated Hartree term is cancelled by a corresponding deformation of the neutralizing background. The simplest such case being represented by the case of the so called deformable jellium. A rather interesting case is discussed in [20].

[10] G. F. GIULIANI and G. VIGNALE, "Quantum Theory of The Electron Liquid", Cambridge University Press, 2005.
[11] S. CHESI and G. F. GIULIANI, Phys. Rev. B **75** (2007), 155305.
[12] S. CHESI and G. F. GIULIANI, Phys. Rev. B **75** (2007), 153306.
[13] P. FEDDERS and P. C. MARTIN, Phys. Rev. **143** (1966), 245.
[14] That the states of Equations (2.3) and (2.4) are solutions of the Hartree-Fock problem was shown in Reference [11].
[15] Notice that in this case, the generalized chirality, as defined for instance in Reference [11], also equals one.
[16] A. W. OVERHAUSER, Phys. Rev. **167** (1968), 691.
[17] S. CHESI and G. F. GIULIANI, Phys. Rev. B **75** (2007), 153306.
[18] http://functions.wolfram.com/08.04.02.0001.01 .
[19] http://functions.wolfram.com/08.05.02.0001.01 .
[20] G. F.GIULIANI and A. W. OVERHAUSER. Phys. Rev. B **20** (1979), 1328.

A brief overview of Giuliani's contribution to many-body aspects of the electron gas

Sudhakar Yarlagadda

Abstract. Among the various important contributions that Gabriele F. Giuliani made in the area of theoretical condensed matter physics, the large body of work (done over a period of a few decades) on the many-body aspects of the electron gas is certainly seminal and has been recognized (very rightly) with an APS fellowship. The present article is a perspective by his first PhD student and attempts to provide highlights of the work with hopefully appropriate references. The main contribution is a detailed analysis of the vertex corrections due to charge-density and spin-density fluctuations in the electron liquid; employing these results to a microscopic determination of the Landau Fermi liquid parameters.

Understanding the many-body aspects of an electron gas (EG) has been a subject of ongoing interest for some decades. While the high-density limit corresponds to an ideal paramagnetic gas, the low-density limit results in a solid. Contrastingly, the intermediate densities are not well characterized owing to the absence of a small parameter to conduct perturbation theory. In the intermediate regime, a noteworthy phenomenological (and approximate) approach is the Landau's theory of Fermi liquids. Landau wrote down an effective Hamiltonian (for low-lying excitations) which yields various properties of the system such as the effective mass of an electron, the effective g-factor of an electron, spin susceptibility, specific heat, etc.

As regards microscopic approaches to the electron-gas problem, Quinn and Ferrel [1] worked out the theory within a random-phase approximation (RPA). Next, Rice [2] (and later others) incorporated the effect of vertex corrections due to charge fluctuations without consistently including the corrections due to spin-density fluctuations; consequently, the approach lead to serious problems.

More detailed analyses that take into account the vertex corrections associated with both charge-density and spin-density fluctuations were carried out for an unpolarized EG by various groups using different approaches [3–5]. First, Kukkonen and Overhauser [3] (KO) proposed an analytic, clear, and simple scheme for taking into account the effects of exchange and correlations in an EG by considering both charge and spin fluctuations. For the case of paramagnetic jellium, KO were the first to exploit fully the many-body local field framework introduced earlier on by Hubbard [7]. The KO theory was the first to provide a derivation of the effective interaction between two electrons in an unpolarized EG. Second, the KO results were later confirmed by Vignale and Singwi (VS) for the paramagnetic EG by means of a diagrammatic technique [4]. Next, Giuliani *et al.* [5,6] proposed a physically transparent approach, the renormalized Hamiltonian approach (RHA), which was a generalization of the RPA-based pioneering theory of Hamann and Overhauser [8]. The RHA involves going beyond RPA by incorporating the effects of vertex corrections associated with charge-density and spin-density fluctuations to account for exchange and correlation effects in the paramagnetic EG. In the RHA, a few electrons in the EG are labeled as test electrons; the remaining part of the EG is treated as a screening medium. Needless to say that the test electrons and the screening medium are not distinct physical entities. Owing to their common nature, exchange processes occur between the test electrons and the screening medium. When the test electrons move, they excite charge-density and spin-density fluctuations in the screening medium; these fluctuations, in turn, screen the interaction between the test electrons. Thus, the screening medium (on the average) mimics the true physical processes. On averaging over the coordinates of the screening medium, an effective renormalized Hamiltonian for the clothed test electrons is obtained. Here, it must be noted that the coupling with the screening medium occurs only through the charge-density and spin-density fluctuations excited by the test electrons.

The diagrammatic analysis of VS was extended to the case of an infinitesimally polarized electron gas by Ng and Singwi [9]. Next, Giuliani *et al.* extended their results for the unpolarized EG (using RHA) to the case of an infinitesimally polarized EG [10]; they derived a quasiparticle pseudo-Hamiltonian in terms of the charge and spin response functions after incorporating the many-body local fields of the system. A major motivating factor for the enterprise was to keep separate track of the two spin occupation numbers so that functional derivatives can be taken with respect to them to evaluate Landau Fermi liquid parameters. Later on, Giuliani *et al.* [11] generalized the elegant and transparent approach of KO to the case of an infinitesimally polarized EG. They derived the effective interaction between two electrons for an infinitesimally polarized degenerate multi-valley system. They obtained the electron self-energy, from the screened interaction, by using the GW approximation [12]. They also show that the obtained self-energy is very similar to that derived by them using RHA [10].

It is *remarkable* that very similar results (*i.e.*, self-energy of an electron, effective interaction between two electrons, etc.) for the EG for both unpolarized and infinitesimally polarized cases were derived by three different approaches thereby lending credibility to the final results.

As regards response functions of an EG, it is worth mentioning that analytical expressions were derived by Giuliani *et al.* for the static many-body, spin-symmetric and spin-antisymmetric local-field factors of a homogeneous two-dimensional electron gas [13, 14]; their expressions reproduce diffusion Monte Carlo data [15] and match the exact asymptotic behaviors at both small and large wave numbers [16].

Lastly, an important enterprise of Giuliani *et al.* was to apply the derived EG theory to calculating quantities of physical interest such as the quasiparticle effective mass, the spin susceptibility enhancement, the effective g-factor, etc. [17, 18]. A quantitative comparison between experiment and calculations of quasiparticle properties (that take into account quasi-2D effects of the electronic wave function and valley degeneracy) for silicon MOSFETs has been successfully carried out earlier for the weak-coupling regime by Giuliani *et al.* in Reference [19]. On the other hand, for the case of strongly interacting electrons, comparison between theory and experiments (showing significant enhancement in effective mass) in a single-valley GaAs/AlGaAs quantum well was also done by Giuliani *et al.* [20]. Surprisingly, the on-shell approximation (meant for weak-coupling) yielded a reasonable fit to the effective mass data unlike the strong-coupling, Dyson-equation approach. Thus, perhaps, more effort is needed to understand the unexplored aspects of many-body local-fields such as their frequency dependence.

All in all, while the contributions of Giuliani *et al.* in the area of many-body EG are significant, the EG problem (given its many-body nature) is far from being fully understood and will continue to be an area of focus for a number of years to come.

Acknowledgements. It is very hard to forget Giuliani's enthusiasm for working on the electron gas. When I write about Giuliani's work (involving me) on the electron gas, I must also acknowledge Overhauser's substantive influence on us.

References

[1] J. J. Quinn and R. A. Ferrell, Phys. Rev. **112** (1958), 812.
[2] T. M. Rice, Ann. Phys. (N.Y.) **31** (1965), 100.

[3] C. A. KUKKONEN and A. W. OVERHAUSER, Phys. Rev. B **20** (1979), 550.
[4] G. VIGNALE and K. S. SINGWI, Phys. Rev. B **32** (1985), 2156.
[5] S. YARLAGADDA and G. F. GIULIANI, Solid State Commun. **69** (1989), 677.
[6] G. F. GIULIANI and G. VIGNALE, "Quantum Theory of the Electron Liquid", Cambridge University Press, Cambridge, 2005.
[7] J. HUBBARD, Proc. R. Soc. London, Ser. A **242** (1957), 539; **243** (1957), 336.
[8] D. R. HAMANN and A. W. OVERHAUSER, Phys. Rev. **143** (1966), 183.
[9] T. K. NG and K. S. SINGWI, Phys. Rev. B **34** (1986), 7738; **34** (1986), 7743.
[10] S. YARLAGADDA and G. F. GIULIANI, Phys. Rev. B **49** (1994), 7887.
[11] S. YARLAGADDA and G. F. GIULIANI, Phys. Rev. B **61** (2000), 12556.
[12] L. HEDIN and S. LUNDQUIST, In: "Solid State Physics", F. Seitz, D. Turnbull, and H. Ehrenreich (eds.), Academic, New York, 1969, Vol. 23.
[13] B. DAVOUDI, M. POLINI, G. F. GIULIANI and M. P. TOSI, Phys. Rev. B **64** (2001), 153101.
[14] B. DAVOUDI, M. POLINI, G. F. GIULIANI and M. P. TOSI, Phys. Rev. B **64** (2001), 233110.
[15] G. SENATORE, S. MORONI and D. M. CEPERLEY, In: "Quantum Monte Carlo Methods in Physics and Chemistry", M. P. Nightingale and C. J. Umrigar (eds.), Kluwer, Dordrecht, 1999.
[16] G. E. SANTORO and G. F. GIULIANI, Phys. Rev. B **37** (1988), 4813.
[17] For a quantitative treatment of many-body enhancement of the spin susceptibility of a two-dimensional electron gas, see S. YARLAGADDA and G. F. GIULIANI, Phys. Rev. B **40** (1989), 5432.
[18] For a quantitative study of the quasiparticle properties of a three-dimensional electron gas, see G. E. SIMION and G. F. GIULIANI, Phys. Rev. B **77** (2008), 035131.
[19] S. YARLAGADDA and G. F. GIULIANI, Phys. Rev. B **49** (1994), 14188.
[20] R. ASGARI, B. DAVOUDI, M. POLINI, G. F. GIULIANI, M. P. TOSI and G. VIGNALE, Phys. Rev. B **71** (2005), 045323.

Gabriele's scientific highlights

PHYSICAL REVIEW B VOLUME 26, NUMBER 8 15 OCTOBER 1982

Lifetime of a quasiparticle in a two-dimensional electron gas

Gabriele F. Giuliani* and John J. Quinn
Department of Physics, Brown University, Providence, Rhode Island 02912
(Received 13 May 1982)

We have investigated the inelastic Coulomb lifetime τ_{ee} of a quasiparticle near to the Fermi surface of a two-dimensional electron gas. Within a perturbative approach based upon the random-phase approximation, we find that at low temperature $1/\tau_{ee}$ behaves like $T^2 \ln T$. Furthermore at small quasiparticle excitation energy, the leading contribution to $1/\tau_{ee}$ is inversely proportional to the electronic density and does not depend upon the electric charge. Although the plasmon frequency goes to zero at long wavelength, plasmon emission contributes to the quasiparticle decay only when the quasiparticle excitation energy exceeds a certain threshold. The threshold becomes a small fraction of the Fermi energy in the high-density limit.

I. INTRODUCTION

The effect of Coulomb interaction on the lifetime of the electronic states close to the Fermi surface is a classic problem in many-body theory. For the ordinary three-dimensional (3D) electron gas, the inverse inelastic lifetime $1/\tau_{ee}$ associated with the electron-electron interaction has been evaluated within a perturbative approach by several authors since the pioneering work of Landau and Pomerantschusk.[1–3] At zero temperature for a quasiparticle state with wave vector p close to the Fermi wave vector p_F, it is found[3] that $1/\tau_{ee} \propto (p-p_F)^2$. Luttinger[4] has established the validity of this result at all the orders in perturbation theory. In a one-dimensional electron gas neutralized by a rigid positive background it has been found that $1/\tau_{ee} \propto |p-p_F|$.[4,5] The corresponding calculation for a two-dimensional (2D) electron gas has been performed by Chaplik.[6] The result is $1/\tau_{ee} \propto (p-p_F)^2 \ln|p-p_F|$.

There has been considerable interest in recent years in the physical properties of 2D metals. Electrons confined in silicon inversion layers and to the GaAs layer of GaAs-$Al_xGa_{1-x}As$ heterojunctions provide a vivid realization of such peculiar systems.[7] The inelastic broadening of the electronic states in these conductors plays a major role in the interpretation of magnetoconductance experiments[8–12] and its bearing upon the localization problem.[13–15] This is discussed in detail by Wheeler.[8] Several authors have invesitgated the Coulomb inelastic lifetime of the electronic states of a 2D metal in the presence of a finite concentration of impurities.[12,16] Their analysis, however, is restricted to the diffusive regime and the results cannot be extrapolated to the pure-metal limit. The aim of this work is to present a detailed and comprehensive investigation of the temperature-dependent Coulomb inelastic broadening in the simple case of a pure 2D electron gas. An interesting feature of the 2D situation is the possibility of plasma modes affecting the Coulomb broadening of the electronic states. In the usual 3D case this phenomenon is inhibited by the large energy associated with plasma oscillations. For a 2D metal however, the plasma frequency goes to zero in the long-wavelength limit[17] and plasmon emission can in principle become an available decay channel also for thermal or low-energy electronic excited states.

The paper is organized as follows. In Sec. II the microscopic theory of the Coulomb inelastic lifetime of a quasiparticle is revisited and specialized to the case of a 2D electron gas. In Sec. III we evaluate $1/\tau_{ee}$ and explicitly establish its asymptotic behavior. Section IV provides some discussion with emphasis on the peculiar temperature and charge dependence of the results. Finally three appendixes complete the paper by providing a discussion of a few technical aspects of the theory.

II. INELASTIC LIFETIME OF A QUASIPARTICLE

Consider a degenerate gas of N electrons in its normal ground state. This can be well described in terms of filled Fermi sea. A quasiparticle is obtained by adding to the system an extra electron which occupies an otherwise empty state characterized by a wave vector $\vec{p}$ and a spin projection σ. In complete analogy a quasihole can be obtained by removing an electron from an otherwise occupied state. At $T=0$ K, if p_F is the Fermi wave vector, necessarily $p \geq p_F$ for a quasiparticle and $p \leq p_F$ for a quasihole.

The ground state for these $N \pm 1$ electrons configurations is of course again a filled Fermi sea with the same Fermi wave vector, apart from corrections of order $1/N$. In the absence of any relaxation mechanism, quasiparticle, and quasihole states are stationary. The mutual Coulomb interaction however, provides a way to redistribute energy and momentum among the electrons and causes a quasiparticle (quasihole) state to decay. This leads to a finite inelastic lifetime $1/\tau_{ee}$ for these electronic states.

For $T=0$ K the situation is readily analyzed via standard time-dependent perturbation theory and $1/\tau_{ee}$ is given by the decay rate of the corresponding plane-wave state. At finite temperature the situation is more complicated and $1/\tau_{ee}$ is defined by the relaxation rate of the occupation number $n_{\vec{p},\sigma}$, as obtained by an approach based on a transport equation of the type[18]

$$\frac{\partial n_{\vec{p},\sigma}}{\partial t} = -\frac{n_{\vec{p},\sigma} - n^0_{\vec{p},\sigma}}{\tau_{ee}} , \tag{1}$$

where $n^0_{\vec{p},\sigma}$ is the distribution function at equilibrium. In both cases $1/\tau_{ee}$ can be evaluated within perturbation theory, with the use of the usual Fermi golden rule,[19]

$$\frac{1}{\tau_{ee}} = \frac{2\pi}{\hbar} \sum_{\vec{k},\vec{q},\sigma'} n^0_{\vec{k},\sigma'}(1-n^0_{\vec{k}-\vec{q},\sigma'})(1-n^0_{\vec{p}+\vec{q},\sigma}) \, | V_c(\vec{p},\vec{q}) |^2 \delta(E_{\vec{p}+\vec{q}} + E_{\vec{k}-\vec{q}} - E_{\vec{p}} - E_{\vec{k}}) , \tag{2}$$

where $V_c(\vec{p},\vec{q})$ is the matrix element of suitable electron-electron interaction potential.

Some discussion is in order as far as the proper choice of V_c is concerned. As pointed out by Quinn and Ferrell,[3] the use in Eq. (2) of the bare Coulomb potential matrix element $v(q)$ for V_c leads to the unphysical result $1/\tau_{ee} = \infty$. Such a difficulty can however be surmounted by allowing for screening effects. This is readily done within the random-phase approximation (RPA).[3,19] Accordingly we choose a dynamically screened interaction of the form

$$V_c(\vec{p},\vec{q}) = \frac{v(q)}{\epsilon(q,(E_{\vec{p}} - E_{\vec{p}+\vec{q}})/\hbar)} , \tag{3}$$

where $\epsilon(q,\omega)$ is the wave vector and frequency-dependent dielectric function of a two-dimensional electron gas.[6,20] ϵ is here evaluated at the frequency $(E_{\vec{p}} - E_{\vec{p}+\vec{q}})/\hbar$ corresponding to the energy transferred to the electron gas by the extra electron (hole) during the scattering. Notice that the use of a dynamical screening (as compared to a static one) makes V_c a complex quantity.

The sum over $\vec{k}$ and σ' appearing in Eq. (2) can be performed and, with the use of the fluctuation and dissipation theorem, expressed in terms of the imaginary part of the susceptibility $\chi^0(q,\omega)$ of a noninteracting electron gas,[19]

$$\sum_{\vec{k},\sigma'} n^0_{\vec{k},\sigma'}(1-n^0_{\vec{k}-\vec{q},\sigma'})\delta(E_{\vec{k}-\vec{q}} - E_{\vec{k}} - \omega) = -\frac{\mathrm{Im}\chi^0(q,\omega)}{S\pi(1-e^{-\hbar\omega/k_BT})} , \tag{4}$$

where S is the total surface and k_B is Boltzmann's constant.[21] Using this result and Eq. (3) in (2), $1/\tau_{ee}$ can be expressed as

$$\frac{1}{\tau_{ee}(\Delta)} = \frac{1}{\hbar S(1+e^{-\Delta/k_BT})} \sum_{\vec{q}} v(q) \int_{-\infty}^{+\infty} \frac{d\omega}{2\pi} \left[\coth\left[\frac{\hbar\omega}{2k_BT}\right] - \tanh\left[\frac{\hbar\omega - \Delta}{2k_BT}\right] \right]$$
$$\times \mathrm{Im}\left[\frac{1}{\epsilon(q,\omega)}\right] \delta\left[\omega - \frac{\Delta + \mu - E_{\vec{p}+\vec{q}}}{\hbar}\right] , \tag{5}$$

where μ is the chemical potential and we have introduced the quantity $\Delta = E_{\vec{p}} - \mu$. Δ is just the excitation energy of the quasiparticle (hole) state. This expression for $1/\tau_{ee}$ applies equally well to the usual three-dimensional case (see Appendix C).[22]

If we use for the single-particle energy $E_{\vec{k}}$ the free-electron value $\hbar^2 k^2/2m$, the angular part of the integration involved in Eq. (5) can be carried out analytically. For a 2D system the result is

$$\int_0^{2\pi} d\phi\, \delta\left(\omega - \frac{\Delta+\mu-E_{\vec{p}+\vec{q}}}{\hbar}\right) = \begin{cases} 0, & \omega > \Omega(q)\,, \\ \dfrac{2}{\{[\Omega(q)-\omega][\Omega(q)+\hbar q^2/m+\omega]\}^{1/2}}\,, & -\Omega(q)-\dfrac{\hbar q^2}{m} \le \omega \le \Omega(q) \\ 0, & \omega < -\Omega(q)-\hbar q^2/m \end{cases} \tag{6}$$

where $\hbar\Omega(q)$ (see Fig. 1) is the maximum value of the energy transfer for a scattering process in which the extra electron (hole) changes its wave vector by q,

$$\Omega(q) = \frac{\hbar p q}{m} - \frac{\hbar q^2}{2m}\,. \tag{7}$$

With (6) in (5) and $v(q)=2\pi e^2/q$ we finally obtain for a 2D electron gas,

$$\frac{1}{\tau_{ee}(\Delta)} = -\frac{e^2}{\pi\hbar(1+e^{-\Delta/k_B T})}\left[\int_{-\infty}^{0} d\omega \int_{-q_-(\omega)}^{q_+(\omega)} dq + \int_0^{(\mu+\Delta)/\hbar} d\omega \int_{q_-(\omega)}^{q_+(\omega)} dq\right] \times \left[\coth\left(\frac{\hbar\omega}{2k_B T}\right) - \tanh\left(\frac{\hbar\omega-\Delta}{2k_B T}\right)\right] \frac{\mathrm{Im}(1/\epsilon(q,\omega))}{\{[\Omega(q)-\omega][\Omega(q)+(\hbar q^2/m)+\omega]\}^{1/2}}\,, \tag{8}$$

where $q_\pm(\omega)$ are the solutions of the equation $\Omega(q)=\omega$, i.e.,

$$q_\pm(\omega) = p\left[1 \pm \left(1-\frac{\hbar\omega}{\mu+\Delta}\right)^{1/2}\right]. \tag{9}$$

Figure 1 illustrates the geometrical constraints imposed by energy and momentum conservation to the decay processes in a 2D electron gas.

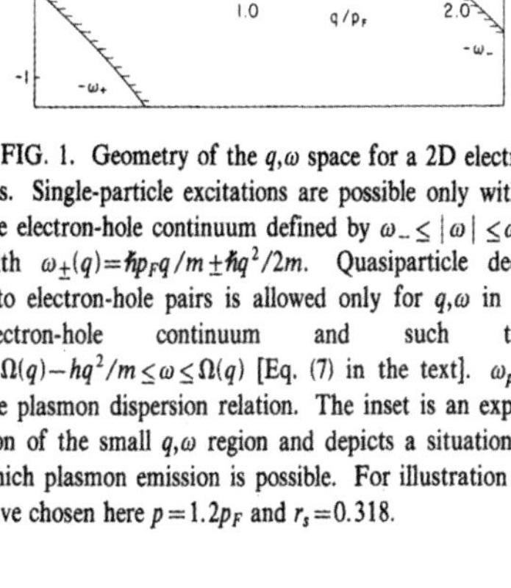

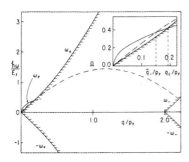

FIG. 1. Geometry of the q,ω space for a 2D electron gas. Single-particle excitations are possible only within the electron-hole continuum defined by $\omega_- \le |\omega| \le \omega_+$, with $\omega_\pm(q)=\hbar p_F q/m \pm \hbar q^2/2m$. Quasiparticle decay into electron-hole pairs is allowed only for q,ω in the electron-hole continuum and such that $-\Omega(q)-hq^2/m \le \omega \le \Omega(q)$ [Eq. (7) in the text]. ω_p is the plasmon dispersion relation. The inset is an expansion of the small q,ω region and depicts a situation in which plasmon emission is possible. For illustration we have chosen here $p=1.2p_F$ and $r_s=0.318$.

A completely equivalent approach to this problem is to evaluate to the lowest order in the screened interaction the self-energy $\Sigma(\vec{p},E_{\vec{p}})$ of the quasiparticle (quasihole). The corresponding diagram is shown in Fig. 2. $1/\tau_{ee}$ is then obtained via[3,19]

$$\frac{1}{\tau_{ee}} = -\frac{2}{\hbar}\,\mathrm{Im}\Sigma(\vec{p},E_{\vec{p}})\,. \tag{10}$$

III. DECAY PROCESSES

We turn now to the analysis of the elementary processes by which a quasiparticle (quasihole) state can decay, as described by the imaginary part of the inverse dielectric function in (8). Within RPA we can divide $\mathrm{Im}(1/\epsilon)$ as follows:

$$\mathrm{Im}\frac{1}{\epsilon(q,\omega)} = \mathrm{Im}\frac{1}{\epsilon(q,\omega)}\bigg|_{e\text{-}h} + \mathrm{Im}\frac{1}{\epsilon(q,\omega)}\bigg|_{\mathrm{pl}}\,. \tag{11}$$

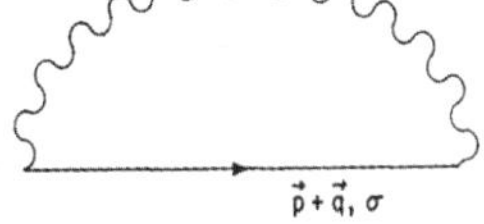

FIG. 2. The simplest self-energy diagram for an electron of wave vector $\vec{p}$ and spin projection σ. The dashed line represents the screened Coulomb potential, Eq. (3).

The first term is associated with single electron-hole pair excitations with wave vector q and energy $\hbar\omega$. The second describes inelastic processes involving excitation of plasma modes. Since in a three-dimensional electron gas the plasma frequency is always finite, for small excitation energies Δ, single electron-hole pair excitations represent the only relevant dissipative processes. In a 2D system however, the plasma frequency (see Appendix A) goes to zero at long wavelength[17] and, as already mentioned above, plasmons became available at small energies. Multipairs excitations are also possible but they lead to a small effect at low excitation energies, and are disregarded in RPA.

A. Decay into single-particle excitations

Within RPA single electron-hole pair excitations are possible only for q and ω inside the electron-hole continuum (see Fig. 1). At low temperature and small excitation energy Δ only the region of small ω is relevant. In this case Eq. (8) can be considerably simplified.

We first notice that because of the singular behavior of the integrand in Eq. (8) for $\omega=\Omega(q)$ [i.e., $q=q_{\pm}(\omega)$], the main contribution to the decay rate at low energies comes from scattering processes involving a small wave-vector change q of the order of $p-p_F$. Accordingly, we write

$$\operatorname{Im}\frac{1}{\epsilon(q,\omega)}\bigg|_{e\text{-}h} = -\frac{\hbar\omega}{2e^2 p_F}\left[1-\left(\frac{m\omega}{\hbar q p_F}\right)^2\right]^{1/2}, \tag{12}$$

where we have made use in (A3) of the small q and ω expansion of the electronic susceptibility [Eqs. (5), (A1), and (A2)].[23] For q and ω outside of the electron-hole continuum, $\operatorname{Im}(1/\epsilon)_{eh}$ is zero.

At $T=0$ K, the frequency integral of Eq. (8) is restricted to the interval 0, $\Delta/\hbar$. In this case, making use of (12) in (8), $1/\tau_{ee}|_{e\text{-}h}$ can be reduced, after some straightforward manipulations, to a single quadrature. A direct inspection allows us to extract the leading contribution. We find

$$\frac{1}{\tau_{ee}(\Delta)}\bigg|_{e\text{-}h} \simeq -\frac{E_f}{4\pi\hbar}\left[\frac{\Delta}{E_f}\right]^2\left[\ln\left[\frac{\Delta}{E_F}\right]-\frac{1}{2}-\ln\left[\frac{2q_{\mathrm{TF}}^{(2)}}{p_F}\right]\right], \quad T=0\ \mathrm{K}, \quad \Delta \ll \frac{\hbar^2 p_F q_{\mathrm{TF}}^{(2)}}{m} \tag{13}$$

a result previously obtained by Chaplik.[6] In Eq. (13) $E_F=\hbar^2 p_F^2/2m$ is the Fermi energy of the electronic system and $q_{\mathrm{TF}}^{(2)}$ is the Thomas-Fermi screening wave vector in 2D, given by $2me^2/\hbar^2$. The result of the numerical integration is shown in Fig. 3.

At finite temperatures the integrals involved in Eq. (8) are not feasible. However, in the region of temperatures much larger than Δ/k_B and much smaller than E_F/k_B, we have been able to evaluate the relevant contribution. The result is

$$\frac{1}{\tau_{ee}(\Delta)}\bigg|_{e\text{-}h} \simeq -\frac{E_F}{2\pi\hbar}\left[\frac{k_B T}{E_F}\right]^2\left[\ln\left[\frac{k_B T}{E_F}\right]-\ln\left[\frac{q_{\mathrm{TF}}^{(2)}}{p_F}\right]-\ln 2-1\right], \quad \Delta \ll k_B T \ll E_F . \tag{14}$$

B. Decay into plasma modes

At zero temperature, the contribution to $\operatorname{Im}(1/\epsilon)$ associated with the collective modes is given in RPA by[10]

$$\operatorname{Im}\frac{1}{\epsilon(q,\omega)}\bigg|_{\mathrm{pl}} = -\pi\left[\frac{\partial \operatorname{Re}\epsilon(q,\omega)}{\partial\omega}\right]^{-1}\delta(\omega-\omega_P(q)), \tag{15}$$

where $\omega_P(q)$ is the plasma dispersion relation, as discussed in Appendix B. This expression is defined only for values of q and ω lying outside of the electron-hole continuum (see Fig. 1). For a 2D electron gas the quantity $\partial \operatorname{Re}\epsilon/\partial\omega$ can be readily evaluated using Eqs. (A3) and (A1).

Inserting Eq. (15) in (8) we find after some straightforward manipulations,

$$\frac{1}{\tau_{ee}(\Delta)}\bigg|_{\mathrm{pl}} = \frac{2e^2}{\hbar}\int_0^{q_c} dq\, \frac{\Theta(\Omega(q)-\omega_+(q))}{\{[\Omega(q)-\omega_P(q)][\Omega(q)+(hq^2/m)+\omega_P(q)]\}^{1/2}}\frac{1}{[\partial \operatorname{Re}\epsilon(q,\omega_P(q))/\partial\omega]}$$
$$\times\int_{\min[\Delta/\hbar,\Omega(q)]}^{\min[\Delta/\hbar,\omega(q)]} d\omega\,\delta(\omega-\omega p(q)), \tag{16}$$

where q_c is the critical wave vector for plasma modes (see Appendix B) and min$[a,b]$ is the minimum between a and b.

Finite contributions to $1/\tau_{ee}$ in Eq. (16) come only from wave vectors q for which the condition $\omega_+(q)\le\omega_P(q)\le\Omega(q)$ is satisfied (see Fig. 1). Furthermore, at zero temperature, the excitation energy Δ must be larger than $\hbar\omega_P(\tilde{q}_-)$, with $\tilde{q}_-$ defined in Eq. (B6). This leads to the existence of a finite excitation-energy threshold Δ_c for the decay into plasmons. By using Eq. (B6) in (B2) we obtain for Δ_c the following equation:

$$\Delta_c=\left[\frac{32me^2E_F(E_F+\Delta_c)^{1/2}}{3\hbar}\right]^{1/2}\cos\left[\frac{\pi}{3}+\frac{1}{3}\arccos\left[\frac{E_F+\tilde{\Delta}(r_s)}{E_F+\Delta_c}\right]^{3/4}\right], \tag{17}$$

where $\tilde{\Delta}$ is given by Eq. (B5). In the high-density limit, Eq. (17) reduces to the simpler form,

$$\Delta_c\simeq\sqrt{2}r_sE_F,\quad r_s\ll 0\,. \tag{18}$$

Here r_s is the average interelectronic distance measured in Bohr radii.

In the general case Eq. (17) must be used. The values of Δ_c as given by Eq. (17) and (18) are compared for small r_s in Fig. 4. At metallic densities Eq. (17) gives quite large values for the excitation energy threshold Δ_c, and the quasiparticle decay into plasma modes is inhibited. In the high-density regime however, Δ_c can be still considered as a small fraction of the Fermi energy. In this case, for small Δ and Δ_c, we can make use in (16) of the approximate form

$$\left.\frac{\partial\mathrm{Re}\epsilon(q,\omega)}{\partial\omega}\right|_{\omega=\omega_P(q)}\simeq\frac{2}{\omega_P(q)}, \tag{19}$$

valid at small q. The result for $1/\tau_{ee}\,|_{\rm pl}$ is

$$\left.\frac{1}{\tau_{ee}(\Delta)}\right|_{\rm pl}\simeq\frac{2e^3m^{1/2}}{\hbar^2}\Theta(\Delta-\Delta_c)(q_m-\tilde{q}_-)^{1/2}, \tag{20}$$

with $q_m=\min[\Delta^2/2e^2E_F,q_c]$ and $\tilde{q}_-$ is given in Eq. (B6). For Δ slightly larger than Δ_c, (20) reduces to

$$\left.\frac{1}{\tau_{ee}(\Delta)}\right|_{\rm pl}\simeq\frac{2\sqrt{2}me^2}{\hbar^3p_F}\Delta_c\left[\frac{\Delta-\Delta_c}{\Delta_c}\right]^{1/2},\quad \Delta\simeq\Delta_c\,. \tag{21}$$

Finally, as Δ exceeds $\hbar\omega_P(q_c)=(2e^2E_Fq_c)^{1/2}$, Eq. (20) can be written as

$$\left.\frac{1}{\tau_{ee}(\Delta)}\right|_{\rm pl}\simeq\frac{2e^4m}{\hbar^3}\left[\frac{\Delta}{E_F}\right]^{1/2},\quad \Delta\ge\hbar\omega_P(q_c)\,. \tag{22}$$

IV. DISCUSSION

In this paper we have calculated within a perturbative approach the temperature-dependent

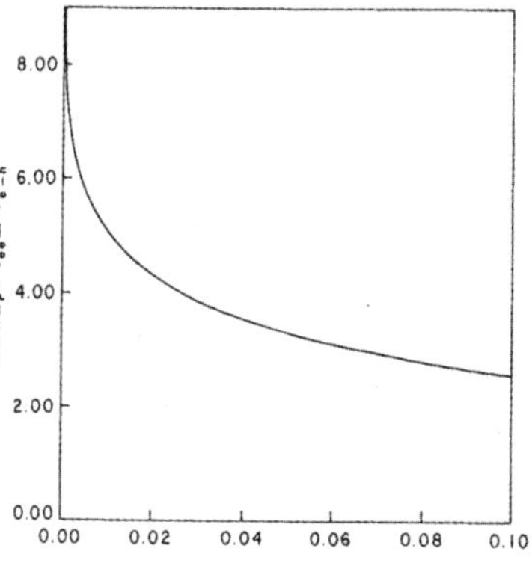

FIG. 3. Plot of $(\tau_{ee}\Delta^2)^{-1}$ measured in units of $4\pi\hbar E_F$ versus $\Delta/4E_F$, as obtained via direct numerical computation. For illustration we have taken here $r_s=2$.

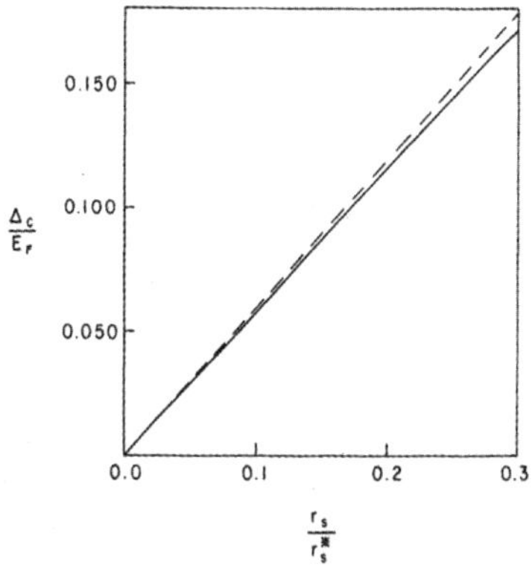

FIG. 4. Plasmon emission threshold Δ_c (measured in units of E_F) versus r_s/r_s^* ($r_s^*=8\sqrt{2}/27$) in the high-density region. Δ_c is obtained solving numerically Eq. (17). The dashed line is the asymptotic formula $\Delta_c/E_F=\sqrt{2}r_s$ [Eq. (18)].

Coulomb inelastic lifetime τ_{ee} of a quasiparticle in a 2D electron gas. Our findings [Eqs. (13) and (20)] complemented by the results of earlier calculations, are schematically summarized in Table I. It is clear that the customary textbook "phase-space argument"[1] leads to the correct answer only in the 3D case.

In a 2D system, τ_{ee} displays an extra logarithmic dependence.[6] This peculiar result stems from the concurrent effects of the planar geometry and the conservation of energy and momentum in the electronic collision processes, as expressed by Eq. (6).[24] This has been overlooked by several previous investigators.

Another interesting feature is the complete independence of τ_{ee} from the electric charge, as manifest in Eqs. (13) and (14). This is just one of the consequences of the analytic dependence of the screened Coulomb potential regarded as a function of electric charge and wave vector.

We have investigated for comparison the dependence on the electric charge e of τ_{ee} in the usual 3D case. In the high-density limit Quinn and Ferrel[3] find that $1/\tau_{ee}$ is simply proportional to e. In the general case however, this dependence is much more involved as shown by the calculation of $1/\tau_{ee}$ for a 3D electron gas presented in Appendix C.

Quite generally, the dependence of $1/\tau_{ee}$ upon e is dictated by which values of the wave-vector transfer q are the most relevant ones in the decay process. In a 3D metal all the q values between zero and $2p_F$ provide a relevant contribution to $1/\tau_{ee}$ leading to the complicated structure of Eq. (C1). For a 2D system the singular behavior displayed in Eq. (6) makes the q values of the order of $p-p_F$ to contribute the leading term at low excitation energy. Since at long wavelength the screened Coulomb potential is independent of e so does $1/\tau_{ee}$.

The situation resembles the dirty-metal case.[14] In the presence of a finite concentration of impurities momentum conservation is relaxed and diffusion dominates the dynamics at low energy. In this case, in all dimensions, the most relevant contributions to the inelastic Coulomb lifetime come from q values of the order $(k_BT/D)^{1/2}$, where D is the diffusion constant. As a consequence $1/\tau_{ee}$ does not depend upon e both in two[16] and three[25] dimensions.

TABLE I. Asymptotic behavior of the inelastic Coulomb lifetime τ_{ee} for $p \to p_F$ in a 3D, 2D, and 1D degenerate electron gas.

	$1/\tau_{ee}$
3D[a]	$(p-p_F)^2$
2D[b]	$(p-p_F)^2\ln\lvert p-p_F\rvert$
1D[c]	$\lvert p-p_F\rvert$

[a]Landau and Pomerantschusk (Ref. 1), Baber (Ref. 2), and Quinn and Ferrell (Ref. 3).
[b]Chaplik (Ref. 6) and this work.
[c]Luttinger (Ref. 4).

In the comparison with experiment the dependence of the leading term in a temperature expansion of $1/\tau_{ee}$ upon the electronic density is usually of interest. For a 3D metal $1/\tau_{ee} \propto A_3T^2$ with A_3 proportional to $n^{-3/2}$. Our analysis of the 2D case gives [see Eq. (14)] $1/\tau_{ee} \propto A_2T^2 \ln T$ with A_2 inversely proportional to n.

As discussed in Sec. III, at $T=0$ K, there exists a finite energy threshold for quasiparticle decay into plasma modes. This threshold is a substantial fraction of the Fermi energy at metallic densities. Such a fraction however decreases as the electronic concentration increases [see Eq. (18)]. At finite temperatures the calculation becomes involved and no clear-cut statement can be made. We expect, however, the existence of a typical temperature threshold T_c of the order r_sE_F/k_B for which the plasma decay mechanism becomes as important as the single-particle processes in the broadening of quasiparticle states. Accordingly, above T_c, $1/\tau_{ee}$ will display an additional contribution proportional to $(T-T_c)^{1/2}$ and then to $T^{1/2}$ as T is further increased.

ACKNOWLEDGMENT

We are grateful to the Office of Naval Research for support of this work.

APPENDIX A: DIELECTRIC FUNCTION OF A 2D ELECTRON GAS

The susceptibility $\chi^0(q\omega)$ of a noninteracting electron gas in 2D is readily evaluated at $T=0$ K, via linear response theory. The result is[20]

$$\mathrm{Re}\chi^0(q,\omega) = -N_0\left[1-\Theta(|x+y|-1)\frac{\mathrm{sgn}(x+y)}{2x}[(x+y)^2-1]^{1/2} - \Theta(|x-y|-1)\frac{\mathrm{sgn}(x-y)}{2x}[(x-y)^2-1]^{1/2}\right], \tag{A1}$$

$$\mathrm{Im}\chi^0(q,\omega)=-\frac{N_0}{2x}\{\Theta(1-|x+y|)[1-(x+y)^2]^{1/2}-\Theta(1-|x-y|)[1-(x-y)^2]^{1/2}\}\ , \tag{A2}$$

where $x=q/2p_F$, $y=m\omega/\hbar q p_F$, and N_0 is the density of states at the Fermi energy. $\Theta(x)$ is the usual step function. Within RPA the dielectric constant $\epsilon(q,\omega)$ is then given by

$$\epsilon(q,\omega)=1-v(q)\chi^0(q,\omega)\ , \tag{A3}$$

with $v(q)=2\pi e^2 q$.

APPENDIX B: PLASMA WAVES IN A 2D ELECTRON GAS

The dispersion relation $\omega_P(q)$ for plasma waves can be readily established within RPA. Making use of the results of Appendix A, the collective mode condition $\epsilon(q,\omega_P(q))=0$ gives at long wavelength,[17,6,20]

$$\omega_P^2(q)\simeq\alpha q+\beta q^2\ , \tag{B1}$$

with $\alpha=2e^2E_F/\hbar^2$ and $\beta=3E_F/2m$. The second term in (B1) is relevant only for $q\geq(4/3\sqrt{2})r_s p_F$, r_s being the average interelectronic distance in Bohr radii. For $r_s\geq 1$, and in any case at low frequencies, the expression

$$\omega_P(q)\simeq(\alpha q)^{1/2} \tag{B2}$$

provides a satisfactory approximation.

Plasma waves are well-defined collective modes only for q less than a critical wave vector q_c defined by $\omega_P(q_c)=\omega_+(q)$, $\omega_+(q)$ being the upper edge of the electron-hole continuum (see Fig. 1). From the condition $\epsilon(q_c,\omega_+(q))=0$ we obtain

$$q_c(r_s)=\frac{8p_F r_s}{3\sqrt{2}}\left[\cos\left(\frac{\phi(r_s)}{3}\right)-\frac{1}{2}\right]\ ,\quad r_s\geq\tilde{r}_s \tag{B3}$$

with

$$\phi(r_s)=\arccos[1+(r_s/2\tilde{r}_s)^{-2}-4(r_s/\tilde{r}_s)^{-1}]^{1/2}$$

and

$$q_c(r_s)=2p_F\left[\left(\frac{r_s}{2}\right)^{2/3}\left\{\left[1-\frac{r_s}{2\tilde{r}_s}+\left(1-\frac{r_s}{\tilde{r}_s}\right)^{1/2}\right]^{1/3}+\left[1-\frac{r_s}{2\tilde{r}_s}-\left(1-\frac{r_s}{\tilde{r}_s}\right)^{1/2}\right]^{1/3}\right\}-\frac{\sqrt{2}}{3}r_s\right]\ ,\quad r_s\leq\tilde{r}_s\ . \tag{B4}$$

In (B3) and (B4) $\tilde{r}_s=27\sqrt{2}/32\simeq 1.19$. For q larger than q_c plasmons suffer Landau damping.

The plasmon frequency $\omega_P(q)$ intersects $\Omega(q)=\hbar p q/m-\hbar q^2/2m$ (see text) only if $\Delta=E-\mu$ is larger than the threshold value $\Delta(r_s)$. With the use of Eq. (B2) we obtain

$$\tilde{\Delta}(r_s)=\left[\left(\frac{r_s}{r_s^*}\right)^{2/3}-1\right]E_F\ , \tag{B5}$$

where $r_s^*=8\sqrt{2}/27\simeq 0.42$. In this case the condition $\omega_P(q)=\Omega(q)$ is satisfied by $\tilde{q}_\pm$m

$$\tilde{q}_\pm(r_s)=\tfrac{8}{3}p\cos^2\left[\frac{\pi}{3}\mp\frac{1}{3}\arccos\left(\frac{E_F+\tilde{\Delta}(r_s)}{E_F+\Delta}\right)^{3/4}\right]\ , \tag{B6}$$

where $\tilde{q}_-\leq\tilde{q}_+$.

APPENDIX C: CHARGE DEPENDENCE OF $1/\tau_{ee}$

In a 3D electron gas the only contribution to $1/\tau_{ee}$ comes from single electron-hole particle excitations. The calculation can be carried out using Eq. (5) and the standard formulas for $\epsilon(q,\omega)$ and $v(q)$.[19] At $T=0$ K we obtain[26]

$$\left.\frac{1}{\tau_{ee}(\Delta)}\right|_{3\mathrm{D}}\simeq\frac{e^2p_F}{32\hbar}\left[\frac{1}{1+(q_{\mathrm{TF}}^{(3)}/2p_F)^2}+\frac{2p_F}{q_{\mathrm{TF}}^{(3)}}\tan^{-1}\left(\frac{2p_F}{q_{\mathrm{TF}}^{(3)}}\right)\right]\left(\frac{\Delta}{E_F}\right)^2\ , \tag{C1}$$

where $q_{\mathrm{TF}}^{(3)}$ is the usual 3D Thomas-Fermi screening wave vector. The extreme RPA limit (i.e., high densities) for $1/\tau_{ee}$, as calculated by Quinn and Ferrell,[3] is recovered in the limit of $q_{\rm TF}^{(3)} \ll p_F$,

$$\left.\frac{1}{\tau_{ee}(\Delta)}\right|_{3D} \simeq \frac{e^2 p_F^2 \pi}{32\hbar q_{\rm TF}^{(3)}}\left[\frac{\Delta}{E_F}\right]^2, \text{ high-density limit .} \tag{C2}$$

Since $q_{\rm TF}^{(3)} \propto e$ we observe that $1/\tau_{ee}\,|_{3D}$ is proportional to e in the high-density limit, whereas in the general case its charge dependence is fairly complicated [see Eq. (C1)].

In obtaining Eq. (C1) we have used for $\epsilon_1(q,\omega)$ the approximate expression $1+(q_{\rm TF}^{(3)}/q)^2$. Had we disregarded the one with respect to $(q_{\rm TF}^{(3)}/q)^2$, $1/\tau_{ee}\,|_{3D}$ would have been charge independent. This is however not justified since in 3D, unlikely in the 2D case, all the wave-vector values between zero and $2p_F$ lead to a contribution of the same order of magnitude to the sum of Eq. (5).

---

*On leave from Scuola Normale Superiore, Pisa, Italy.

[1]L. Landau and I. Pomerantschusk, Phys. Z. Sowjetunion 10, 649 (1936).

[2]W. G. Baber, Proc. R. Soc. London Ser. A 158, 383 (1937).

[3]J. J. Quinn and R. A. Ferrell, Phys. Rev. 112, 812 (1958).

[4]J. M. Luttinger, Phys. Rev. 121, 942 (1961).

[5]If the neutralizing background is deformable a one-dimensional electron gas suffers a Peierls instability. For the implication of this phenomenon on the inelastic Coulomb lifetime see for instance, M. Kaveh, J. Phys. C 13, L611 (1980).

[6]A. V. Chaplik, Zh. Eksp. Teor. Fiz. 60, 1845 (1971) [Sov. Phys.—JETP 33, 997 (1971)].

[7]See, for instance, the Proceedings of the International Conference on the Electronic Properties of Two-Dimensional Systems [Surf. Sci. 58, (1976); 73, (1978); 98, (1980); 113, (1982)].

[8]R. G. Wheeler, Phys. Rev. B 24, 4645 (1981).

[9]Y. Kawaguchi and S. Kawaji, Surf. Sci. 113, 505 (1982); J. Phys. Soc. Jpn. 48, 699 (1980); 49, 983 (1980).

[10]R. G. Wheeler, K. K. Choi, and A. Goel, Surf. Sci. 113, 523 (1982).

[11]K. K. Choi and R. G. Wheeler, Bull. Am. Phys. Soc. 27, 249 (1982), and private communication.

[12]M. J. Uren, R. A. Davies, M. Kaveh, and M. Pepper, J. Phys. C 14, L395 (1981).

[13]S. Hikami, A. I. Larkin, and Y. Nagaoka, Prog. Theor. Phys. 63, 707 (1980).

[14]B. L. Al'tshuler, D. E. Khmel'mitzkii, A. I. Larkin, and P. A. Lee, Phys. Rev. B 22, 5142 (1980).

[15]For a recent review see, for instance, H. Fukuyama, Surf. Sci. 113, 489 (1982).

[16]E. Abrahams, P. W. Anderson, P. A. Lee, and T. V. Ramakrishnan, Phys. Rev. B 24, 6783 (1981).

[17]R. H. Ritchie, Phys. Rev. 106, 874 (1957); see also R. A. Ferrell, *ibid.* 111, 1214 (1958).

[18]See for instance the discussion contained in Refs. 1, 2, and 19 and in J. Appel and A. W. Overhauser, Phys. Rev. B 18, 758 (1978).

[19]D. Pines and P. Nozières, *The Theory of Quantum Liquids* (Benjamin, New York, 1966).

[20]F. Stern, Phys. Rev. Lett. 18, 546 (1967).

[21]Here, within our perturbative approach, the distribution function $n^0_{\vec{p},\sigma}$ are assumed to be given by the usual equilibrium Fermi-Dirac expression.

[22]Use has been made here of Eq. (A3) valid within RPA.

[23]This expression for $\mathrm{Im}(1/\epsilon)$ is valid for $\omega - \hbar q p_F/m \gg \hbar q^2\omega/m$.

[24]Notice that the extra logarithmic dependence of τ_{ee} discussed in this paper is completely unrelated to the one proposed in Ref. 16. The latter results from the diffusive processes dominating the dynamics of a dirty metal at low energy.

[25]A. Schmid, Z. Phys. 271, 251 (1974).

[26]This formula is similar to Eq. (6.15) in R. H. Ritchie, Phys. Rev. 114, 644 (1959).

Quantization of the Hall conductance in a two-dimensional electron gas

Gabriele F. Giuliani,* J. J. Quinn, and S. C. Ying
Department of Physics, Brown University, Providence, Rhode Island 02912

(Received 18 February 1983)

We present a microscopic theory of the Hall conductance in a two-dimensional electron gas. Our approach is based on a single-particle picture and explicitly accounts for the effects of a random impurity potential. Within the geometry introduced by Laughlin a general expression is derived from which it is possible to evaluate the Hall conductance in terms of the properties of the electronic spectrum at the Fermi energy for any value of the magnetic field. When the chemical potential lies between the bulk extended states of well-defined neighboring Landau bands, the Hall conductance is quantized in integral multiples of e^2/h, even in the presence of a large density of localized states. Within our model the exactness of this quantization depends on the shape of the confining potential, the thickness of the sample, and the magnetic field.

I. INTRODUCTION

One of the most interesting properties of the two-dimensional electron gas which can occur at a semiconductor interface is the quantization of the Hall conductance.[1] At very low temperature T and high magnetic field strength B, the Hall conductance σ_{xy} as a function of the Landau-level filling factor $\nu=n_S(eB/hc)^{-1}$, n_S being the number of electrons per unit area, is characterized by flat steps[2] at integral multiples of the fundamental value e^2/h. In those regions of concentration n_S in which σ_{xy} has the quantized value, σ_{xx} is essentially equal to zero (it would presumably be zero at zero temperature). This effect was first indicated by Ando, Matsumoto, and Uemura in a study of the effects of impurity centers on the properties of an otherwise noninteracting two-dimensional electron gas.[3]

This result suggests very strongly that if we think of the density of states associated with a particular Landau level as broadened by impurity scattering, extended states exist only very close to the center of the Landau level and localized states exist everywhere else. Establishing the validity of this picture from microscopic theory remains a fundamental unsolved problem. Work in this direction has been recently carried out by several authors.[4–6]

The quantized Hall effect has received particular attention within the single-particle picture in which a prominent role is played by the electronic states which are localized by the impurity random potential.[7–9] Within the same framework Kazarinov and Luryi have presented an argument based on quantum percolation theory.[10] Thouless and co-workers have investigated how the effect is influenced by the presence of a periodic substrate potential.[11]

The observation of a sizable cyclotron resonance shift in Si inversion layers[12] and of additional quantized steps in the Hall resistance at $3h/e^2$ and probably $3h/2e^2$ in GaAs-AlGaAs heterojunctions in the extreme quantum limit[13] casts, however, some doubts on the validity of the single-particle picture. The normal state of a two-dimensional electron gas in the presence of a magnetic field is inherently unstable with respect to a many-body charge-density-wave or Wigner-lattice type of ground state.[14] It is not clear, however, what happens in the presence of an impurity random potential and what the magnetotransport of such an exotic ground state would be. The implication of such many-body effects on the Hall conductance of an ideal two-dimensional electron gas have recently received a great deal of attention.[15–17]

In order to establish the relevance of the many-body effects in the quantized Hall-effect problem a complete and reliable theory based on the single-particle picture must be first at hand which can treat exactly the problem associated with the impurity random potential for any value of the external magnetic field.

Laughlin has presented an elegant argument which attempts to demonstrate that the quantization is due to the long-range phase-rigidity characteristic of a supercurrent, and that it can be derived from gauge invariance and the existence of a mobility gap.[8] He does this by considering the response of a two-dimensional metallic ribbon to a change in the flux threading the ribbon. Because changing the flux threading the ribbon is certainly not a simple gauge transformation, the terminology of Laughlin's argument is inappropriate.[18] Furthermore, his argument contains the implicit assumption that the only consequence of adding an integral number of flux quanta hc/e is to repopulate the current-carrying states. This assumption is obviously valid for the ideal system, but it is not so obvious in the presence of disorder when localized states can exist at the Fermi level.

In this paper we investigate, within a single-particle picture, the eigenfunctions, eigenvalues, and distribution functions of the electrons in a two-dimensional metallic ribbon using a cyclindrical coordinate system appropriate to the geometry of the problem. In Sec. II we consider the ideal system, free of impurities, and establish the notation. There it is stressed that it is strictly necessary to mix orbitals belonging to different Landau levels in order to be able to describe localized states. In Sec. III we introduce an effective random potential and discuss the structure of the electric spectrum. In particular we prove that in a rib-

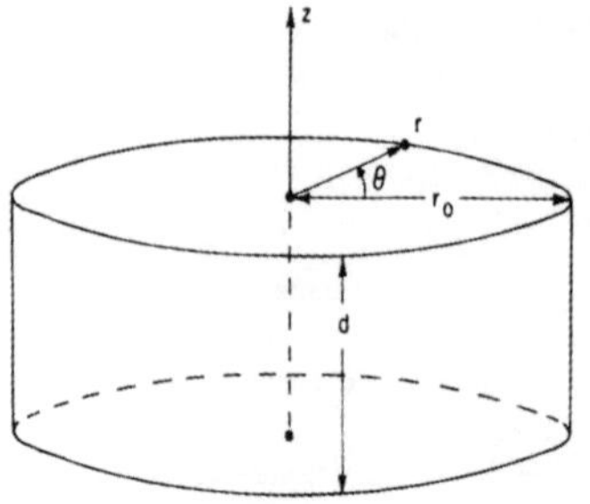

FIG. 1. Geometrical arrangement for the ribbon geometry used throughout the paper.

bon of finite radius the spectrum of the extended states (if any) is countable. This analysis is used in deriving a virtually exact expression for the Hall current which explicitly displays the dependence of this quantity on the temperature and the shape of the confining potential. The conditions for observing a quantization of the Hall conductance are discussed. Finally, Sec. IV contains a discussion of the relevant physical questions and the conclusions.

II. IDEAL METALLIC RIBBON

In this section we consider an idealized model in which the system of electrons is confined to an impurity free ribbon of radius r_0 as shown in Fig. 1. Cylindrical coordinates (r,θ,z) will be used to describe the motion of the electrons. The electrons are restricted to the radius $r=r_0$ by some confining potential, but they are free to move between the edges $(0\le z\le d)$ and around $(0\le\theta\le 2\pi)$ the ribbon. For large values of r_0 any small section of the ribbon will be indistinguishable from a similar small section of a standard Hall bar, in that a current I will flow in the θ direction and a voltage ΔV will be present across the ribbon. The ratio of I to ΔV will define the Hall conductance.

In order to mimic the behavior of a standard Hall bar, we want a magnetic field B which is everywhere perpendicular to the ribbon. We introduce a vector potential $\vec{A}=(A_r,A_\theta,A_z)$ in the cylindrical coordinate system in which $\hat{r}$, $\hat{\theta}$, and $\hat{z}$ are unit vectors, and choose $A_r=A_z=0$ and $A_\theta=-(Bz+A_1)$. Taking the curl of $\vec{A}$ leads to a magnetic field

$$\vec{B}=B\hat{r}-(Bz+A_1)\hat{z}/r\ .$$

Now $B\hat{r}$ is exactly radial magnetic field we want. The other term $r^{-1}(Bz+A_1)\hat{z}$ is extra. It does not bother us because the electrons are strictly confined to $r=r_0$ by a potential of the type

$$V(r)=0\begin{cases}0 & \text{if } r=r_0\\ \infty & \text{otherwise}\ .\end{cases} \tag{1}$$

Therefore, the Lorentz force assoicated with B_z will have no effect on the classical motion of the electrons. Of course, the constant vector potential A_1 just leads to "trapped flux" inside the ribbon which is independent of the magnetic field $B\hat{r}$, but does not affect the classical motion of the electrons.

The flux through the ribbon at a plane z is given by

$$\Phi(z)=\int_{\text{at }z}\vec{B}\cdot d\vec{a}=\oint_{\text{at }z}\vec{A}\cdot d\vec{l}=-2\pi r_0(Bz+A_1)\ . \tag{2}$$

We define $\phi=-2\pi r_0A_1$ as the trapped flux associated with A_1, and we introduce the set z_l of values of z defined by the equation

$$-2\pi r_0Bz_l+\phi=-l\phi_0\ , \tag{3}$$

where $\phi_0=hc/e$ is the flux quantum and l is an integer. Because A_θ depends on z, the flux passing through the circle defined by the intersection of the cylinder $r=r_0$ and the plane $z=$const depends upon z. The plane $z=z_l$ defines the circle through which $-l$ flux quanta pass.

Now let us look at the Hamiltonian. To start we will neglect boundary effects at $z=0$ and d. Because the electrons are confined to the radius $r=r_0$, the radial coordinate does not enter the Hamiltonian. We have

$$\begin{aligned}H_0&=\frac{1}{2m}p_z^2+\frac{1}{2m}\left[\frac{p_\theta}{r_0}+\frac{e}{c}A_\theta\right]^2,\\&=\frac{-\hbar^2}{2m}\frac{\partial^2}{\partial z^2}+\frac{1}{2m}\left[\frac{-i\hbar}{r_0}\frac{\partial}{\partial\theta}+\frac{e}{c}A_\theta\right]^2,\end{aligned} \tag{4}$$

where for simplicity only the spatial degrees of freedom have been retained. Because A_θ is independent of θ we can assume that the eigenfunctions of H_0 are of the form

$$\psi(z,\theta)=e^{il\theta}u(z)\ , \tag{5}$$

where l is an integer. Substituting this form into the Schrödinger equation gives the equation

$$\left[\frac{p_z^2}{2m}+\tfrac{1}{2}m\omega_c^2(z-z_l)^2\right]u(z)=Eu(z)\ . \tag{6}$$

Equation (6) is just the Schrödinger equation for a simple harmonic oscillator of frequency $\omega_c=eB/mc$ centered at $z=z_l$; thus, the eigenfunctions and eigenvalues of H_0 can be written

$$\begin{aligned}\psi_{nl}(z,\theta)&=\frac{e^{il\theta}}{\sqrt{2\pi}}u_n(z-z_l)\ ,\\E_{nl}&=\hbar\omega_c(n+\tfrac{1}{2})\ ,\end{aligned} \tag{7}$$

where $u_n(z)$ is the nth harmonic-oscillator eigenfunction. The quantum numbers n and l have here the following meaning. n labels a set of well-defined Landau levels whose energy spacing is given by $\hbar\omega_c$. l represents the angular momentum of the state. For a given n states with different values of l are degenerate. If the system has unit area the degeneracy of each Landau level is $(2\pi a_B^2)^{-1}$ where $a_B=(c\hbar/eB)^{1/2}$ is the magnetic length. This degeneracy is lifted by any potential term depending on the coordinate z added to the single-particle Hamiltonian (4).

Note that in (7) the "orbit-center" coordinate z_l is exactly the value given by Eq. (3); this value was determined by requiring the circle defined by $r=r_0, z=z_l$ to enclose $-l$ flux quanta. Changing the trapped flux by $\Delta\phi$ simply shifts each orbit center by $\Delta z_l=-\Delta\phi/2\pi r_0 B$. Any change $\Delta\phi$ is allowed; the electron orbits simply adjust themselves by shifting their centers to still enclose an integral number of flux quanta. In the absence of a potential which lifts the degeneracy, the eigenvalue E_{nl}, given by Eq. (7), is unchanged.

If we introduce a constant electric field $\vec{E}$ in the z direction (across the ribbon), there is an additional term eEz in the Hamiltonian. The eigenvalues and eigenfunctions of the ideal system in the presence of $\vec{E}$ are

$$\psi_{nl}(z,\theta)=\frac{e^{il\theta}}{\sqrt{2\pi}}u_n\left[z-z_l+\frac{v_D}{\omega_c}\right],$$
$$E_{nl}=\hbar\omega_c(n+\tfrac{1}{2})+eEz_l-\tfrac{1}{2}mv_D^2, \quad (8)$$

where the drift velocity v_D is defined as $v_D=cE/B$. If $\vec{E}\to 0$ these equations reduce to Eqs. (7). Clearly n and l are still good quantum numbers but the degeneracy with respect to l has been lifted. Notice that in the presence of an homogeneous applied electric field the orbitals $\psi_{nl}(z,\theta)$ are centered at z_l-v_D/ω_c, as compared to z_l in the field-free case.

It is worth mentioning here that the operator $\dot{\theta}=\partial H_0/\partial p_\theta$ can be written

$$(mr_0^2)^{-1}[p_\theta+(er_0/c)A_\theta],$$

where $A_\theta=-(Bz+A_1)$. Because p_θ is a constant of motion with value $\hbar l$, the Hall current I_H^{nl} carried by an electron in state $|nl\rangle$ is

$$I_H^{nl}=-e\langle nl|\dot{\theta}r_0|nl\rangle=e\omega_c\langle nl|(z-z_l)|nl\rangle=-ev_D. \quad (9)$$

It is straightforward to verify that

$$\langle nl|\dot{z}|nl\rangle=0 \quad (10)$$

so that no current is carried in the direction of the applied electric field. In (9) and (10), $|nl\rangle$ is the Dirac notation for the state whose wave function is $\psi_{nl}(z,\theta)$. In the absence of an external electric field no current flows in the system. If an electric field is present in the z direction, as in (8), all the electronic states $|nl\rangle$ carry the same Hall current, $-ev_D$, along $\hat{\theta}$. Equations (9) and (10) lead to the values $\sigma_{xx}=0$ for the longitudinal conductivity and

$$\sigma_{xy}=-(e^2N/h)(dr_0/a_B^2)^{-1},$$

where N is the total number of electrons.

It is interesting to realize that exactly the same results are obtained for any wave function $\phi_{nl}(z,\theta)$ of the type

$$\phi_{n\alpha}(z,\theta)=\sum_l c_{nl}^{\alpha}\psi_{nl}(z,\theta). \quad (11)$$

This implies that it is strictly necessary to allow for the mixing of different Landau levels in order to describe a non-current-currying state in a magnetic field.[19]

A different and elegant procedure to evaluate the Hall current in this geometry has been proposed by Laughlin.[8] The approach is based on the following formula:

$$I_H=-c\frac{\Delta E_T}{\Delta\phi}, \quad (12)$$

where E_T is the total energy of the system and ΔE_T is the change in E_T caused by a change in flux $\Delta\phi$. This result can be obtained by noting that the current-density operator is

$$j_\theta^{\text{op}}=-er_0\dot{\theta}=-er_0\frac{\partial H_0}{\partial p_\theta}=c\frac{\partial H_0}{\partial A_1}.$$

Taking the expectation value of j_θ^{op} and summing over occupied states leads to Eq. (12). We can calculate E_T using perturbation theory. Let us write

$$H=H_0+\Delta H, \quad (13)$$

where

$$\Delta H=(e/c)v_\theta\Delta A_1=(e\,\Delta\phi/2\pi c)\dot{\theta}$$

is the change in the Hamiltonian caused by a small change in A_1 (or in the flux $\Delta\phi=-2\pi r_0\Delta A_1$). Let us think of ΔA_1 as varying in time as $\exp(i\omega t)$, where ω is a very low frequency. Then the perturbation ΔH causes a change in the single-particle density matrix $\Delta\rho=\rho-\rho_0$, where ρ_0 is the density matrix in the absence of ΔH. The use of linear-response theory gives

$$\langle nl|\Delta\rho|n'l'\rangle=\frac{f_0(\epsilon_{n'l'})-f_0(\epsilon_{nl})}{\epsilon_{n'l'}-\epsilon_{nl}-\hbar\omega}\langle nl|\Delta H|n'l'\rangle, \quad (14)$$

where $f_0(\epsilon)$ is the usual Fermi-occupation function. The change in the total energy can be written

$$\Delta E=\text{Tr}(\rho H-\rho_0 H_0). \quad (15)$$

Keeping terms linear in the small perturbation gives

$$\Delta E_T=\sum_{n,l}[f_0(\epsilon_{nl})\langle nl|\Delta H|nl\rangle+\epsilon_{nl}\langle nl|\Delta\rho|nl\rangle]. \quad (16)$$

It is clear from Eq. (14) that the second term vanishes; therefore

$$\Delta E_T=+\frac{e\,\Delta\phi}{2\pi c}\sum_{n,l}f_0(\epsilon_{nl})\langle nl|\dot{\theta}|nl\rangle=\frac{eNv_D\Delta\phi}{2\pi r_0 c}, \quad (17)$$

where we have made use of Eq. (9). Also $\sum_{n,l}f_0(\epsilon_{nl})=N$ since, as noticed above, every state $|nl\rangle$ carries the same current. Another way to obtain this result is to notice that a change $\Delta\phi$ in the threading flux displaces each orbit center by the same amount, $\Delta z_l=-\Delta\phi/B$. From Eqs. (8) this leads to a change $eE\Delta\phi/B$ in the single-particle energy. Summing over all the occupied states the expression (17) for ΔE_T is recovered.[20]

Inserting (17) in (12) we get again

$$I_H=-eNv_D=-\frac{ecN}{B}E. \quad (18)$$

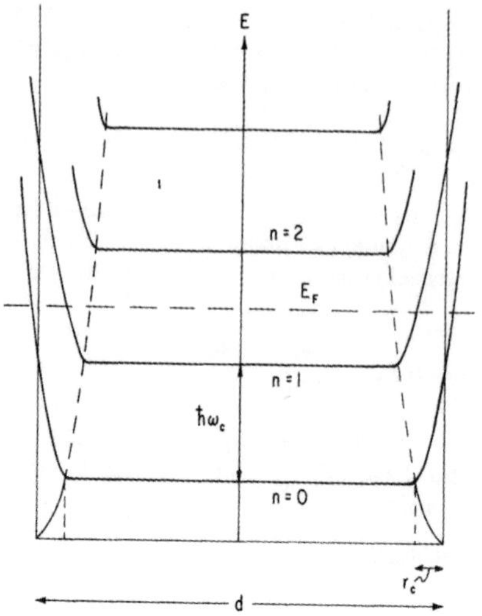

FIG. 2. Schematic energy spectrum for an ideal two-dimensional electron gas confined on a ribbon of width d. In the bulk region the spectrum consists of Landau levels separated by an energy $\hbar\omega_c$. r_c represents the cyclotron orbit radius for the $n=0$ Landau level.

In the case in which an integral number n of Landau levels are fully occupied we can write $N=nr_0d/a_B^2$ and get

$$I_H=-n\frac{e^2}{h}Ed\ , \tag{19}$$

which amounts to the well-known expression for the quantized Hall conductivity

$$\sigma_{xy}=-n\frac{e^2}{h}\ ,\quad n=0,1,2,3,\ldots\ . \tag{20}$$

It is obvious that Eqs. (18)–(20) do not provide a theory for the quantized Hall effect since within this extremely simplified model σ_{xy} is just a monotonic function of both B and the number of electrons N. The quantized expression (20) is recovered only for a set of values of N (or B) of zero measure.

Consider next the problem associated with the edge of the system. The electrons in our metallic ribbon must be confined to the region $\theta\leq z\leq d$ by some potential. If we assume that there is an infinite potential barrier at these positions, then the Landau-level energies will increase as the orbit center comes within the cyclotron radius of the wall. The energy as a function of z_l, the position of the orbit center, is sketched in Fig. 2 for the ideal case with $\vec{E}=0$. Note that the $n=0$ level with $z_l=0$ and d has the energy $(\frac{3}{2})h\omega_c$, the value of the $n=1$ level in the bulk. The reason for this is, of course, that for the potential appropriate to these orbit centers, only the odd eigenfunctions of the full harmonic-oscillator potential inside the ribbon are solutions to the Schrödinger equation. Note that for orbit centers outside the ribbon the energy continues to increase approximately quadratically. These states represent "skipping orbits" or edge states. The very presence of the edge states leads to a smooth variation of the Fermi energy between the bulk Landau levels as they provide a nonvanishing density of states in such regions. The role and the physical properties of the edge states of a two-dimensional electron gas in a Corbino disk have been discussed by Halperin.[21] His treatment can be easily repeated for our cylindrical geometry and all his conclusions apply also in the present case.

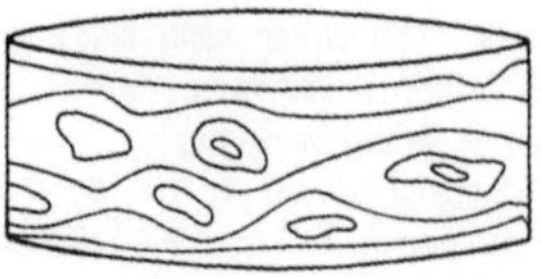

FIG. 3. Schematic of the classical orbits for extended and localized states in the ribbon geometry. Extended states are associated with classical orbits which encircle any flux threading the ribbon.

III. METALLIC RIBBON WITH IMPURITIES

A. Structure of the energy spectrum

Before attempting to evaluate the Hall current a discussion on the nature and the structure of the spectrum of the system in this case is in order. We start with the Hamiltonian

$$H=H_0+ezE+V(z,\theta)\ , \tag{21}$$

where H_0 is given by Eqs. (4) and where $V(z,\theta)$ is the random potential caused by the impurities. As V explicitly depends upon z and θ in general; n and l are not good quantum numbers, which is to say, the set of wave functions ψ_{nl} of Eqs. (8) does not represent a set of eigenstates of H. Accordingly we will introduce a new index (or set of indices) α in order to label the wave functions ψ_α and eigenvalues E_α of the system.

The precise nature of the electronic states of a system in the presence of a random potential has not yet been elucidated. In the absence of an externally applied magnetic field it is believed on the basis of scaling arguments that all the states must be localized.[22] However, when a uniform magnetic field is applied to the system the situation is different and it has been argued by several authors that extended states exist at the center of each Landau level.[4–6,21] We will assume here that this is the case.[23] Accordingly we will divide the states ψ_α into extended and localized states. The particular geometry of our system allows us a quite natural distinction between localized and extended states. Following an analysis similar to that used by Kazarinov and Luryi,[10] we can define the extended states as the ones associated with classical electronic orbits which circle the entire ribbon, whereas the localized states do not, as schematically pictured in Fig. 3. More precisely a state $\psi_{\alpha e}$ is extended if for any value of θ between 0 and

2π there exists a value of z between 0 and d such that $|\psi_{ae}|^2$ goes to zero such as r_0^{-1} as r_0, the radius of the ribbon, goes to infinity. A state ψ_{aL} will instead be localized at z^*,θ^* if $|\psi_{aL}(z^*,\theta^*)|^2$ is of the order r_0^{-1}, whereas for any value of z, $|\psi_{aL}(z,\theta^*+\pi)|^2$ goes exponentially to zero with r_0 as r_0 is made to grow.

We shall now give an argument to show that even in the presence of impurities for a ribbon of finite size the energy eigenvalues of the extended states are isolated and therefore constitute a countable set. Let ψ_{ae}^{ϕ} and E_{ae}^{ϕ} be the wave function and the energy of a given extended eigenstate ae of H^{ϕ}, which is the Hamiltonian of the system, Eq. (21), where the "trapped flux" associated with A_1 is ϕ [see Eqs. (2) and (3)]. Suppose now that the flux is adiabatically changed to $\phi+\Delta\phi$, with $\Delta\phi<\phi_0$. Since ψ_{ae} is associated with an orbit linked to the flux change an electron in this state will experience an induced emf and will respond to the perturbation. Accordingly ψ_{ae}^{ϕ} and E_{ae}^{ϕ} will be mapped into $\psi_{ae}^{\phi+\Delta\phi}$ and $E_{ae}^{\phi+\Delta\phi}$ which in principle can be obtained by solving the Schrödinger equation for $H^{\phi+\Delta\phi}$.[24] In the ideal case, where $V=0$, the result of such a flux change is readily established. If we start from a state $\phi_{nl}^{\phi},E_{nl}^{\phi}$ as given in Eqs. (8), it is easy to establish that two possible solutions of the problem are possible,

$$\psi_{n,l}^{\phi},E_{nl}^{\phi}\rightarrow\psi_{nl}^{\phi+\Delta\phi}=e^{i\eta\theta}\psi_{n,l}^{\phi},E_{nl}^{\phi+\Delta\phi}=E_{nl}^{\phi}\ , \tag{22}$$

corresponding to an increment of angular momentum by $\hbar\eta$, or

$$\psi_{nl}^{\phi},E_{nl}^{\phi}\rightarrow\psi_{nl}^{\phi+\Delta\phi}=\tilde{\psi}_{nl}^{\phi},E_{nl}^{\phi+\Delta\phi}=E_{n}^{\phi+\Delta\phi}-\frac{\eta h v_D}{2\pi r_0}\ . \tag{23}$$

Here

$$\tilde{\psi}_{nl}^{\phi}=(2\pi)^{-1/2}\exp(il\theta)u_n\left[\frac{z-z_l+v_D}{\omega_c+\eta a_B^2/r_0}\right]$$

corresponds to a rigid displacement of the orbit center by exactly the amount required to maintain an integral number of flux quanta threading its orbits. In (22) and (23), $\eta=\Delta\phi/\phi_0$. Because the wave function must be single valued, it is clear that for $\eta<1$, the solution (22) is not acceptable and the response of the system will be characterized by the orbit-shifting process of Eq. (23). Notice that this implies that E_{nl}^{ϕ} does not belong to the spectrum of $H^{\phi+\Delta\phi}$. An analogous phenomenon occurs in the system in the presence of impurities. Although $\exp(i\eta\theta)\psi_{ae}^{\phi}$ is an eigenfunction of $H^{\phi+\Delta\phi}$ with eigenvalue E_{ae}^{ϕ}, such a solution is not acceptable for nonintegral η since it is not a single-valued function of θ. Thus, E_{ae}^{ϕ} does not belong to the spectrum of $H^{\phi+\Delta\phi}$.[25] Therefore, the energy of the state ae will change with ϕ. Furthermore, due to the random nature of $V(z,\theta)$, it is reasonable to expect that for very small values of η, $E_{ae}^{\phi+\Delta\phi}$ will be smaller than E_{ae}^{ϕ} by an amount linear in η. This process is exemplified in Fig. 4(a). It is obvious that here the change in ψ_{ae}^{ϕ} will be much more complicated than a simple rigid shift of the center of the orbit since the potential $V(\theta,z)$ has a complicated local structure. Now, since $\psi_{ae}^{\phi+\Delta\phi}$ in an eigenfunction of $H^{\phi+\Delta\phi}$, the wave function $\exp(-i\eta\theta)\psi_{ae}^{\phi+\Delta\phi}$ is an eigenfunction of H^{ϕ} with eigenenergy $E_{ae}^{\phi+\Delta\phi}$. The fact

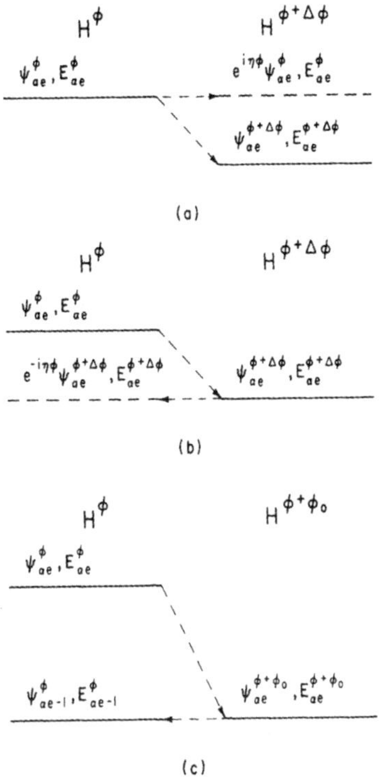

FIG. 4. Evolution of an extended state level E_{ae} (with wave function ψ_{ae}) upon an adiabatic change $\Delta\phi$ of the threading flux. $\eta=\Delta\phi/\phi_0$. (a) $\eta<1$, $\exp(i\eta\theta)\psi_{ae}^{\phi}$ is not an acceptable wave function; (b) $\eta<1$, $\exp(-i\eta\theta)\psi_{ae}^{\phi+\Delta\phi}$ is not an acceptable wave function and E_{ae} is isolated; (c) $\eta=1$, $E_{ae}^{\phi+\phi_0}=E_{ae-1}^{\phi}$ as $\psi_{ae}^{\phi+\phi_0}=\exp(i\theta)\psi_{ae-1}^{\phi}$.

that $\psi_{ae}^{\phi+\Delta\phi}$ is an acceptable (i.e., single-valued) solution implies that $\exp(-i\eta\theta)\psi_{ae}^{\phi+\Delta\phi}$ is not and that the eigenvalue $E_{ae}^{\phi+\Delta\phi}$ does not belong to the spectrum of H^{ϕ}. Since the flux change $\Delta\phi$ is completely arbitrary we conclude that for a ribbon of finite size the eigenvalues E_{ae}^{ϕ} of H^{ϕ} are isolated and the spectrum of the extended states of the system is countable. This situation is represented in Fig. 4(b).

If $\eta=1$, i.e., $\Delta\phi=\phi_0$, then $\exp(-i\theta)\psi_{ae}^{\phi+\phi_0}$ is an acceptable solution of H^{ϕ} with eigenenergy $E_{ae}^{\phi+\phi_0}$, which by definition is nothing other than ψ_{ae-1}^{ϕ} [see Fig. 4(c)]. The spectrum of $H^{\phi+m\phi_0}$, m being any integer, is the same as the one of H^{ϕ}; the wave functions simply differ by the phase factor $\exp(im\theta)$. We have established that by changing the trapped flux ϕ by one flux quantum each extended state ψ_{ae} is mapped into its nearest-neighboring ψ_{ae-1}, with

$$E_{ae}^{\phi+\phi_0}=E_{ae-1}^{\phi}\ . \tag{24}$$

This expression will be useful later.

It is obvious that formally the same argument can be

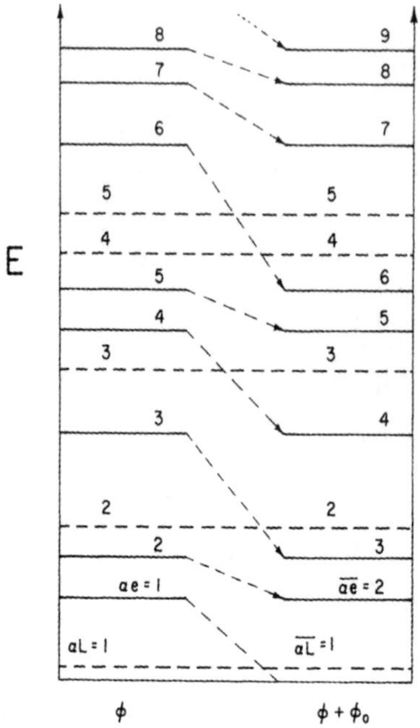

FIG. 5. Schematic illustration of how the spectrum of the system is modified by an adiabatic change of the flux threading the ribbon. αL and $\overline{\alpha L}$ label the localized states of H^{ϕ} and $H^{\phi+\phi_0}$; αe and $\overline{\alpha e}$ are the corresponding labels for the extended states. After a whole quantum of flux ϕ_0 has been added, the states αe are mapped into $\overline{\alpha e}=\alpha e-1$. The localized states are unaffected by the flux change.

carried out for a localized state $\psi_{\alpha L}$. However, the nature of a localized wave function is such that this electronic state will not be significantly modified by any change of trapped flux. This amounts to the fact that it is only for the localized states that a change in the trapped flux ϕ has the same effect of a gauge transformation. The situation for both extended and localized states is schematically pictured in Fig. 5.

If the external magnetic field is large enough the spectrum of the extended states of the system will still be characterized by a Landau-level structure. Each Landau level will, however, be broadened by the effect of the random potential. In such a case the label αe can be resolved into a Landau band index ν and a generalized orbit-position quantum number λ. With this notation Eq. (24) can be written as

$$E_{\nu\lambda}^{\phi+\phi_0}=E_{\nu,\lambda-1}^{\phi}\,, \tag{25}$$

where it has been recognized that for a macroscopic system ϕ_0 is a very small change of flux and cannot possibly induce a change of the Landau band index ν.

A word is in order here about the occupation of the extended states while the trapped flux ϕ is changed. If the step-by-step change in ϕ is adiabatic, we can make use of time-dependent perturbation theory as done in Sec. II. By a direct inspection of Eq. (14) (valid also in the presence of the potential V), it is easily realized that in this situation the occupation of the states does not change. If an electron occupies an extended state $\psi_{\nu\lambda}^{\phi}$ after a full quantum of flux ϕ_0 has been added to ϕ, the state $\psi_{\nu\lambda}^{\phi+\phi_0}$, i.e., $\psi_{\nu,\lambda-1}^{\phi}$ will be occupied. This is the transfer process originally suggested by Laughlin.[8] However, notice that for the extended states the flux change by no means can be thought of as a gauge transformation.

B. Hall current

Within our single-particle picture the Hall current I_H can be evaluated as follows:

$$I_H=\mathrm{Tr}(\rho j_\theta^{\mathrm{op}})=\sum_\alpha f(E_\alpha)\langle\alpha\,|\,j_\theta^{\mathrm{op}}\,|\,\alpha\rangle\,, \tag{26}$$

where the sum runs over the possible states of the system. $f(E_\alpha)$, ρ, and j_θ^{op} have been defined in Sec. II. Making use of the relation

$$j_\theta^{\mathrm{op}}=-c\frac{\partial H}{\partial A_0}=c\frac{\partial H}{\partial\phi}\,,$$

(26) can be written as

$$I_H=c\sum_{\alpha e} f(E_{\alpha e})\frac{\partial E_{\alpha e}}{\partial\phi}\,, \tag{27}$$

where we have restricted the sum to the extended states only because as discussed in Sec. III, $\partial E_{\alpha L}/\partial\phi$ is zero. Notice that in absence of extended states the Hall current is zero.

In a macroscopic system we can expand Eq. (24) in powers of ϕ_0 and get

$$\frac{\partial E_{\alpha e}}{\partial\phi}\simeq\frac{1}{\phi_0}\frac{\partial E_{\alpha e}}{\partial\alpha e}\,, \tag{28}$$

where we have also made use of the fact that the extended states are closely spaced. Corrections to Eq. (28) are readily shown to vanish with the inverse of the ribbon radius. By using (28) in (27) we obtain

$$I_H=-\frac{c}{\phi_0}\sum_{\alpha e} f(E_{\alpha e})\frac{\partial E_{\alpha e}}{\partial\alpha e}\,, \tag{29}$$

after integrating by parts we find

$$I_H=\frac{c}{\phi_0}\sum_{\alpha e}\frac{\partial f(E_{\alpha e})}{\partial E_{\alpha e}}\frac{\partial E_{\alpha e}}{\partial\alpha e}E_{\alpha e}\,. \tag{30}$$

Here we have used

$$\frac{\partial f(E_{\alpha e})}{\partial\alpha e}=\left[\frac{\partial f(E_{\alpha e})}{\partial E_{\alpha e}}\right]\left[\frac{\partial E_{\alpha e}}{\partial\alpha e}\right].$$

At low temperatures the derivative of the Fermi function is essentially the negative of a δ function which picks out the value of $E_{\alpha e}$ for which $E_{\alpha e}=\zeta$, the chemical potential. Because of the possible presence of an electric field along the z direction, ζ can, in general, vary with position and each extended state will experience some average value

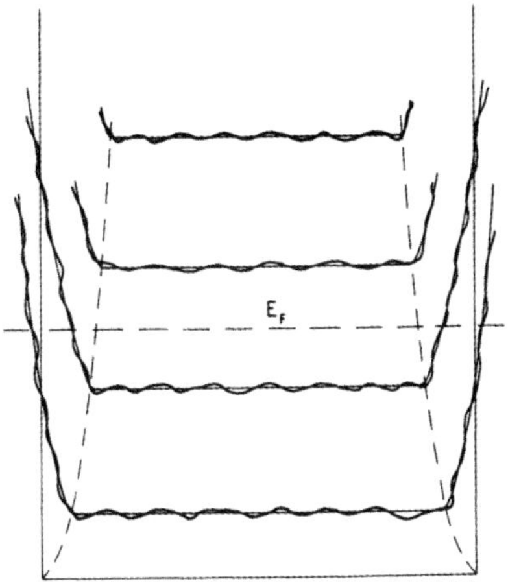

FIG. 6. Schematic energy spectrum for the extended states of a two-dimensional electron gas in the presence of an effective random potential $V(z,\theta)$ at a fixed value of θ. Here the strength of $V(z,\theta)$ is assumed to be much smaller than $\hbar\omega_c$ so that the Landau levels still provide an approximately good description scheme. Localized states levels (not shown) can be present everywhere in the energy range.

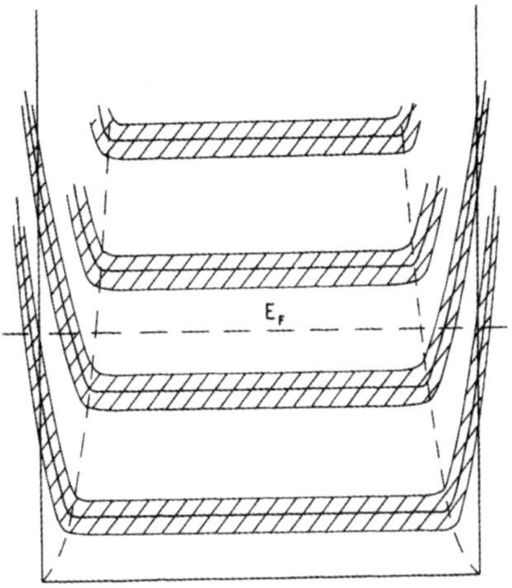

FIG. 7. When curves like those of Fig. 6 are drawn for all values of θ, the Landau levels are broadened into Landau bands labeled with the index ν, as shown schematically.

$\zeta_{\alpha e}$. We introduce the set of values αe_i for which $E_{\alpha e}=\zeta_{\alpha e}$, the local value of the chemical potential, and make use of the result

$$\delta(f(x))=\sum_i \delta(x-x_i)\left|\frac{\partial f}{\partial x}\right|^{-1}$$

to obtain

$$I_H=-\frac{e}{h}\sum_{\alpha e}\sum_i \delta(\alpha e-\alpha e_i)\mathrm{sgn}\left[\frac{\partial E_{\alpha e}}{\partial \alpha e}\right]\zeta_{\alpha e}\ . \tag{31}$$

When the Landau bands are well defined, Eq. (31) can be specialized to

$$I_H=-\frac{e}{h}\sum_{i,\nu}\mathrm{sgn}\left[\frac{\partial E_{\nu\lambda}}{\partial\lambda}\right]_{\lambda_i}\zeta_{\lambda_i}\ , \tag{32}$$

where we have made use of the indices ν and λ introduced in the preceding section. In (32) the sum over ν extends to all the Landau bands which cross the chemical potential at some value of λ.

First, let us apply Eq. (32) to the ideal case with boundaries for which the energy diagram of Fig. 2 applies. When the chemical potential lies between two Landau levels, say, n and $n+1$, then the only crossings occur close to the boundaries. In the absence of an applied electric field there is no current unless the local values of the Fermi energy at the two edges differ by ΔE_F. In this case Eq. (32) gives $I_H=-ne\Delta E_F/h$ which is the current carried by the edge states discussed by Halperin. If an electric field is applied along z with the use of (32) we obtain

$$I_H=-n\frac{e^2}{h}\Delta V\ , \tag{33}$$

where ΔV is the potential drop across the ribbon and n is the number of occupied bulk Landau bands. Equation (33) is a generalization of Eq. (19) to the case in which the electric field need not be homogeneous. In absence of impurities Eq. (33) is valid only for a very restricted range of values of N once B is fixed (or of B when N is fixed). The reason is that the chemical potential changes very rapidly between the Landau levels because the density of edge states is relatively small. In this case the width of the plateaus described by Eq. (33) would be a surface effect heavily dependent on the size of the sample.

The situation does not change in an essential way when the effect of the random potential is considered. In this case, the diagram of Fig. 2 no longer applies since l is no longer a good quantum number. However, for a particular value of θ we can still draw energy levels as a function of the orbit center (taken as the average value of z for that particular value of θ for the particular eigenstate in question). Then, instead of Fig. 2, something like the picture shown in Fig. 6 results. The wavy lines in Fig. 6 result from the particular distribution of impurities close to the value of θ for which the curves are drawn. If we repeatedly draw the equivalent of Fig. 6 for a large number of values of θ between 0 and 2π, we obtain a picture like that shown in Fig. 7. Again, if an electric field is applied and ζ lies between the broadened Landau bands in the bulk, the value (33) for the Hall current is obtained, in spite of the presence of not current carrying states.

Strictly speaking, I_H turns out to be slightly smaller than (33). In the evaluation of the chemical potential differences we have assumed the value $-e\Delta V$ for each Landau band. However, it takes the full energy $-e\Delta V$ to transfer an electron in the $n=0$ Landau level across the ribbon, it takes somewhat less than $-e\Delta V$ to transfer an electron in a higher level across the ribbon.

We make one final argument which gives somewhat different insight into the mechanism of the Hall quantization. Let us once more think of slowly changing, in time, the trapped flux ϕ by an amount $\Delta\phi \propto \exp(i\omega t)$.

According to Faraday's law the rate of change of ϕ gives rise to the induced emf, $\vec{F}=-c^{-1}(\partial\phi/\partial t)\hat{\theta}$. If we assume that $\sigma_{xx}=0$, then the only response the system can make to this emf is a current flow I_z perpendicular to $\vec{F}$,

$$I_z=\sigma_{xy}\left[-\frac{1}{c}\frac{\partial\phi}{\partial t}\right]. \tag{34}$$

If we integrate this to obtain the charge transfer associated with a flux change $\Delta\phi$ we find

$$Q=\int I_z dt=-\frac{1}{c}\sigma_{xy}\Delta\phi . \tag{35}$$

But due to the fact shown above that for $\Delta\phi=\phi_0$ each orbit center moves exactly one step into the position of its neighboring orbit center, we know that $Q=en$, where n is the number of filled Landau levels. Thus we obtain

$$en=\sigma_{xy}(-c^{-1})\frac{hc}{e}, \tag{36}$$

and we again find $\sigma_{xy}=-n(e^2/h)$. The point to be emphasized is that, for the topology of the metallic ribbon, the change in flux *is not* a gauge transformation. There is a real emf associated with the rate of change of flux. The electrons sense this emf and their response leads to a flow of charge across the sample.[26]

IV. DISCUSSION

In this paper we have provided a microscopic theory of the Hall conductance for a two-dimensional electron gas in the presence of an impurity random potential. Our main result is Eq. (32) where the value of the Hall current I_H is related to the electronic spectrum at the Fermi level.

In previous works on quantized Hall conductance, the results are first derived for free electrons and then qualitative arguments are given why the impurities cannot destroy the quantized nature of the Hall conductivity. No explicit calculations were done for the disordered system. In addition, the question on the accuracy of the quantization was not satisfactorily addressed.

We have analyzed the electronic level structure in the presence of impurities. We show explicitly that for the extended states, when the trapped flux is adiabatically changed, the spectrum shifts in a way analogous to the behavior of free electrons and maps into itself when the change in flux is precisely one quantum. For the localized states, the change of the flux does not shift the level and amounts to just a gauge transformation. These are exact results that enable us to *derive* the quantized Hall conductance for electrons in the presence of impurities and the result can be considered as a nontrivial generalization to the many-impurity case of Prange's simple result concerning the single-impurity problem.[17] It is also worth mentioning that the analysis of Sec. III provides a firm theoretical ground for justifying and assessing the limits of the picture put forward by Laughlin.[8] Our general expression (32) for the Hall conductance can also be used in the case in which a weak periodic substrate potential acts on the electrons.[11] Making use of it one easily finds that if the Fermi level lies between two of the magnetic subbands in which each Landau band is split by the periodic substrate potential, the Hall conductance is still an integer multiple of e^2/h in agreement with the conclusions of Thouless and co-workers.[11]

The other major point in our theory is the careful inclusion of the edge effects. We find that whenever the energy of an extended state crosses the local Fermi level ζ, a term $\pm e^2\zeta/h$ is contributed to I_H, the sign being the same as that of $\partial E_{\alpha e}/\partial\alpha e$. When the Fermi level lies between two Landau bands, for instance, n and $n+1$, the only crossings of this type occur at the edges and I_H is found to assume the quantized value $-ne^2\Delta V/h$, ΔV being the voltage drop across the sample. As discussed at the end of Sec. III, however, the local values of ζ at the two crossing points at the edges of a given Landau band differs in general by an amount smaller than ΔV. The correction is found to be dependent on the ratio of the corresponding skipping-orbit radius to the size of the sample. If our picture is correct the exactness of the quantization of I_H must depend upon the specific shape of the potential confining the electron gas in the plane. In the case of a shallow confining potential the Hall conductance will not be quantized as in Eq. (33). In the Appendix we present a model calculation in which this phenomenon is explicitly demonstrated.

It must be stressed here that the Hall current is not carried by the edge states only but is typically a bulk phenomenon. The restriction of I_H to the nature of the confining boundary potential is the result of a great number of cancellations of bulk contributions. This situation is reminiscent of the Landau diamagnetism in which a similar phenomenon occurs. Finally, as clear from Eq. (27), temperature effects will also cause I_H to deviate from the quantized values.

The Hall conductance will maintain its quantized value as long as the Fermi level remains in the localized region of the density of states between two Landau bands. The width of such plateaus depends, therefore, on the capacity of the impurity potential to localize the electronic states. The presence of extended states in the bulk of the sample is necessary for our model to give a finite value for I_H.[21] When the Fermi level is within one of the Landau bands, i.e., within one of the shaded regions of Fig. 7, Eq. (33) need not be valid because several crossings of the local chemical potential by extended states can occur, leading via (32) to a completely different value of I_H. This situation is schematically exemplified in Fig. 8 where the Hall conductivity is plotted against the fractional occupation number a_B^2N/r_0d.

Our approach is based on a single-particle picture and no explicit reference is made to the electron-electron interaction. It must be understood, however, that the present theory implicitly contains some of the effects asso-

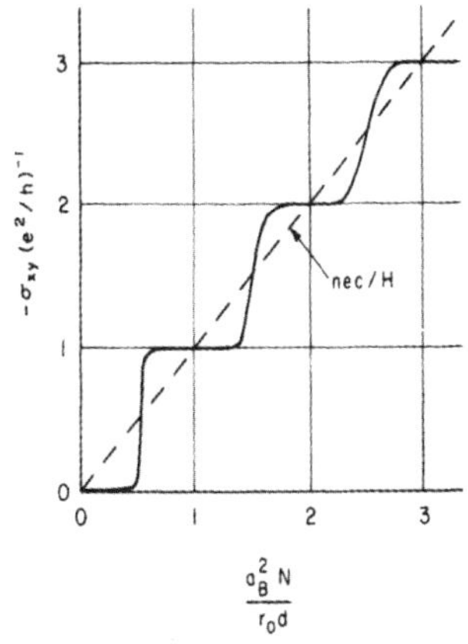

FIG. 8. Behavior of the Hall conductivity σ_{xy} in the quantized region. The plateau regions correspond to a situation in which the chemical potential lies between two Landau bands and localized states are being filled. Corrections associated with edge effects discussed in the text are neglected here.

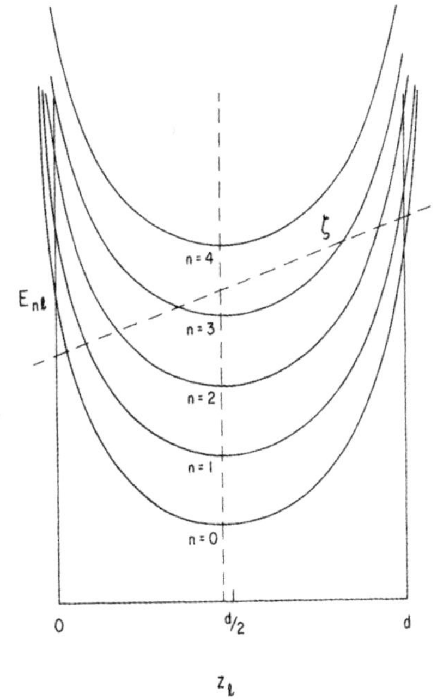

FIG. 9. Schematic energy spectrum of a two-dimensional electron gas confined on a ribbon by a harmonic potential. The Landau levels are parabolas. Energy is plotted vs $z_l=a_B^2 l/r_0$. Various Landau levels are separated by an energy $\hbar\tilde{\omega}_c$ [Eq. (A5)]. Crossing points between the parabolas and the local chemical potential ζ (dashed line) determines the Hall current I_H via Eq. (32). "Edge corrections" are very large in this case and σ_{xy} is not quantized.

ciated with this interaction. In particular, the random potential-energy term $V(z,\theta)$ of Eq. (21) can be thought of as the total effective potential seen by a single quasiparticle including the Hartree-Fock contributions within the normal state. Many-body effects[15–17] possibly responsible for an instability of the normal state[14] have been neglected and are currently under investigation.

ACKNOWLEDGMENT

We are grateful to the U. S. Office of Naval Research for partial support of this work.

APPENDIX: HALL CONDUCTANCE OF AN IDEAL TWO-DIMENSIONAL ELECTRON GAS CONFINED BY A HARMONIC POTENTIAL

In Sec. III we have shown that within the single-particle picture the occurrence of the quantized Hall effect crucially depends on the shape of the confining potential. If the latter is too shallow the Hall conductance will not be expressed by Eq. (33). In particular, for a realistic system, an edge correction will arise which is of the order of the ratio of a_B to the sample size. For the sake of illustration, we explicitly demonstrate this phenomenon here, making use of the exactly solvable model of an ideal system confined by a harmonic potential. Within the geometrical setup discussed in the text, we write the following Hamiltonian:

$$H_\Omega=\frac{p_z^2}{2m}+\frac{1}{2m}\left[\frac{p_\theta}{r_0}-m\omega_c z\right]^2+eE_z z+\tfrac{1}{2}m\Omega^2\left[z-\frac{d}{2}\right]^2, \qquad \text{(A1)}$$

where we have chosen $A_1=0$. The last term represents a confining potential. H_Ω can be readily solved exactly in the same way as done for H_0 in Sec. II, as p_θ is still a conserved quantity. The problem reduces to the following Schrödinger equation for the z-dependent part of the wave function:

$$\left[\frac{p_z^2}{2m}+\frac{m\tilde{\omega}_c^2}{2}(z-\tilde{z}_l)^2+\Delta E_l-E\right]u(z)=0, \qquad \text{(A2)}$$

where the integer l has the same meaning as in Eq. (5) of the text. In (A2) we have defined

$$\tilde{\omega}_c=(\omega_c^2+\Omega^2)^{1/2}, \qquad \text{(A3)}$$

$$\tilde{z}_l=\frac{1}{1+(\Omega/\omega)^2}\left[z_l+\left(\frac{\Omega}{\omega_c}\right)^2\left(\frac{d}{2}-\frac{v_D}{\omega_c}\right)\right], \qquad \text{(A4)}$$

$$\Delta E_l=\frac{m\omega_c^2}{2[1+(\omega_c/\Omega)^2]}\left[z_l-\frac{d}{2}+\left(\frac{\omega_c}{\Omega}\right)^2\frac{v_D}{\omega_c}\right]^2+\text{const}, \qquad \text{(A5)}$$

with $z_l=a_B^2 l/r_0$. The eigenfunctions and eigenvalues of H_Ω can still be classified by making use of the quantum numbers n and l as in the free case. We have

$$\psi_{nl}^{\Omega}(z,\theta)=\frac{1}{\sqrt{2\pi}}e^{il\theta}u_n(z-\tilde{z}_l)\,,$$
$$E_{nl}^{\Omega}=\hbar\tilde{\omega}_c(n+\tfrac{1}{2})+\Delta E_l\,. \tag{A6}$$

Notice that as $\Omega\to 0$ these equations reduce to Eqs. (8).

The spectrum of the system is composed by a series of parabolic Landau levels separated by an energy $\hbar\tilde{\omega}_c$ as described in Eqs. (A6). This spectrum is represented in Fig. 9.

The Hall current carried by each state ψ_{nl}^{Ω} is easily evaluated and is given by

$$I_H^{nl}=-\frac{e\omega_c}{1+(\omega_c/\Omega)^2}\left[z_l-\frac{d}{2}+\left(\frac{\omega_c}{\Omega}\right)^2\frac{v_D}{\omega_c}\right]. \tag{A7}$$

Notice that in this case even in the absence of an electric field every state carries a finite current proportional to $z_l-d/2$. This is in contrast with the picture discussed by Halperin in which only the eigenstates carry a finite current in this field as the localizing potential is flat in the bulk region.[21]

Making use of the Eq. (A7), it is straightforward to evaluate the total Hall current. The result for $T\simeq 0$ K is given by the general expression Eq. (32) obtained in the text. Looking at Fig. 9 it is now obvious that I_H is not an integer multiple of e^2/h since the value of the local chemical potential cannot be approximated by either zero or $e\Delta V$ as in the case of a sharp confining potential such as, for instance, the one represented in Fig. 2.

---

*On leave from the Scuola Normale Superiore, Pisa, Italy.

[1] K. von Klitzing, G. Dorda, and M. Pepper, Phys. Rev. Lett. 45, 494 (1980).

[2] M. A. Paalanen, D. C. Tsui, and A. C. Gossard, Phys. Rev. B 25, 5566 (1982).

[3] T. Ando, Y. Matsumoto, and Y. Uemura, J. Phys. Soc. Jpn. 39, 279 (1975).

[4] D. J. Thouless, J. Phys. C 14, 3475 (1981).

[5] Y. Ono, J. Phys. Soc. Jpn. 51, 2055 (1982).

[6] See also the recent review by T. Ando, in *Anderson Localization*, edited by Y. Nagaoka and H. Fukuyama (Springer, New York, 1982), p. 176.

[7] R. E. Prange, Phys. Rev. B 23, 4802 (1981).

[8] R. B. Laughlin, Phys. Rev. B 23, 5632 (1981); Surf. Sci. 113, 22 (1982).

[9] R. E. Prange and R. Joynt, Phys. Rev. B 25, 2943 (1982).

[10] R. F. Kazarinov and S. Luryi, Phys. Rev. B 25, 7626 (1982).

[11] D. J. Thouless, M. Kohmoto, M. P. Nightingale, and M. den Nijs, Phys. Rev. Lett. 49, 405 (1982).

[12] B. A. Wilson, S. J. Allen, Jr., and D. C. Tsui, Phys. Rev. Lett. 44, 479 (1980).

[13] D. C. Tsui, H. L. Störmer, and A. C. Gossard, Phys. Rev. Lett. 48, 1559 (1982).

[14] H. Fukuyama, P. M. Platzman, and P. W. Anderson, Phys. Rev. B 19, 5211 (1979).

[15] H. Fukuyama and P. M. Platzman, Phys. Rev. B 25, 2934 (1982).

[16] P. Streda, J. Phys. C 15, L717 (1982).

[17] E. Tosatti and M. Parrinello, Lett. Nuovo Cimento 36, 289 (1983).

[18] J. J. Quinn and B. D. McCombe, Comments Solid State Phys. 10, 139 (1982).

[19] This conclusion has been also verified in Ref. 7.

[20] If we take $\Delta\phi=\phi_0$, then this is actually the Laughlin argument; for that case $z_l+\Delta z_l=z_{l+1}$, and each state moves to its neighboring orbit. The net effect is to transfer one electron per occupied Landau level across the ribbon giving $\Delta E_T=\nu e\,\Delta V$, where ν is the number of occupied Landau level and $\Delta V=dE$ in the potential drop associated with $\vec{E}$. Using Eq. (12), $I_H=-c\eta e\,\Delta V/\phi_0$ and we are led once more to results (19) and (20).

[21] B. I. Halperin, Phys. Rev. B 25, 2185 (1982).

[22] E. Abrahams, P. W. Anderson, D. C. Licciardello, and T. V. Ramakrishnan, Phys. Rev. Lett. 42, 673 (1979).

[23] A weaker assumption is actually sufficient since we can still allow all the states to be localized but have some of them with a localization length of the order of the size of the ribbon (see Ref. 5).

[24] As according to our definition extended and localized states are topologically distinct it is reasonable to assume that an adiabatic change of the trapped flux does not modify the nature of the given state.

[25] We have assumed here that the applied electric field and the random potential lift any possible degeneracy of the Hamiltonian operator. The accidental degeneracy with an extended state close to one of the edges does not have any bearing on the argument because of the spatial separation.

[26] This argument has been first put forward in Ref. 18. A similar reasoning due to Takemori can be found in Ref. 6.

PHYSICAL REVIEW B VOLUME 29, NUMBER 4 15 FEBRUARY 1984

Effects of diffusion on the plasma oscillations of a two-dimensional electron gas

Gabriele F. Giuliani* and J. J. Quinn
Brown University, Providence, Rhode Island 02912
(Received 22 November 1983)

In the presence of a finite elastic mean free path the frequency of the plasma excitations of a two-dimensional electron gas is shown to be significantly affected in the low-density region by the diffusive nature of the electronic dynamics. This remarkable phenomenon, absent in the familiar three-dimensional case, is peculiar to the electronic systems of lower dimensionality and appears to have been observed in silicon inversion layers.

In a normal three-dimensional metal the plasma frequency and its dispersion are not significantly affected by the presence of a small impurity concentration. In a two-dimensional electron gas, however, the plasma frequency goes to zero with the wave vector[1] and the diffusive nature of the electronic dynamics leads at low density to a sizable and remarkable effect on the plasmon dispersion relation in the long-wavelength limit.

Plasma excitations of the quasi-two-dimensional electron gas in the inversion layer of silicon metal-oxide-semiconductor field-effect transistor (MOSFET) systems have been observed and intensively investigated experimentally.[2–6] The measurements are typically performed via far-infrared transmission[3] and emission[4] techniques, the coupling between the plasmons and the radiation field being provided by an artificial grating overlay. Thermal excitation of two-dimensional plasmons has also been reported.[6]

In the high-density regime, in which the electronic dynamics is essentially metallic in character, the observed resonance position of the plasma modes is well described by a simple analysis based on Drude theory.[7,3] At low densities, however, the measured plasma frequency significantly deviates from the value predicted by this simple approach as the plasmon appears to soften considerably.

In this Rapid Communication we discuss a microscopic theory of the plasma modes of a two-dimensional electron gas in which the finite elastic lifetime of the electron momentum eigenstates is explicitly accounted for. This allows us to describe the effects of the random impurity potential on the plasma oscillations in the crossover from the metallic to the diffusive regime of the electronic dynamics, which takes place as the electron density is lowered.

In the presence of a random distribution of impurities the dielectric properties of an electron gas still can be described in terms of a dielectric function $\epsilon(q,\omega)$. The dispersion relation for the plasma modes of this system is obtained by the usual requirement $\epsilon(q,\omega)=0$. Within the random-phase approximation (RPA) this condition acquires the customary form $1-v(q)\chi^0(q,\omega)=0$. Here $v(q)$ is the Fourier transform of the electron-electron interaction potential and $\chi^0(q,\omega)$ is the density response function of the corresponding noninteracting system.[8]

In the presence of a small density of impurities the density response function of a two-dimensional noninteracting electron gas can be evaluated by means of standard perturbation theory.[9] Keeping diagrams of the ladder type only [Fig. 1(a)], and assuming a delta function electron-impurity interaction, we obtain at zero temperatures[10]

$$\chi^0(q,\omega)=-N_0\left\{1-\frac{\omega}{[(\omega+i/\tau)^2-2Dq^2/\tau]^{1/2}-i/\tau}\right\}, \quad (1)$$

$$q \ll p_F,\quad 0<\omega \ll E_F/\hbar ,$$

where $N_0=m^*/\pi\hbar^2$ is the density of states at the Fermi energy E_F, and $D=v_F^2\tau/2$ is the diffusion constant in two dimensions. m^*, p_F, and v_F are, respectively, the electron effective mass, the Fermi wave vector, and velocity. Here τ is the elastic lifetime of the momentum eigenstates in the presence of impurities. This expression for $\chi^0(q,\omega)$ is extremely useful since it correctly describes the electronic density response for any value of the elastic lifetime. For very large values of τ, Eq. (1) reduces to the correct expression for the susceptibility of a noninteracting two-dimensional electron gas[11] in which the dynamics is purely metallic. For finite τ, in the limit of small ω and q, $\chi^0(q,\omega)$ displays the pole structure characteristic of the diffusive regime.[12]

For vanishing frequencies the expression (1) for the density response is inadequate as the relevant class of diagrams is in this case represented by the maximally crossed graphs [Fig. 1(b)] which describe the onset of Anderson localization.[13] The importance of this set of diagrams can be estimated by evaluating their contribution to $\chi^0(q,\omega)$ for $q=0$. We find that this is proportional to $N_0\hbar\omega\ln\omega\tau/iE_F$ which corresponds to the well-known correction to the longitudinal conductivity in the weak localization regime.[14] This effect is important, and eventually dominating, only for very small values of $\omega\tau$. As in the actual experimental situation the value of $\omega\tau$ is typically of order one, it can be assumed that Eq. (1) provides a good approximation to the density response function in this regime.

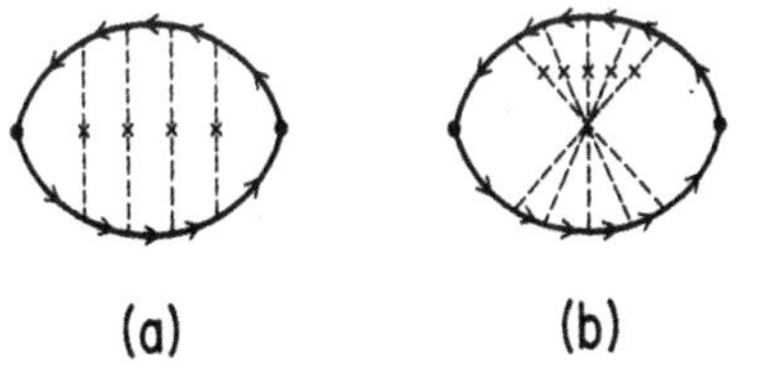

FIG. 1. Diagrams contributing to $\chi^0(q,\omega)$ in the electron-impurity interaction perturbation theory: (a) typical ladder diagram; (b) typical maximally crossed diagram.

Making use of Eq. (1) in the dispersion relation and taking $v(q)=2\pi e^2/\epsilon_0 q$, ϵ_0 being a background dielectric constant, the frequency of the plasmon $\Omega(q,\tau)$ is found to be

$$\Omega(q,\tau)=\frac{1+x}{2+x}\left[\left[x(2+x)v_F^2k^2-\frac{1}{\tau^2}\right]^{1/2}-\frac{i}{\tau}\right], \quad (2)$$

where $x=q/k$, and we have defined the two-dimensional Thomas-Fermi screening wave vector k as $2\pi e^2 N_0/\epsilon_0$. $\Omega(q,\tau)$ displays several interacting features: (i) The plasma frequency acquires a wave-vector-dependent imaginary part which corresponds to a plasmon lifetime roughly comparable to 2τ; (ii) plasma oscillations are softened at small wave vectors as $\Omega(q,\tau)$ is lower than the corresponding plasma frequency in the pure case[7,11] which, in our notation, is approximately given by $(v_F^2kq/2)^{1/2}$; (iii) plasma modes exist only for wave vectors q larger than a critical value q^* given by

$$q^*=k\left\{\left[1+\left(\frac{1}{v_Fk\tau}\right)^2\right]^{1/2}-1\right\} . \quad (3)$$

As q approaches q^* from above, the real part of $\Omega(q,\tau)$ tends to zero proportionally to $(q-q^*)^{1/2}$. Clearly a region of values of q larger than q^* exists such that $\mathrm{Re}(\Omega)$ is comparable to $\mathrm{Im}(\Omega)$. In this case the plasma excitation is not well defined and loses its meaning.

Notice that for $\tau\to\infty$, Eq. (2) recovers the result previously obtained by Stern for the case of a pure two-dimensional metal, up to second order in q.[11]

It is easy to realize that the plasmon softening occurring for small q is a direct consequence of the quasidiffusive nature of the electronic dynamics in this regime. A direct inspection of Eq. (2) shows that the magnitude of this effect is strongly dependent on the value of the perturbation theory expansion parameter $\hbar/\tau E_F=\hbar N_0/\tau n$, and therefore on the electronic density n. The lower the density the larger is the effect. This phenomenon is shown in Fig. 2.

Making use of Eq. (2) it is also possible to calculate the heat capacity associated with the excitation of two-dimensional plasma modes. In doing so, care must be exercised in dealing with the plasmon width.[15]

Our theory finds a natural application in the interpretations of the available data on the plasmons in silicon inversion layers. In this case, the theory developed above requires some minor modifications in view of the specific structure of the system at hand. In particular, the potential $v(q)$ must be properly chosen in order to account for the oxide layer of dielectric constant ϵ_{ox} and thickness d, and the metallic gate. Following Chaplik a suitable choice is provided here by[7]

$$v(q)=(4\pi e^2/q)(\epsilon_0+\epsilon_{ox}\coth qd)^{-1} .$$

The plasmon frequency can be obtained in this case in a straightforward way making use of Eq. (1) by following the procedure outlined above.[16] With particular reference to the data of Ref. 3, in which the first observation of two-dimensional intrasubband plasmons in the inversion layer of a (100) p-type Si MOSFET system was reported, we find that the present theory leads to a good fit of the plasmon resonance position throughout the whole density range assuming a single value for the electronic effective mass. A comparison of our theory and experiment is shown in Fig. 3.

It must be stressed here, however, that particular care must be taken in the choice of the elastic lifetime τ. We have not attempted here to evaluate τ from first principles but have regarded it as a parameter. If the value of τ as given by a dc conductivity measurement (i.e., the transport elastic lifetime τ_{tr}) is used, the agreement with the experimental findings is lost as the predicted resonance position becomes close to the one obtained for $\tau\to\infty$. We find that a value of τ from three to four times smaller than τ_{tr} is

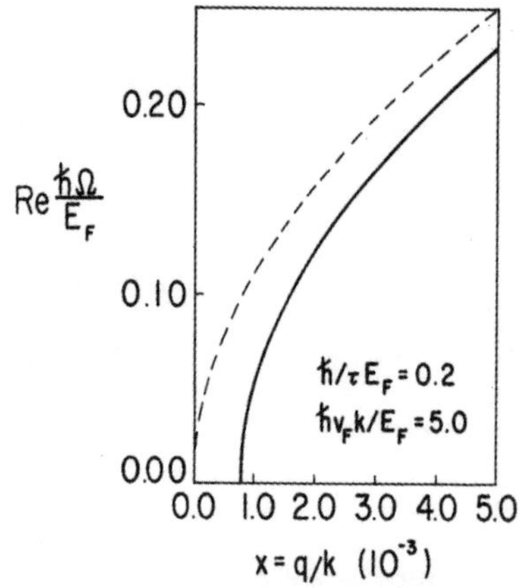

FIG. 2. Two-dimensional plasmon dispersion relation Ω vs q for finite τ as obtained from Eq. (2) (full line). The result for the pure metallic limit (dashed line) is shown for comparison. For illustration we have chosen $\hbar/\tau E_F=0.2$ and $\hbar v_F k/E_F=5.0$.

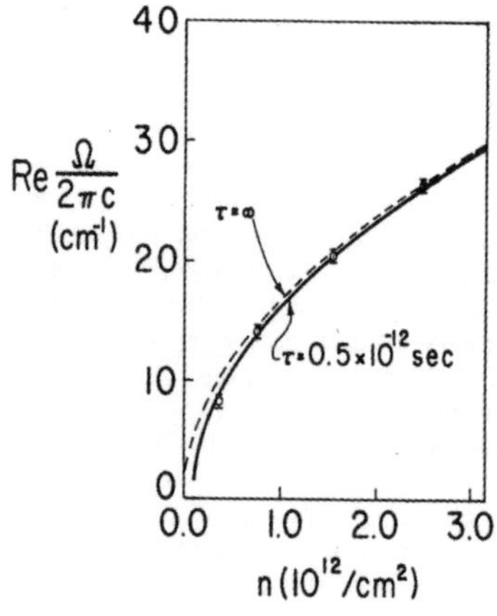

FIG. 3. Comparison of the present theory (full line) with the experimental data taken from Ref. 3, in a plot of $\Omega/2\pi c$ (in cm^{-1}) vs electronic density (in 10^{12} cm^{-2}) at a fixed value of the excitation wave vector $q=1.78\times10^4$ cm^{-1}. The dashed line represents the result one would obtain in the absence of impurities. As indicated, the value $\tau=0.5\times10^{-12}$ sec has been used. Here the values of the parameters are proper to a (100) p-type Si inversion layer system with $\epsilon_0=12$, $\epsilon_{ox}=4$, $d=0.14\times10^{-4}$ cm, $m^*=0.2$ m.

necessary to reach agreement with the data. In general, τ is expected to be shorter than τ_{tr} but is altogether unknown so that no firm conclusion on this matter can be drawn at this stage.[17]

In conclusion, we have discussed a microscopic theory of the plasma oscillations in a two-dimensional electron gas in the presence of a small impurity concentration. At variance with the usual three-dimensional situation the onset of a diffusive electronic dynamics has remarkable effects on the plasmon dispersion relation. In particular we predict the existence of a wave-vector cutoff for the existence of such modes. Evidence of these phenomena appears to surface in the existing data but it is manifest that further experimental exploration is necessary in order to clarify the relevance of the present analysis.

We would like to thank S. J. Allen, Jr., D. C. Tsui, and R. A. Logan for allowing us to reproduce in our Fig. 3 their data of Ref. 3. One of us (G.F.G.) kindly acknowledges the stimulating discussions on the subject with S. C. Ying and T. N. Theis. This work was supported by the National Science Foundation under Grant No. DMR 812069.

*On leave from the Scuola Normale Superiore, Pisa, Italy.

[1]R. H. Ritchie, Phys. Rev. 106, 874 (1957); R. A. Ferrell, *ibid.* 111, 1214 (1958).

[2]For a review up to 1980, see T. N. Theis, Surf. Sci. 98, 515 (1980).

[3]S. J. Allen, Jr., D. C. Tsui, and R. A. Logan, Phys. Rev. Lett. 38, 980 (1977).

[4]D. C. Tsui, E. Gornik, and R. A. Logan, Solid State Commun. 35, 875 (1980).

[5]D. Heitmann, J. P. Kotthaus, and E. G. Mohr, Solid State Commun. 44, 715 (1982).

[6]R. A. Hoepfel, E. Vass, and E. Gornik, Phys. Rev. Lett. 49, 1667 (1982).

[7]A. V. Chaplik, Zh. Eksp. Teor. Fiz. 62, 746 (1972) [Sov. Phys. JETP 35, 395 (1972)].

[8]Our aim here is to discuss the plasmon dispersion for small values of the ratio q/p_F. For q/p_F of order one it has been shown that the RPA must be improved upon. For work in this direction, see, for instance, A. K. Rajagopal, Phys. Rev. B 15, 4264 (1977), and references quoted therein.

[9]A. A. Abrikosov, L. P. Gorkov, and I. E. Dzyaloshinskii, *Methods of Quantum Field Theory in Statistical Physics* (Prentice-Hall, Englewood Cliffs, NJ, 1969).

[10]G. F. Giuliani and J. J. Quinn, in *Proceedings of the Fifth International Conference on Electronic Properties of Two-Dimensional Systems, Oxford, England,* edited by R. J. Nicholas *et al.* [Surf. Sci. (to be published)].

[11]F. Stern, Phys. Rev. Lett. 18, 546 (1967).

[12]A shortcoming of the assumption of a delta function electron-impurity potential is that the elastic lifetime and the transport lifetime which enters the mobility are here the same.

[13]E. Abrahams, P. W. Anderson, D. C. Licciardello, and T. V. Ramakrishnan, Phys. Rev. Lett. 42, 673 (1979).

[14]See, for instance, the review by H. Fukuyama, Surf. Sci. 113, 489 (1982).

[15]A simple discussion of the problem of calculating the heat capacity associated with damped modes can be found in M. Danino and A. W. Overhauser, Phys. Rev. B 26, 1569 (1982).

[16]Further details will be reported at a later date in G. F. Giuliani and J. J. Quinn (unpublished).

[17]A discussion relevant to this point can be found in the paper by C. S. Ting, S. C. Ying, and J. J. Quinn, Phys. Rev. B 16, 5394 (1977).

Spin-polarization instability in a tilted magnetic field of a two-dimensional electron gas with filled Landau levels

Gabriele F. Giuliani
Department of Physics, Purdue University, West Lafayette, Indiana 47907

J. J. Quinn
Department of Physics, Brown University, Providence, Rhode Island 02912
(Received 14 December 1984)

We have investigated the stability of a two-dimensional electron gas with two filled Landau levels (of opposite spin) in the high-field limit. The Zeeman energy can be increased by adding a component of the magnetic field parallel to the surface. The lowest-lying excitations can be described in terms of singlet and triplet excitons. Taking interaction effects into account, we have found that at a critical Zeeman energy (smaller than the cyclotron energy), there is a first-order transition to a fully spin-polarized state in which two Landau levels of equal spin are filled. Exotic intermediate states of the spin-density-wave type have been found not to occur in the simple case of nondegenerate bands.

INTRODUCTION

In the presence of a sufficiently large magnetic field the carriers of a two-dimensional electron gas will populate only the lowest Landau, lowest spin level. The occurrence of the fractional quantum Hall effect[1] when this level is partially occupied has led to a great deal of interest in the effect of electron-electron interactions[2,3] on the properties of the system. The fractional quantum Hall regime is particularly challenging because all of the single-particle states are degenerate in the absence of electron-electron interactions. If the cyclotron energy $\hbar\omega_c$ is sufficiently large, then the Coulomb energy e^2/l, where $l=(\hbar c/eB)^{1/2}$ is the magnetic length, is the only relevant energy scale in the problem. Because of this, no small parameter exists with which one can construct a perturbation expansion.

In the situation where Landau levels are either completely filled or completely empty, the situation can be considerably simpler. The low-lying excitations consist of electron-hole (e-h) pairs, and there are two types of e-h pair excitations. The singlet e-h pair occurs when a carrier is promoted to a higher Landau level without a change in spin; the triplet excitation involves the promotion of a carrier to the opposite spin state of either the same Landau level or a higher one. Because the excited electron and the hole left behind in the lower energy level interact with one another, they can form a bound state (exciton). The binding energy of singlet and triplet excitons in two-dimensional systems in a strong magnetic field has been studied by a number of authors.[4] The ratio of the Coulomb energy to the cyclotron energy can act as the small parameter for a perturbation expansion.

The present work is motivated by the consideration of systems in which the spin splitting is of the same order of magnitude as the cyclotron splitting. Then if both spin states of a given Landau level are fully occupied and the next Landau level is empty, the lowest energy excitation results from promoting an electron from the occupied upper spin state of the filled Landau level to the lower spin state of the next Landau level. In the absence of electron-electron interactions the energy of such an excitation would be $\epsilon\equiv\hbar(\omega_c-\omega_s)$, where ω_s is the spin-resonance frequency. Because the cyclotron frequency ω_c depends on the component of magnetic field normal to the surface, while the spin-resonance frequency depends on the magnitude of $\mathbf{B}$, this excitation can be made arbitrarily small (in fact, it can be negative in which case the lower spin states of both Landau levels are occupied while the upper spin states are empty). When electron-hole interactions are included, the triplet exciton binding energy can exceed ϵ and one might expect a spin-density-wave (SDW) instability.[5] What actually occurs is that before ϵ becomes smaller than the triplet exciton binding energy, the electron-electron interactions cause a paramagnetic to ferromagnetic phase transition.[6]

EXCITONS

Before proceeding to investigate the phase transition, it is worth reviewing the evaluation of the singlet and triplet excitons.

The energy of the excitonic excitations of a two-dimensional electron gas in a strong perpendicular magnetic field can be evaluated in closed form making use of the formalism developed in Ref. 3 in connection with the fractional quantum Hall-effect problem.

The Hamiltonian for a two-dimensional electron gas in the presence of a dc magnetic field $\mathbf{B}$ can be written as $H=H_0+H_I$, where

$$H_0=\sum_{n,k,\sigma}[\hbar\omega_c(n+\tfrac{1}{2})+\sigma\epsilon_Z]c^\dagger_{n,k,\sigma}c_{n,k,\sigma}\,, \qquad (1)$$

and

$$H_I=\sum_{\substack{n,n',m,m',\\ k,k',q,\\ \sigma,\sigma'}} V_{nm,n'm'}(k',k;q) \times c^\dagger_{n,k+q,\sigma}c^\dagger_{n',k'-q,\sigma'}c_{m',k',\sigma'}c_{m,k,\sigma} \ . \qquad (2)$$

In these equations σ takes on the values ± 1, and ϵ_Z is the Zeeman energy ($2\epsilon_Z=\hbar\omega_s$). The operator $c^\dagger_{n,k,\sigma}$ creates an electron of spin $\sigma=\frac{1}{2}$ in the single-particle state,

$$\psi_{n,k}(r)=L^{-1/2}e^{iky}u_n(x+l^2k) \ . \qquad (3)$$

Here L is the length of the sample, $k=(2\pi/L)$ times an integer, $u_n(x)$ is the nth eigenfunction of the simple harmonic oscillator, and $l=(\hbar c/eB_z)^{1/2}$ is the magnetic length. For any given Landau-level index n there are $N_L=L^2/2\pi l^2$ such states labeled by the wave vector k. These N_L states are degenerate solutions of the noninteracting problem. The matrix element $V_{nm,n'm'}(k',k;q)$ is given by

$$V_{nm,n'm'}(k',k;q)=\int d^2\mathbf{r}\,d^2\mathbf{r}'\psi^*_{n,k+q}(r)\psi^*_{n',k'-q}(r')V(|\mathbf{r}-\mathbf{r}'|)\psi_{m',k'}(r')\psi_{m,k}(r) \ , \qquad (4)$$

where $V(|\mathbf{r}-\mathbf{r}'|)$ is the electron-electron interaction. The sum appearing in Eq. (2) is over all values of n, n', m, m', k, k', q, σ, and σ'. In writing down these equations we have taken the two-dimensional electron gas to lie in the plane $z=0$, and have used the Landau gauge $\mathbf{A}=(0,B_zx,0)$ for the vector potential causing the normal component of the magnetic field.

For the sake of simplicity we concentrate on the situation in which the filling factor ν, defined as the ratio of the number of electrons N to the Landau-level degeneracy N_L, is equal to two, so that the two spin states of the $n=0$ Landau level are the only occupied states. This situation is sketched in Fig. 1. The elementary excitations which we consider are the singlet exciton and the triplet exciton. The former is generated by promoting an electron from the filled $n=0$ Landau level to the same spin state of the $n=1$ Landau level and then "turning on" the many-particle interactions. The latter is generated when the spin of the electron is flipped in the process of promotion to the next Landau level.

Because we consider the cyclotron energy $\hbar\omega_c$ to be much larger than the Coulomb energy e^2l^{-1}, a simple perturbation theory can be constructed in powers of $e^2l^{-1}/\hbar\omega_c$. To first order in this parameter one need only consider intermediate states containing a single exciton. Then, the exciton energy consists of three parts:[4] (i) the "kinetic" energy, i.e., the excitation energy in the absence of electron-electron interactions, (ii) the exchange energy of the particle and hole, and (iii) the electron-hole binding energy. For the singlet exciton the kinetic energy is $\hbar\omega_c$, while for the triplet it is $\hbar(\omega_c\mp\omega_s)$.

The exchange-matrix element E_x^{nm} of an electron in the nth Landau level interacting with an electron of the same spin in the mth Landau level is $-v_{nm}(lq,0)$ where

$$v_{nm}(lq,l(p'-q-q))=V_{nn,mm}(p',p;q) \ .$$

Useful expressions are

$$v_{00}(x,y)=\frac{e^2}{L}\int_{-\infty}^{\infty}dz(z^2+x^2)^{-1/2}\exp[-\tfrac{1}{2}(z^2-2izy+x^2)] \ , \qquad (5)$$

$$v_{10}(x,y)=\frac{e^2}{L}\int_{-\infty}^{\infty}dz\,(z^2+x^2)^{-1/2}\left[1-\frac{x^2}{2}-\frac{z^2}{2}\right]\exp-[\tfrac{1}{2}(z^2-2izy+x^2)] \ , \qquad (6)$$

$$v_{11}(x,y)=\frac{e^2}{L}\int_{-\infty}^{\infty}dz\,(z^2+x^2)^{-1/2}\left[1-\frac{x^2}{2}-\frac{z^2}{2}\right]^2\exp[-\tfrac{1}{2}(z^2-2izy+x^2)] \ . \qquad (7)$$

In particular if both electrons are in the $n=0$ level we have

$$E_x^{00}(q)=-v_{00}(lq,0) =\frac{-e^2}{L}e^{-(ql/2)^2}K_0(q^2l^2/2) \ , \qquad (8)$$

where K_0 is a modified Bessel function.[7] If the two electrons are one in the $n=0$ and one in the $n=1$ level the result is

$$E_x^{01}(q)=-v_{01}(lq,0) =\frac{-2\sqrt{\pi}e^2}{L}e^{-(ql/2)^2}W_{1/2,1/2}(q^2l^2/2) \ , \qquad (9)$$

where $W_{1/2,1/2}$ is a Whittaker function.[7] The exchange energy of an electron in the nth Landau level interacting with electrons of the same spin in the filled mth Landau level is

$$\epsilon_x^{nm}=-\sum_p v_{nm}(p,0) \ . \qquad (10)$$

The exchange energy ϵ_x^{00} is equal to $(\pi/2)^{1/2}e^2l^{-1}$, and $\epsilon_x^{10}=\frac{1}{2}\epsilon_x^{00}=\frac{2}{3}\epsilon_x^{11}$. Promoting an electron from the full $n=0$ to the otherwise empty $n=1$ Landau level gives rise to a net change $\epsilon_x^{00}-\epsilon_x^{01}=\frac{1}{2}\epsilon_x^{00}$ in exchange energy.

The electron-hole binding energy results from solving the following integral equation for the electron-hole vertex function $\Gamma_{n\sigma,m\sigma'}(p'-p,q,\omega)$ in terms of the irreducible vertex function $\gamma_{n\sigma,m\sigma'}(p'-p,q)$, and the Green's functions $G^{e,h}_{n,\sigma}(p,\omega)$,

$$\Gamma(p'-p,q,\omega)=\gamma(p'-p,q)+i\sum_{p'',\omega''}\gamma(p''-p,q)\Gamma(p''-p',q,\omega)G^h(p'',\omega'')G^e(p''+q,\omega+\omega'') . \tag{11}$$

This integral equation is shown schematically in Fig. 2(a). The corresponding expansion for $\gamma_{n\sigma,m\sigma'}$ is graphically represented in Fig. 2(b). In Eq. (11) appropriate Landau-level and spin indices are understood. It is clear that within the Landau gauge the various contributions to $\gamma_{n\sigma,m\sigma'}$ are given by the matrix elements $v_{nm}(lp,lq)$ introduced above.

The fact that we are restricting the calculation to the single-exciton approximation results in considerable simplification. For an electron and hole of opposite sign Eq. (11) generates only the ladder graphs because the unscreened interaction is instantaneous. For an electron and a hole of the same spin the first diagram of Fig. 2(b) corresponds to the RPA for $\gamma_{n\sigma,m\sigma'}(p'-p,q)$. For the case of the triplet exciton the RPA diagram does not occur and $\gamma(p'-p,q)$ is in the present case equal to $v_{10}(p'-p,q)$. The single-particle Green's functions are given by

$$G_h^{e,h}(p,\omega)=(\omega-\epsilon_{n,\sigma}^{e,h}\pm i\delta)^{-1} , \tag{12}$$

where $\epsilon_{n,\sigma}^{e,h}$ is the spin-dependent Hartree-Fock single-particle or single-hole energy in the nth Landau level including the exchange contribution and the $\pm i\delta$ refers to electron or hole states. The solution to Eq. (11) can readily be obtained by introducing $\widetilde{\Gamma}(k,q,\omega)$ the Fourier transform of $\Gamma(p'-p,q,\omega)$ with respect to the variable $p'-p$. The poles of $\widetilde{\Gamma}(k,q,\omega)$ are the exciton energies.[3] It is easily found that if the electron and the hole are, respectively, in the nth and mth Landau level the resulting exciton energy can be simply written as

$$E_{t,s}^{nm}(R^2)=\Delta_{n\sigma,m\sigma'}-\widetilde{\gamma}_{n\sigma,m\sigma'}(R^2) , \tag{13}$$

where $\Delta_{n\sigma,m\sigma'}=\epsilon_{n,\sigma}^{e}-\epsilon_{m,\sigma'}^{h}$ is the energy of the noninteracting electron-hole pair, and $\widetilde{\gamma}_{n\sigma,m\sigma'}$ is the Fourier transform with respect to the variable $p'-p$ of the appropriate irreducible vertex function. The labels t and s stand for triplet and singlet. In Eq. (13) we have made explicit that $\widetilde{\gamma}_{n\sigma,m\sigma'}$ (and therefore $E_{t,s}^{nm}$) is solely a function of $R^2=l^2(k^2+q^2)$, a quantity that can be interpreted in terms of the exciton size.[3,6] It is interesting to notice that this property is an explicit proof of the independence of our analysis upon the particular choice of the gauge [Eq. (3)].

The exciton energies of Eq. (13) can be readily evaluated in closed form and the results agree with those given by various authors.[4]

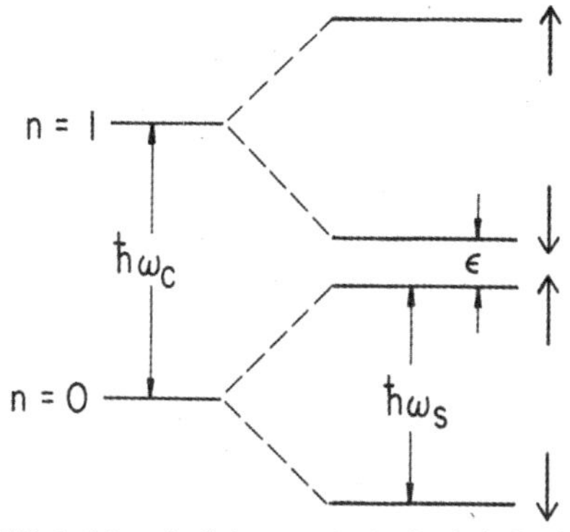

FIG. 1. Schematic of the energy levels: $\hbar\omega_c$ is the Landau-level separation, whereas $\hbar\omega_s$ is the spin splitting. Here $\hbar\omega_c$ and $\hbar\omega_s$ are comparable in magnitude.

SDW VERSUS FERROMAGNETIC INSTABILITY

The case of particular interest in this work is that of the triplet exciton whose "kinetic" energy is $\epsilon\equiv\hbar(\omega_c-\omega_s)$. We find that the energy of this exciton as a function of R^2 is given by[6]

$$E_t^{01}(R^2)=\epsilon+[\tfrac{1}{2}-\mu(R^2)]\epsilon_x^{00} , \tag{14}$$

with

$$\mu(x)=\tfrac{1}{2}e^{-x}[(1+2x)I_0(x)-2xI_1(x)] . \tag{15}$$

In this equation $I_n(x)$ is the modified Bessel function of order n. It is clear that $E_t^{01}(R^2)$ becomes negative when $\epsilon/\epsilon_x^{00}<\mu(R^2)-\frac{1}{2}$. The maximum value of $\mu(x)$ is $\mu_{max}\simeq 0.573$, so this corresponds to a positive value of $\epsilon\equiv\hbar(\omega_c-\omega_s)$. If this inequality is satisfied, the binding energy of the triplet exciton is larger than the sum of the "kinetic" and exchange energies, and an instability must occur.

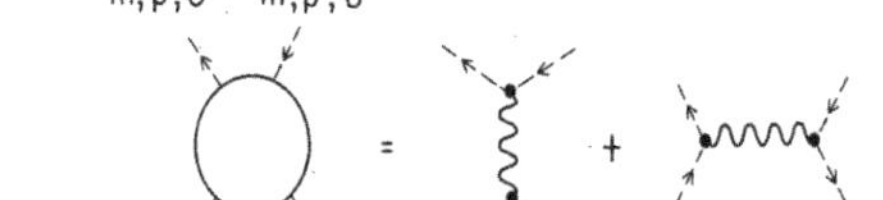

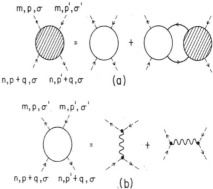

FIG. 2. (a) Diagrammatic representation of the Bethe-Salpeter equation for the vertex function of interacting electron-hole pairs of a two-dimensional electron gas in a magnetic field. Here we use the asymmetric Landau gauge representation in which the noninteracting electronic states are labeled by means of an integer Landau index n, one of the components of the wave vector, and the spin projection. (b) Perturbative contributions to the electron-hole irreducible vertex function to be used in the Bethe-Salpeter equation. The second term is absent in the triplet case.

At first glance one might expect a spin-density-wave instability[5] with the value of the SDW wave vector $Q_{\rm SDW}$ determined by the location of the maximum of $\mu(x)$, i.e., $Q_{\rm SDW}\simeq 1.2l^{-1}$. In order to investigate the behavior of the system for values of e^2l^{-1} large enough to cause this instability, we introduce new operators which are linear combinations of $c_{nk\sigma}$ for $|nk\sigma\rangle$ equal to $|0,k,\downarrow\rangle$ and $|l,k+Q,\downarrow\rangle$. Because these are the only two Landau levels which are modified in the new ground state, we make the following simplification in notation: c_k stands for $c_{0k\uparrow}$ and a_k stands for $c_{1k\downarrow}$. The Hamiltonian can be written $H=H_0+V$ where

$$H_0=N_L(\tfrac{1}{2}\hbar\omega_c-\epsilon_Z)+\sum_k[(\tfrac{1}{2}\hbar\omega_c+\epsilon_Z)c_k^\dagger c_k+(\tfrac{3}{2}\hbar\omega_c-\epsilon_Z)a_k^\dagger a_k] \quad (16)$$

and

$$V=-\tfrac{1}{2}N_L\epsilon_x^{00}-\epsilon_x^{10}\sum_k a_k^\dagger a_k+\tfrac{1}{2}\sum_{pp'q}\{v_{00}(q,p'-p-q)c_{p+q}^\dagger c_{p'-q}^\dagger c_{p'}c_p+v_{11}(q,p'-p-q)a_{p+q}^\dagger a_{p'-q}^\dagger a_{p'}a_p$$
$$+v_{10}(q,p'-p-q)[c_{p+q}^\dagger a_{p'-q}^\dagger a_{p'}c_p+a_{p+q}^\dagger c_{p'q}^\dagger c_{p'}a_p]\}\ . \quad (17)$$

The three terms in H_0 correspond to the kinetic energies of the $0\downarrow$ Landau level, the $0\uparrow$, and the $1\downarrow$ levels. The potential energy has five terms: The first is the exchange energy of the electrons in the $0\downarrow$ level interacting among themselves. The second is the exchange energy due to the particles in the $1\downarrow$ level interacting with those in the $0\downarrow$ level. The final three terms are the interactions of the $0\uparrow$ particles among themselves, the $1\downarrow$ particles among themselves, and finally the interaction of the $0\uparrow$ and the $1\downarrow$ particles with one another. In writing down this approximation we assume that the $0\downarrow$ level is full (contains N_L electron) and always remains full. The $1\uparrow$ level and all higher levels are empty and always remain empty. Only the $0\uparrow$ and $1\downarrow$ levels enter the dynamics.

We make a Bogoliubov-Valatin transformation to new operators[8] α_p and β_k defined by

$$c_p=\cos\theta_p\alpha_p+\sin\theta_p\beta_p\ , \quad (18)$$

$$a_{p+Q}=-\sin\theta_p\alpha_p+\cos\theta_p\beta_p\ . \quad (19)$$

We express the Hamiltonian in terms of the operators α_p and β_p and their Hermitian conjugates.[9] We then apply the Hartree-Fock approximation assuming that the linear combination corresponding to the state α_p is the lower energy state and therefore the occupied state. That is, we assume that $\langle\alpha_p^\dagger\alpha_p\rangle=N_p$ is finite while $\langle\beta_p^\dagger\beta_p\rangle=0$ where the angular brackets denote ground-state expectation value. After assuming that $v_{nm}(0,p'-p)=0$ due to charge neutrality we find that

$$\langle H\rangle_{\rm HF}=N_L(\tfrac{1}{2}\hbar\omega_c-\epsilon_Z-\tfrac{1}{2}\epsilon_x^{00})+(\tfrac{1}{2}\hbar\omega_c+\epsilon_Z)\sum_k\cos^2\theta_k+(\tfrac{3}{2}\hbar\omega_c-\epsilon_Z-\epsilon_x^{10})\sum_k\sin^2\theta_k$$
$$-\tfrac{1}{4}\sum_{k,k'}v_{10}(k'-k,Q)\sin2\theta_k\sin(2\theta_{k'})-\tfrac{1}{2}\sum_{k,k'}v_{00}(k'-k,0)\cos^2\theta_k\cos^2\theta_{k'}$$
$$-\tfrac{1}{2}\sum_{k,k'}v_{11}(k'-k,0)\sin^2\theta_k\sin^2\theta_{k'}\ . \quad (20)$$

Because all the single-particle states are degenerate in the absence of electron-electron interactions, we expect that with periodic boundary conditions $\cos\theta_k$ must be independent of k. In that case Eq. (20) simplifies to

$$N_L^{-1}\langle H\rangle_{\rm HF}=\tfrac{1}{2}\hbar\omega_c-\epsilon_Z-\tfrac{1}{2}\epsilon_x^{00}+(\tfrac{1}{2}\hbar\omega_c+\epsilon_z)\cos^2\theta+(\tfrac{3}{2}\hbar\omega_c-\epsilon_Z-\epsilon_x^{10})\sin^2\theta$$
$$-\tfrac{1}{4}\epsilon_x^{00}\mu(l^2Q^2)\sin^2(2\theta)-\tfrac{1}{2}\epsilon_x^{00}\cos^4\theta-\tfrac{3}{8}\epsilon_x^{00}\sin^4\theta\ . \quad (21)$$

The extreme of $\langle H\rangle_{\rm HF}$ as a function of θ must satisfy the equation

$$\sin(2\theta)(a-b\sin^2\theta)=0\ , \quad (22)$$

where

$$a=\epsilon+\epsilon_x^{00}[\tfrac{1}{2}-\mu(l^2Q^2)] \quad (23)$$

and

$$b=2\epsilon_x^{00}[\tfrac{7}{8}-\mu(l^2Q^2)]\ . \quad (24)$$

There are three possible solutions to Eq. (22): $\theta=0$, $\theta=\pi/2$, and $\theta=\theta^*$ where $\sin^2\theta^*=a/b$. The solution $\theta=0$ corresponds to $\alpha_k=c_k$ and gives the paramagnetic state. This state is a stable Hartree-Fock solution $(\partial^2\langle H\rangle_{\rm HF}/\partial^2\theta>0)$ if $\epsilon/\epsilon_x^{00}>\mu_{\max}-\tfrac{1}{2}\approx 0.073$. This is exactly the condition we found for the triplet exciton instability, so that our starting paramagnetic state is a stable Hartree-Fock solution when the triplet exciton energy $E_t^{01}(R^2)$ is positive.

The solution $\theta=\pi/2$ corresponds to $\alpha_p=a_{p+Q}$ giving a ferromagnetic ground state (i.e., the $n=0$ and $n=1$ spin-down states are both occupied while the $0\uparrow$ state is empty). This extremum corresponds to a stable Hartree-Fock solution if $\epsilon/\epsilon_x^{00}<\tfrac{5}{4}-\mu_{\max}\sim 0.667$. The energy per particle (remember we have $2N_L$ particles) is

$$E_{\text{para}}=\tfrac{1}{2}(\hbar\omega_c-\epsilon_x^{00})\,, \tag{25}$$

$$E_{\text{ferro}}=\hbar\omega_c-\epsilon_Z-\tfrac{11}{16}\epsilon_x^{00}\,. \tag{26}$$

These two energies are equal at

$$\epsilon/\epsilon_x^{00}\equiv(\hbar\omega_c-2\epsilon_Z)/\epsilon_x^{00}=\tfrac{3}{8}\,.$$

What about the solution $\theta=\theta^*$ which corresponds to a spin-density-wave state? This solution occurs when $|a|<|b|$ and a and b have the same sign. These conditions are satisfied if $\mu_{\max}-\frac{1}{2}<\epsilon/\epsilon_x^{00}<\frac{5}{4}-\mu_{\max}$, the region where both the paramagnetic and ferromagnetic solutions are minima as functions of θ. This means that θ^* is always a maximum energy solution and hence unstable.

DISCUSSION

As shown in the previous section, in this simple situation the SDW state we expected never occurs.[6,10] When the energy of the triplet exciton of the paramagnetic state vanishes, the paramagnetic state becomes unstable. However, before that occurs a paramagnetic to ferromagnetic phase transition will preempt such an excitonic instability.[11] It is apparent that if we had started with the stable ferromagnetic state and calculated the energy of the "triplet" exciton resulting from promoting a $1\downarrow$ electron to an unoccupied $0\uparrow$ state, we would find that the energy of the exciton vanished when $\epsilon/\epsilon_x^{00}>\frac{5}{4}-\mu_{\max}$. This would signal the instability of the ferromagnetic state.

The paramagnetic to ferromagnetic transition occurs at $\epsilon/\epsilon_x^{00}=\frac{3}{8}$. This can be seen simply by writing the total energy. For the paramagnetic state a $0\downarrow$ particle has energy $\frac{1}{2}\hbar\omega_c-\epsilon_Z-\epsilon_x^{00}$ while $0\uparrow$ particle has energy $\frac{1}{2}\hbar\omega_c+\epsilon_Z-\epsilon_x^{00}$. The total Hartree-Fock energy is the sum of the "kinetic" and half the exchange energies of the individual particles

$$E_P=N_L\left[\frac{\hbar\omega_c}{2}+\epsilon_Z-\tfrac{1}{2}\epsilon_x^{00}\right]+N_L\left[\frac{\hbar\omega_c}{2}-\epsilon_Z-\tfrac{1}{2}\epsilon_x^{00}\right],$$

so that $E_P=N_L(\hbar\omega_c-\epsilon_x^{00})$. For the ferromagnetic state a $0\downarrow$ particle has energy $\frac{1}{2}\hbar\omega_c-\epsilon_Z-\epsilon_x^{00}-\epsilon_x^{01}$, while a $1\downarrow$ particle has energy $\frac{3}{2}\hbar\omega_c-\epsilon_Z-\epsilon_x^{11}-\epsilon_x^{01}$. Adding the kinetic and half the exchange energies gives

$$\begin{aligned}E_F&=N_L(2\hbar\omega_c-2\epsilon_Z-\tfrac{1}{2}\epsilon_x^{00}-\tfrac{1}{2}\epsilon_x^{11}-\epsilon_x^{01})\\&=N_L(2\hbar\omega_c-2\epsilon_Z-\tfrac{11}{16}\epsilon_x^{00})\,.\end{aligned}$$

By equating these we see that $E_F=E_P$ at $\epsilon/\epsilon_x^{00}=\frac{3}{8}$, just as we showed after Eq. (26).

It might be possible to observe the transition discussed in this paper by measuring the magnetic susceptibility in a field whose z component is held fixed (to keep filling factor $\nu=2$) and whose component parallel to the surface is varied. The de Haas–van Alphen effect has recently been studied in two-dimensional systems,[12] so that the magnetization itself is large enough to be detected. Structure in cyclotron and spin resonance should reflect the singlet and triplet exciton spectrum through the memory function[13] or its spin equivalent. Because the exciton energies are different for the paramagnetic and ferromagnetic states, the phase transition might also be observed by this technique.

ACKNOWLEDGMENTS

This work was supported in part by the National Science Foundation Grant No. DMR-81-2069. J.J.Q. would like to acknowledge the hospitality of colleagues at the IBM Thomas J. Watson Research Center, where some of the initial work on the problem was started. G.F.G. wishes to acknowledge the hospitality of the Aspen Center for Physics where part of this work was carried out.

---

[1]D. C. Tsui, H. L. Stormer, and A. C. Gossard, Phys. Rev. Lett. **48**, 1559 (1982).

[2]R. B. Laughlin, Phys. Rev. Lett. **50**, 1395 (1983); Surf. Sci. **142**, 163 (1984); R. Tao and D. J. Thouless, Phys. Rev. B **28**, 1142 (1983); B. I. Halperin, Helv. Phys. Acta **56**, 75 (1983); Phys. Rev. Lett. **52**, 2390 (1984); F. D. M. Haldane, *ibid.* **51**, 605 (1983); J. J. Quinn, Bull. Am. Phys. Soc. **29**, 428 (1984); G. F. Giuliani and J. J. Quinn, *ibid.* **29**, 428 (1984).

[3]G. F. Giuliani and J. J. Quinn, Phys. Rev. B **31**, 3451 (1985).

[4]Yu. A. Bychkov, S. V. Iordanskii, and G. M. Eliashberg, Pis'ma Zh. Eksp. Teor. Fiz. **33**, 152 (1981) [Sov. Phys.—JETP Lett. **33**, 143 (1981)], and references quoted therein; C. Kallin and B. I. Halperin, Phys. Rev. B **30**, 5655 (1984); G. F. Giuliani and J. J. Quinn (unpublished).

[5]A. W. Overhauser, Phys. Rev. Lett. **4**, 462 (1960); Phys. Rev. **128**, 1437 (1962).

[6]G. F. Giuliani and J. J. Quinn, Solid State Commun. (to be published).

[7]I. S. Gradshtein and I. M. Ryzhik, *Tables of Integrals, Series and Products* (Academic, London, 1981).

[8]N. N. Bogoliubov, J. Exptl. Theoret. Phys. (U.S.S.R.) **34**, 58 (1958) [Sov. Phys.–JETP **7**, 41 (1958)]; J. G. Valatin, Nuovo Cimento **7**, 843 (1958).

[9]The most appropriate transformation is in this case of the excitonic type, in which the "new" operators are a linear superposition of the "old" ones which spans the entire range of the wave-vector label. It turns out, however, that in a *translationally invariant* system the simple transformation of Eqs. (18) and (19) is sufficient to describe the phenomenon.

[10]It should be noticed that in the case of silicon the present analysis does not apply in a straightforward fashion as the valley degeneracy, not accounted for in our analysis, might play a relevant role. A study of this specific problem is currently being carried out.

[11]See, for instance, the review by B. I. Halperin and T. M. Rice, in *Solid State Physics,* edited by H. Ehrenreich, F. Seitz, and D. Turnbull (Academic, New York, 1968), Vol. 21, p. 115.

[12]F. F. Fang and P. J. Stiles, Phys. Rev. **28**, 6962 (1983); Surf. Sci. **142**, 290 (1984); T. Haavasoia, H. L. Stormer, D. J. Bishop, V. Narayanamurti, A. C. Gossard, and W. Weigmann, *ibid.* **142**, 294 (1984).

[13]C. S. Ting, S. C. Ying, and J. J. Quinn, Phys. Rev. Lett. **37**, 215 (1976); Phys. Rev. B **16**, 5394 (1977); C. Kallin and B. I. Halperin, Phys. Rev. B **31**, 3635 (1985).

PHYSICAL REVIEW B VOLUME 37, NUMBER 2 15 JANUARY 1988-I

Acoustic plasmons in a conducting double layer

Giuseppe E. Santoro* and Gabriele F. Giuliani
Department of Physics, Purdue University, West Lafayette, Indiana 47907
(Received 20 July 1987)

We have investigated the acoustic plasma branch present in the longitudinal spectrum of two spatially separated parallel quasi-two-dimensional conducting layers. Our approach is based on the dielectric theory and is completely analytical within the random-phase approximation. By means of a systematic analysis we have obtained several exact results concerning the plasma dispersion relation. In particular, we have derived an exact expression for c_p the acoustic plasmon group velocity in terms of the effective masses, densities and geometrical parameters of the heterostructure. We find that when the two layers are identical the system always admits a branch of acoustic plasmons as undamped modes for any finite value of the distance between the layers.

INTRODUCTION

Momentous advances in growth techniques presently allow the fabrication of materials consisting of alternating layers of two or more semiconductors. At the various interfaces electronic or hole layers can be trapped whose low-energy dynamics is, for all practical purposes, quasi-two-dimensional.[1] The current growth techniques can actually be exploited to tailor the properties of a heterostructure to specific dynamical requirements. In particular, it is in principle possible to design heterostructures with a customized excitation spectrum.

Semiconducting electronic heterostructures have recently provided a valuable testing ground for the study of plasma excitations in various geometrical configurations.[2–4] Semiconducting superlattices are a typical example of such a class of materials. In particular, they are the only known electronic systems in which plasmons have been directly observed.[3] Recently the possibility of acoustic surface plasma modes in certain semiconducting superlattices has also been investigated.[5,6]

In this paper we focus our attention on a peculiar electronic system comprised of two spatially separated parallel quasi-two-dimensional conducting layers (Fig. 1). This problem is of current experimental interest because the system at hand is a model for double-quantum-well heterojunctions and single inversion layers with more than one populated subband.[7–9]

The collective plasma excitation spectrum in double-quantum-well electronic structures has been first investigated by numerical means.[7] A more transparent analytic treatment based on the dielectric approach has been later presented which explicitly showed the existence of two plasmon branches of which one is characterized in the electrostatic limit by an acoustic dispersion relation at long wavelength.[8] The properties of such an acoustic branch have also been studied in the context of a single electronic inversion layer with two populated subbands.[9] Recently the possibility of the plasmon mechanism for superconductivity in a double well in which the effective interaction is mediated by both plasmon branches has also been studied.[10]

The present analysis focuses on the condition for the existence as undamped modes of the acoustic plasma branch in terms of the relevant physical parameters of the heterostructure. Although our theory is based on the same formalism used in previous studies, our main results significantly differ from the ones previously reported. In what follows we will develop a systematic and analytic approach which allows us to precisely characterize and examine the plasma dispersion relation in terms of the geometry, the Fermi velocities, and the electronic effective masses. In particular, for the case of identical layers we will explicitly derive in closed form a simple and exact expression for the plasmon group velocity. Finally, we will prove that in the same situation the system always admits a branch of acoustic plasmons as undamped modes for any finite value of the layer separation.

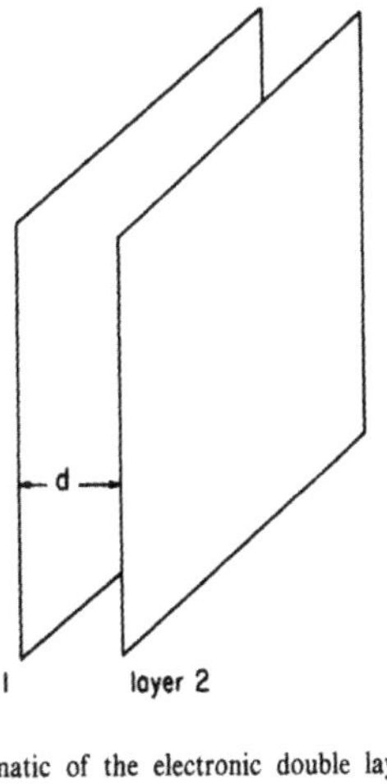

FIG. 1. Schematic of the electronic double layer system studied in the text. The layers are parallel and separated by a distance d.

DISPERSION RELATION

At long wavelength the collective plasma oscillations of a single (i.e., isolated) two-dimensional electronic layer have a dispersion relation given by[11]

$$\omega_0(q) \approx (2\pi n e^2 q/\epsilon_0 m^*)^{1/2} , \quad (1)$$

where q is the magnitude of the (two-dimensional) in-plane wave vector, n and m^* the electronic density and effective mass, and ϵ_0 is the average dielectric constant of the medium. Within the random-phase approximation (RPA), two-dimensional plasma excitations are therefore undamped for $q \approx 0$ since $\omega(q)$ lies outside the continuum of the electron-hole pair excitations. In what follows we shall refer to this type of mode as an *optical plasmon.*

Consider next two parallel conducting quasi-two-dimensional layers separated by a finite distance d (see Fig. 1). Because of the lack of translational invariance the collective modes for the system at hand are given by the zeros of the determinant of the dielectric tensor $\epsilon_{ij}(q,\omega)$.[8] The RPA expression for $\epsilon_{ij}(q,\omega)$ in this system is given by

$$\epsilon_{ij}(q,\omega) = \delta_{ij} - V_{ij}(q)\chi^0_{jj}(q,\omega), \quad i,j=1,2 , \quad (2)$$

where $\chi^0_{jj}(q,\omega)$ is the noninteracting charge susceptibility of the jth layer and $V_{ij}(q)$ stands for the Coulomb interaction vertex between two electrons, respectively, in the ith and jth layer. In obtaining Eq. (2) we have assumed that there is no overlap between electronic wave functions on different layers. If for simplicity sake we limit our analysis strictly to literally two-dimensional electronic layers the matrix elements V_{ij}'s are simply given by $V_{11}(q) = V_{22}(q) = 2\pi e^2/\epsilon_0 q$, and $V_{12}(q) = V_{21}(q) = e^{-qd} 2\pi e^2/\epsilon_0 q$. Making use of Eq. (2) the dispersion relation is readily obtained as given by

$$D(q,\omega) = 1 - V_{11}(q)[\chi^0_{11}(q,\omega) + \chi^0_{22}(q,\omega)] + V^2_{11}(q)(1-e^{-2qd})\chi^0_{11}(q,\omega)\chi^0_{22}(q,\omega) = 0 , \quad (3)$$

where the layer susceptibilities are accordingly evaluated for literally two-dimensional electrons.[11] Clearly this result is independent of the sign of the charge of the carriers on each layer. Equation (3) must then be then explicitly solved for ω as a function of q.

As shown in Ref. 8 the longitudinal spectrum of the total system is comprised of an optical plasmon branch [of the type of Eq. (1)], plus a new branch whose dispersion relation at long wavelength is of the type $\omega(q) \approx c_p q$, i.e., an *acoustic plasmon.* A schematic of this spectrum is shown in Fig. 2. Although it is rather simple to arrive at a rough characterization of the spectrum some care must be taken in evaluating the acoustic plasmon group velocity c_p. The physical origin of the difficulty lies in the fact that in calculating the energy associated with the plasma oscillations a cancellation due to the screening of the long-range part of the Coulomb interaction occurs in the case of the acoustic branch.

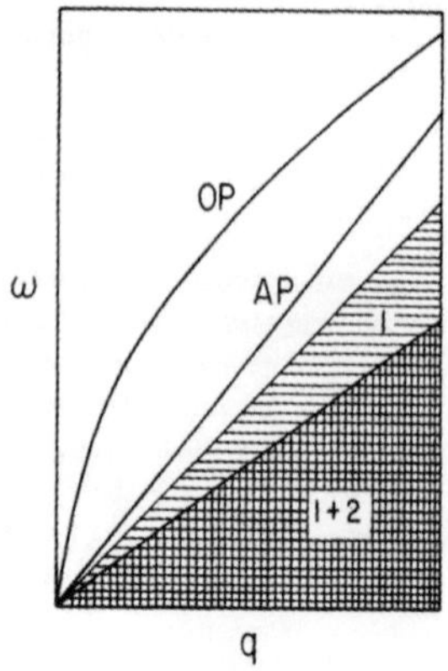

FIG. 2. Schematic of the long-wavelength region of the longitudinal spectrum ω vs q for the electronic double layer system studied in the text. The two shaded regions labeled 1 and 1 + 2 are the electron-hole pair continua for the two layers. In region 1 only pairs in layer 1 can be excited, in region 1 + 2 pairs in both layers can be excited. The two continuous thick lines represent, respectively, the optical (OP) and the acoustic branch (AP) of the plasmon spectrum. In this case the acoustic plasmon group velocity is taken to be larger than v_{F1}, the largest of the two Fermi velocities.

As well known within the RPA a bulk acoustic plasma branch is well defined only if it lies outside the electron-hole pair continua of the two charge components (see dashed regions in Fig. 2), i.e., the region of the q,ω plane in which the imaginary part of the susceptibilities $\chi^0_{ii}(q,\omega)$ is different from zero. If $\omega(q)$ lies inside one of the continua the plasma mode can decay into electron-hole pairs and is Landau damped. Accordingly, in RPA the condition for the existence of the acoustic branch as an undamped mode in the long-wavelength limit is then simply $c_p \geq v_{F1}$, where v_{F1} is by definition the largest of v_{F1} and v_{F2}, the Fermi velocities in the two layers. Our aim is to determine the precise condition for the validity of the above inequality.[12]

In order to obtain an exact expression for the plasmon group velocity c_p in the region of Fig. 2 in which $q \approx 0$ and $\mathrm{Im}\{\chi^0_{ii}(q,\omega)\} = 0$, we proceed as follows. We first introduce for $q \approx 0$ the power expansion

$$\omega(q) = c_p q + c_2 q^2 + c_3 q^3 + \cdots , \quad (4)$$

for the plasmon dispersion relation, and define a function $F(q)$ as

$$F(q) = D(q, c_p q + C_2 q^2 + C_3 q^3 + \cdots) , \quad (5)$$

where D is defined in Eq. (3). For $q \approx 0$, $F(q)$ can in turn be written in terms of the power expansion (in fact a Laurent-Taylor expansion)

$$F(q)=f_{-1}q^{-1}+f_0+f_1q+f_2q^2+\cdots , \quad (6)$$

where the f_i's are suitable coefficients which are derived with the use of Stern's formulas for the $\chi^0_{jj}(q,\omega)$'s (Ref. 11) in Eqs. (3) and (5). The mode condition Eq. (3) is then satisfied by requiring that all the coefficients f_i's vanish independently. As can be readily found f_{-1} depends on c_p only and by equating its expression to zero we arrive after some algebra at the following equation:

$$2k_2d-(1+2k_2d)[1-(v_{F2}/c_p)^2]^{1/2}-(m_2^*/m_1^*)(1+2k_1d)[1-(v_{F1}/c_p)^2]^{1/2} +[1+(m_2^*/m_1^*)+2k_2d][1-(v_{F1}/c_p)^2]^{1/2}[1-(v_{F2}/c_p)^2]^{1/2}\equiv 0 , \quad (7)$$

where $k_i=2m_i^*e^2/\epsilon_0\hbar^2$ is the Thomas-Fermi wave vector of the ith layer. Equation (7) is the sought condition which determines in the general case the plasmon group velocity c_p in terms of the effective masses, densities, and geometrical parameters of the heterostructure.

It should be noted that the present procedure is quite general in character and allows to obtain the dispersion relation at all orders in q. In fact, as it turns out the coefficients of the higher-order terms in the expansion for $\omega(q)$ can in principle be systematically evaluated simply by solving an equation in which only the coefficients of the lower-order terms appear. For instance, the value of c_2 can be readily obtained in terms of c_p as determined in Eq. (7) by requiring that f_0 in Eq. (6) be zero. This leads to the following equation:

$$c_2=\frac{(w_1w_2)^2[w_1w_2-2k_1k_2d^2(c_p-w_1)(c_p-w_2)]}{k_1v_{F1}^2w_2^2[2k_2d(c_p-w_2)-w_2]+k_2v_{F2}^2w_1^2[2k_1d(c_p-w_1)-w_1]} , \quad (8)$$

where we have defined $w_i=(c_p^2-v_{Fi}^2)^{1/2}$.

ACOUSTIC PLASMON: EXISTENCE CONDITION

By definition, the critical value d_c of d is the value of the layers separation for which $c_p=v_{F1}$. By using this value of c_p in Eq. (7) we obtain

$$d_c=\frac{1}{2k_2}\frac{(v_{F1}^2-v_{F2}^2)^{1/2}}{v_{F1}-(v_{F1}^2-v_{F2}^2)^{1/2}} . \quad (9)$$

For larger values of the distance, the branch lies outside the electron-hole continua. We immediately notice that our exact, and indeed very simple, expression for d_c clearly predicts a critical distance equal to zero in the particular case in which the two Fermi velocities are identical. Accordingly, in such a situation an undamped acoustic plasmon branch *always* exists for all finite of the interlayer distance d. This is at variance with the results of Refs. 8 and 9.[13] Furthermore, when $v_{F1}=v_{F2}=v_F$ and $m_1^*=m_2^*$, Eq. (7) has an exact simple closed form solution, given by

$$c_p=\frac{v_F(1+kd)}{(1+2kd)^{1/2}} , \quad (10)$$

where $k=k_1=k_2$. As illustrated in Fig. 3, c_p increases from its critical value v_{F1} at first quadratically and then (asymptotically) like a square root with kd.

We also want to stress the fact that for any finite value of d greater then d_c there *always* exists an acoustic branch. The plasma modes in each layer are decoupled strictly only for $d=\infty$.[14]

With a procedure similar to that used to obtain Eq. (7) we have also studied the solution for the acoustic mode in the region which is outside the electron-hole continuum of the layer with smaller Fermi velocity, but inside the continuum of the electron-hole excitations of the layer with greater Fermi velocity (the shaded region labeled 1 in Fig. 2). In this region $\mathrm{Im}\{\chi^0_{11}(q,\omega)\}$ is different from zero. The solution for $\omega(q)$ is then necessarily complex, and the acoustic branch is damped. One can verify that when c_p approaches (from below) the limiting

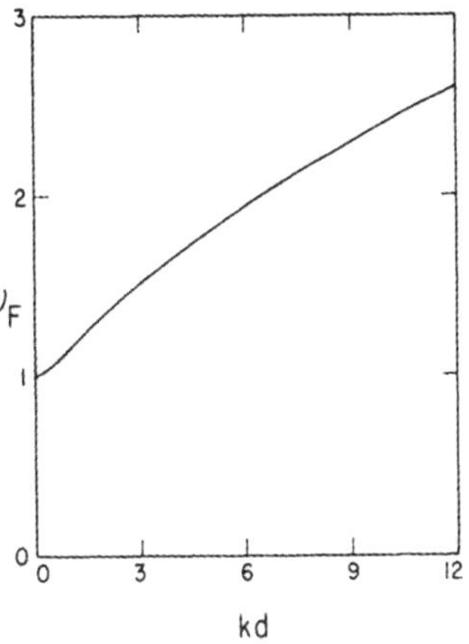

FIG. 3. Plot of the ratio of c_p, the plasmon group velocity, to v_F the Fermi velocity vs kd, the product of the Thomas-Fermi wave vector and the layers distance, for the case in which the two layers are identical. Notice that for small kd, the value of c_p increases quadratically from its critical value v_f. For large kd, c_p increases like $(kd)^{1/2}$

value v_{F1} the imaginary part of $\omega(q)$ is much smaller than the real part and is exactly zero for $c_p = v_{F1}$. The latter situation occurs when $d = d_c$. The two solutions for $c_p(d)$ obtained in this way perfectly match. This is in contrast with the conclusions reached in Ref. 8. There the RPA is held responsible for a series of odd results that are instead a consequence of an incorrect analysis.

It must be mentioned here that Landau damping of the acoustic branch is still possible via intersubband excitations. This will, of course, have an energy threshold close in value to the subband separation and will therefore be negligible in the long-wavelength limit.[15]

In conclusion, we have characterized the acoustic part of the longitudinal spectrum of an electronic double layer system within the RPA. By means of a systematic analytical approach we have obtained several exact results concerning the plasma dispersion relation. In particular, we have derived an exact expression for c_p, the acoustic plasmon group velocity, and d_c, the critical distance for their existence, in terms of the effective masses, densities, and geometrical parameters of the structure. Finally, an interesting and exact result is that when the two layers are identical, for any finite value of their distance, the system always admits a branch of acoustic plasmons as undamped modes.

A similar analysis can be carried out also for the optical plasmon. In such a case, however, the energy is mainly determined by the long-range part of the Coulomb interaction which makes the problem rather trivial.

ACKNOWLEDGMENTS

This work was partially supported by Nation Science Foundation (Materials Research Laboratory Program) Grant No. DMR-84-18453 at Purdue University. One of us (G.S.) acknowledges the support of the Purdue Research Foundation, and of the Scuola Normale Superiore.

---

*On leave from the Scuola Normale Superiore, Pisa, Italy.

[1]See, for instance, *Electronic Properties of Two-Dimensional Systems*, Proceedings of the Yamada Conference XIII, edited by T. Ando (Yamada Science Foundation, Japan, 1986).

[2]S. J. Allen, Jr., D. C. Tsui, and R. A. Logan, Phys. Rev. Lett. **38**, 980 (1977); D. C. Tsui, E. Gornik, and R. A. Logan, Solid State Commun. **35**, 875 (1980).

[3]D. Olego, A. Pinczuk, A. C. Gossard, and W. Wiegmann, Phys. Rev. B **25**, 7867 (1982); R. Sooryakumar, A. Pinczuk, A. C. Gossard, and W. Wiegmann, *ibid.* **31**, 2578 (1985); A. Pinczuk, M. G. Lamont, and A. C. Gossard Phys. Rev. Lett. **56**, 2092 (1986).

[4]G. Fasol, N. Mestres, H. P. Hughes, A. Fisher, and K. Ploog, Phys. Rev. Lett. **56**, 2517 (1986).

[5]R. A. Mayanovic, G. F. Giuliani, and J. J. Quinn, Phys. Rev. B **33**, 8390 (1986); G. Qin, G. F. Giuliani, and J. J. Quinn, *ibid.* **28**, 6144 (1983).

[6]For a review see, G. F. Giuliani, P. Hawrylak, and J. J. Quinn, Phys. Scr. **35**, 946 (1987).

[7]A. Equiluz, T. K. Lee, J. J. Quinn, and K. W. Chin, Phys. Rev. B **11**, 4989 (1975).

[8]S. Das Sarma and A. Madhukar, Phys. Rev. B **23**, 805 (1981).

[9]Y. Takada, J. Phys. Soc. Jpn. **43**, 1627 (1977).

[10]G. F. Giuliani, Surf. Sci. (to be published).

[11]F. Stern, Phys. Rev. Lett. **18**, 546 (1967).

[12]Multipair excitations beyond the RPA will in general always lead to a small damping of the plasmon but will be neglected here for simplicity.

[13]For the case of the double quantum well $GaAs/Al_xGa_{1-x}As$ with equal Fermi velocities, the authors of Ref. 8 arrived at the (incorrect) result $d_c \approx 150$ Å. It is straightforward to see that the assorted results for d_c obtained by the authors of Refs. 8 and 9 can be simply traced to the use of invalid approximations.

[14]Here, care must be taken in correctly handling the limiting procedures involved. The solution with two decoupled optical plasmons given in Ref. 8 never exists for finite albeit large d.

[15]This phenomenon is similar to the absence of Landau damping for surface plasmons in semiconducting superlattices first discussed in G. F. Giuliani and J. J. Quinn, Phys. Rev. Lett. **51**, 919 (1981).

PHYSICAL REVIEW B VOLUME 40, NUMBER 8 15 SEPTEMBER 1989-I

Spin susceptibility in a two-dimensional electron gas

Sudhakar Yarlagadda and Gabriele F. Giuliani
Department of Physics, Purdue University, West Lafayette, Indiana 47907
(Received 28 November 1988)

The problem of the many-body enhancement of the static spin susceptibility at long wavelengths and its relation to the quasiparticle effective mass is investigated for a normal electron gas in two-dimensional space as a function of the electronic density. We start from a discussion of the results of the simple Hartree-Fock approximation for various interaction potentials and proceed to develop a complete theory. We find that the effects of the electron-electron interaction are significantly larger than in the familiar three-dimensional case. Our approach is based on a new self-consistent scheme which goes beyond the simple random-phase approximation by explicitly allowing for charge- and spin-fluctuation-induced vertex corrections of the Hubbard type. We show that when the latter are neglected, the many-body enhancement of the spin susceptibility can be cast in a remarkably simple and elegant analytic form.

I. INTRODUCTION

The enhancement of the paramagnetic spin susceptibility χ_S of an electron gas (EG) due to the electron-electron interaction is a classic many-body problem. Pioneering work on the subject goes all the way back to Bloch, Wigner, and Sampson, and Seitz.[1] The challenging aspect of this problem is the fact that at metallic densities the concomitant effects of exchange and correlation are both large but opposite in sign and eventually lead to a value for χ_S sensibly larger than the free-electron Pauli value. Both for the three-dimensional (3D) and the two-dimensional (2D) case within the Hartree-Fock (HF) approximation, in which only the exchange contribution is retained, χ_S actually diverges for values of the parameter r_s, the average electron distance in Bohr radii, respectively equal to $(9\pi/4)^{1/3}\approx 6.03$ and $\pi/2^{1/2}\approx 2.22$. The situation is further complicated by the appearance of the more exotic instabilities of the spin-density-wave type.[2,3] At metallic densities the effect of the correlations is to rid the spin susceptibility of these instabilities. Ultimately the correct many-body enhancement results from a delicate balance of the two contributions.

The value of χ_S is physically accessible through a number of different experimental techniques. The situation is particularly favorable since the effects of the electron-phonon interaction can in general be safely neglected. This makes the spin-susceptibility problem an especially valuable testing ground for many-body theories.

For the case of a 3D EG there are reported data from conduction-spin-resonance, spin-wave, Knight-shift, and total-susceptibility measurements.[4] Several theoretical methods have been employed for the solution of this problem for the 3D EG.[3,5–9] In general it is believed that there exists a reasonably good agreement between theory and experiment, although Fig. 6 of Ref. 4 may raise some doubts since the various theories arguably seem to cover all the conceivable (as well the inconceivable), experimental results. More specifically there exists a theoretical "consensus curve" of χ_S versus r_s, which not only appears to be supported by the experimental results, but has been reproduced within a few percent via a variety of diverse many-body techniques, ranging from self-consistent-field approaches,[3,9] microscopic Landau Fermi-liquid analyses,[6,7] to full fledged perturbative-theoretic calculations.[5,7,8] In all cases the amount of analytic and numerical work necessary to reach the final answer is considerable.

The major problem with the familiar three-dimensional metals is that for obvious reasons the density dependence of the many-body enhancement of χ_S can only be approximately measured by looking at different materials. This makes it difficult to clearly discern the sought phenomenon amid band-structure effects whose relevance varies from metal to metal.

For the case of a 2D EG the study of the many-body enhancement of the static paramagnetic spin susceptibility has a decisive advantage in that currently available quasi-two-dimensional electronic systems, notably the Si-inversion-layer structures, are characterized by the rather interesting possibility of varying the carrier density and other intrinsic parameters within the same sample.[10] This offers the remarkable possibility of measuring χ_S for a range of density values while keeping constant other uninteresting (albeit not necessarily irrelevant) factors.

In this case an experimental determination of χ_S can be achieved by concomitantly measuring, by magneto-transport techniques, both the quasiparticle effective mass[11,12] and the anomalous Landé g factor.[13,14]

The purpose of the present paper is to provide a theory of the many-body enhancement of the static paramagnetic spin susceptibility in a normal 2D EG at long wavelengths and discuss the density dependence of this remarkable phenomenon as well as its relation to the quasi-particle mass renormalization.

A natural starting point for our analysis is provided by a discussion of the Hartree-Fock theory, which sets the stage for a more complete and reliable approach. As it will be shown however, the inclusion of correlation effects is necessary. Janak was the first to attempt the study of the effects of screening in his study of the effects of the electronic interactions on the Landé g factor.[15] His theory suffered, however, from serious shortcomings and was limited to a static approach to screening. Work along similar lines can be found in Refs. 16–18.

The effect of correlations can more satisfactorily be included by employing, as a first approximation, an approach due to Hamann and Overhauser and based on the dynamically screened exchange approximation.[3] It is quite satisfying, although hitherto unnoticed, that in this case the many-body enhancement of the spin susceptibility can be cast in a simple and elegant analytic form.

In order to go beyond this useful, but necessarily simplified, approach we then proceed to develop a complete theory based on the Landau theory of the Fermi liquids and a new self-consistent scheme in which the effects of charge- and spin-fluctuation-induced vertex corrections are accounted for following the procedure first suggested, in its most elementary form, by Hubbard.[19] An extensive investigation of the relevance of these corrections in realistic situations has been discussed elsewhere.[20]

The present paper is structured as follows: In Sec. II we discuss several interesting results concerning the application of the Hartree-Fock theory to the case of various interaction potentials; in Sec. III we develop the formalism for the dynamically screened exchange approach and relate the many-body enhancement for the spin susceptibility to the quasiparticle effective mass; in Sec. IV we introduce the generalized Hubbard many-body local fields which account for charge- and spin-fluctuation-induced vertex corrections; in Sec. V we develop a general theory for the spin susceptibility by relating such a quantity to the quasiparticle effective interaction; finally, in Sec. VI we discuss our results and provide some conclusions.

II. HARTREE-FOCK THEORY

We start at first with an analysis of the Hartree-Fock (HF) theory. Within this approximate scheme one can readily obtain the static spin susceptibility $\chi_S(\mathbf{q},\omega)$ by following the procedure of Wolff.[21] One finds that $\chi_S(\mathbf{q},0)$ can be expressed as

$$\chi_S(\mathbf{q},0)=2\mu_B^2\sum_{\mathbf{p}}\frac{n_{\mathbf{p}-\mathbf{q}/2}-n_{\mathbf{p}+\mathbf{q}/2}}{E_{\mathbf{p}+\mathbf{q}/2}-E_{\mathbf{p}-\mathbf{q}/2}}u(\mathbf{p}), \quad (1)$$

where μ_B is the Bohr magneton, $n_{\mathbf{p}}$ is the momentum-space occupation number, and $E_{\mathbf{p}}$, the quasiparticle energy, is given by the familiar expression

$$E_{\mathbf{p}}=\frac{p^2}{2m}-\sum_{|\mathbf{p}'|<p_F}v(\mathbf{p}-\mathbf{p}'). \quad (2)$$

In Eq. (1) $u(\mathbf{p})$ is the solution of the integral equation

$$u(\mathbf{k})=1+\sum_{\mathbf{p}}\frac{n_{\mathbf{p}-\mathbf{q}/2}-n_{\mathbf{p}+\mathbf{q}/2}}{E_{\mathbf{p}+\mathbf{q}/2}-E_{\mathbf{p}-\mathbf{q}/2}}v(\mathbf{k}-\mathbf{p})u(\mathbf{p}), \quad (3)$$

$v(\mathbf{q})$ being the Fourier transform of the appropriate interaction potential. In the long-wavelength limit this equation is easily solved to give for χ_S the static value of the susceptibility, $\chi_S(\mathbf{q}\to 0,0)$, the following expression,

$$\frac{\chi_S}{\chi_P}=\frac{m^*}{m}u(p_F), \quad (4)$$

In Eq. (4) χ_P is the Pauli susceptibility given by $\mu_B^2N(0)$, where $N(0)=m/\pi\hbar^2$ is the density of states for a noninteracting electron gas in two dimensions. In Eq. (4) m^* is the quasiparticle effective mass defined in terms of the derivative of the quasiparticle energy evaluated at the Fermi wave vector p_F via the relation

$$v_{\mathrm{qp}}=\frac{p_F}{m^*}=\left(\frac{\partial E(\mathbf{p})}{\partial p}\right)_{p_F}. \quad (5)$$

For illustrative purposes we will first examine the simple case of a short-range (local) interaction which we model here via a delta-function potential so that $v(\mathbf{q})=C$, where C is a constant. In this case there is no correction to the bare mass and the susceptibility can be readily expressed in analytic form as follows:

$$\frac{\chi_S^{(\delta)}(z)}{\chi_P}=\frac{[1-(1-z^{-2})^{1/2}\Theta(1-z^{-2})]}{1-\frac{C}{2\mu_B^2}[1-(1-z^{-2})^{1/2}\Theta(1-z^{-2})]}, \quad (6)$$

where $z=q/2p_F$. It is clear that in this case the ratio is independent of the electronic density. In Fig. 1 we plot $\chi_S^{(\delta)}$ as a function of z for different values of the interaction strength C. We note that for $z<1$, $\chi_S^{(\delta)}$ is a constant

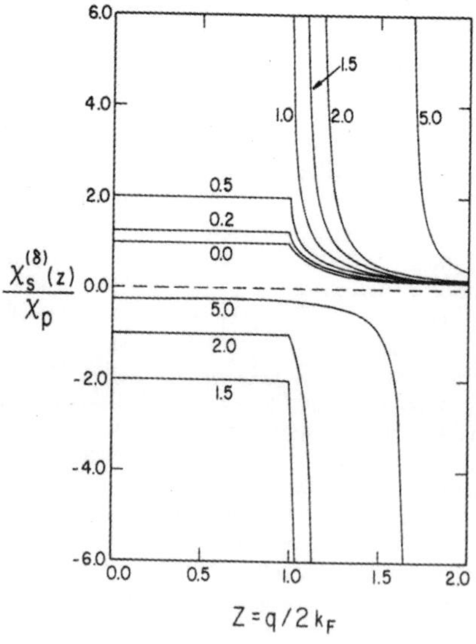

FIG. 1. Enhanced spin susceptibility $\chi_S^{(\delta)}(z)/\chi_P$ vs $z=q/2p_F$, given by Eq. (6) in text, for a delta-function interaction of strength C. The curves are for different values of $C/(2\mu_B^2)$.

whose value diverges as C tends to the critical value $2\mu_B^2$. For C larger than this value $\chi_S^{(\delta)}$ changes sign and decreases in magnitude. For $z>1$ the situation is more interesting. For C less than $2\mu_B^2$, $\chi_S^{(\delta)}$ is a monotonically decreasing function of z. When C exceeds this value, however, $\chi_S^{(\delta)}$ becomes negative (a sign of incipient ferromagnetic instability), and displays a singularity for $z=[1-(1-2\mu_B^2/C)^2]^{-1/2}$. This behavior is quialitatively not dissimilar from the one encountered in the 3D case first discussed by Wolff.[21]

We will consider next the case of a screened interaction potential which we write in the general form

$$v(\mathbf{q})=\frac{2\pi e^2}{q+2\alpha p_F}=\frac{2\pi e^2}{q+\dfrac{2^{3/2}\alpha}{r_s a_B}}, \tag{7}$$

where, as mentioned above, r_s is the mean electronic separation measured in units of the Bohr radius a_B, and is related to the electronic density via the relation $r_s=(\pi a_B{}^2 n)^{-1/2}$. In Eq. (7) α is a positive adjustable parameter which controls the range of the interaction. In the long-wavelength limit, using Eqs. (1), (3), and (5), one can obtain the following simple analytic formula for the HF susceptibility for this case:

$$\frac{\chi_S}{\chi_P}=\frac{m^*}{m}\left\{1-\frac{m^*}{m}\left[\frac{r_s}{2^{1/2}\pi}\frac{\Theta(\alpha^2-1)}{(\alpha^2-1)^{1/2}}\left[\frac{\pi}{2}-\tan^{-1}\frac{1}{(\alpha^2-1)^{1/2}}\right]-\frac{r_s}{2^{3/2}\pi}\frac{\Theta(1-\alpha^2)}{(1-\alpha^2)^{1/2}}\ln\left|\frac{1-(1-\alpha^2)^{1/2}}{1+(1-\alpha^2)^{1/2}}\right|\right]\right\}^{-1}, \tag{8}$$

where $\Theta(x)$ is the familiar step function and the effective mass m^* can be explicitly obtained from Eqs. (2), (5), and (7) and is given by

$$\frac{m}{m^*}=1-\frac{2^{1/2}r_s}{\pi}+\frac{\alpha r_s}{2^{1/2}}-\frac{(2\alpha^2-1)r_s}{2^{1/2}\pi}\frac{\Theta(\alpha^2-1)}{(\alpha^2-1)^{1/2}}\left[\frac{\pi}{2}-\tan^{-1}\frac{1}{(\alpha^2-1)^{1/2}}\right]$$
$$+\frac{(2\alpha^2-1)r_s}{2^{3/2}\pi}\frac{\Theta(1-\alpha^2)}{(1-\alpha^2)^{1/2}}\ln\left|\frac{1-(1-\alpha^2)^{1/2}}{1+(1-\alpha^2)^{1/2}}\right|. \tag{9}$$

A noteworthy feature of Eq. (8) is the manifest possibility of a polarization instability signaled by a diverging spin response. This situation can be realized when the denominator of Eq. (8) vanishes. This in turn occurs when, for a given α, r_s acquires the following critical value

$$r_s^*=2^{1/2}\left[\frac{2}{\pi}-\alpha+\frac{2\alpha^2}{\pi}\frac{\Theta(\alpha^2-1)}{(\alpha^2-1)^{1/2}}\left[\frac{\pi}{2}-\tan^{-1}\frac{1}{(\alpha^2-1)^{1/2}}\right]-\frac{\alpha^2}{\pi}\frac{\Theta(1-\alpha^2)}{(1-\alpha^2)^{1/2}}\ln\left|\frac{1-(1-\alpha^2)^{1/2}}{1+(1-\alpha^2)^{1/2}}\right|\right]^{-1}. \tag{10}$$

A plot of the above expression is displayed in Fig. 2. The critical value r_s^* increases almost linearly with α.

A specific case of interest is that of Thomas-Fermi screening, characterized by the condition $\alpha=r_s/2^{1/2}$. In this case, at variance with the corresponding 3D situation, the screening length is independent of the electron density. As can be readily verified, for this choice of α, the divergence does not occur and the spin susceptibility is a well-behaved simple monotonic function of r_s^*, which is displayed in Fig. 3 by the curve labeled TF. Notice that in this case the many-body enhancement of the spin response is rather small.

Finally for $\alpha=0$ one recovers the HF result for the Coulomb interaction. In this case, upon making use of Eqs. (8) and (9), one readily obtains the following expression for the HF spin susceptibility:

$$\frac{\chi_S}{\chi_P}=\frac{1}{1-\dfrac{\pi r_s}{2^{1/2}}}. \tag{11}$$

Care must be taken in this limit since the ratio m/m^* diverges here logarithmically for vanishing α, i.e.,

$$\frac{m}{m^*}\sim-\frac{r_s}{2^{1/2}\pi}\ln\frac{\alpha}{2}\quad\text{as }\alpha\to 0, \tag{12}$$

As shown in Eq. (11) above, a differential instability occurs for Coulomb interactions at $r_s^*=\pi/2^{1/2}\approx 2.22$. This is displayed by the curve labeled HF in Fig. 3. It should be mentioned at this point, however, that within HF such an instability is preempted by a sudden transition to a ferromagnetic ground state. As is readily found,

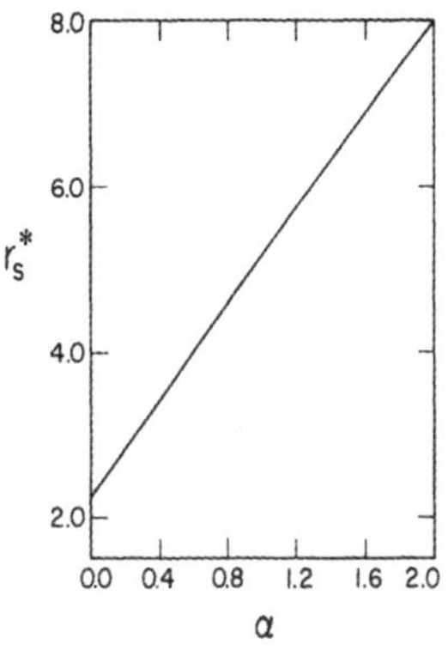

FIG. 2. Plot of r_s^* the critical value of the average electronic separation, given by Eq. (10) in text, vs the screening parameter α. The curve displays the divergence condition of spin susceptibility in the Hartree-Fock approximation with screened interaction.

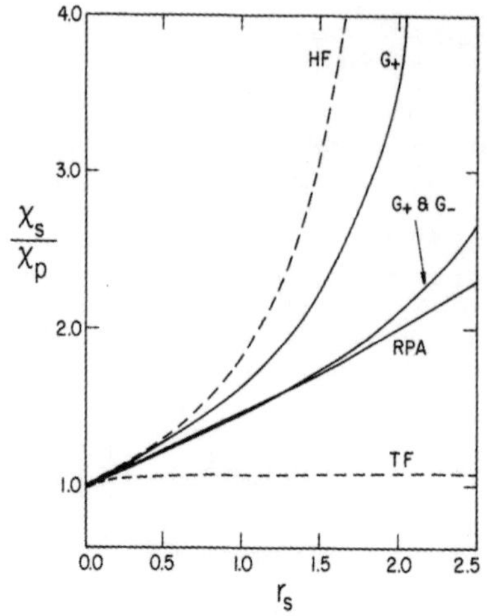

FIG. 3. Plot of the many-body susceptibility enhancement χ_S/χ_P vs the density parameter r_s. The solid curves labeled G_+ & G_-, G_+, and RPA correspond to the following three cases: (i) our full theory, (ii) no spin fluctuations, and (iii) no vertex corrections, respectively. The dashed curves labeled HF and TF correspond, respectively, to Hartree-Fock approximation and Hartree-Fock approximation with Thomas-Fermi screening. The meaning of these curves is explained in the text.

such a phenomenon is characterized by the Bloch condition $r_s \gtrsim 3\pi 2^{1/2}/16(2^{1/2}-1) \approx 2.01$.

As well-known correlations do in general change the nature of the ground state and make the ferromagnetic phase energetically unfavorable, it is interesting to mention, however, how such a situation is drastically modified by the presence of a large quantizing magnetic field when a number of Landau levels are completely filled. In this case the energy separation between the Landau levels will in general lead to a quenching of the correlations, thereby restoring the (at times perhaps more interesting) HF scenario.[22]

III. GENERALIZED HARTREE-FOCK THEORY: THE HAMANN-OVERHAUSER APPROACH

In this section we will derive a simple formula for the susceptibility based on the dynamically screened exchange approach of Hamann and Overhauser which is known to lead to the correct result for the spin problem in 3D.[3] Here, and in what follows, we will focus our analysis on the case of the Coulomb interaction. The gist of the approach is as follows. One starts with the derivation of a suitable pseudo-Hamiltonian in which only the quasiparticle degrees of freedom of the electron gas appear explicitly. As first discussed in Ref. 3 this can be achieved by introducing an appropriate canonical transformation designed to eliminate, (more appropriately average out), the collective part of the spectrum of the system. Such a pseudo-Hamiltonian can be written as

$$H_{\rm qp}=\sum_{{\bf p},\sigma} E_{\bf p}\, :a^\dagger_{{\bf p},\sigma}a_{{\bf p},\sigma}: +\tfrac{1}{2}\sum_{{\bf p},{\bf p}',{\bf q},\sigma,\sigma'} {\rm Re}[\Lambda_C({\bf q},\epsilon_{\bf p}-\epsilon_{{\bf p}-{\bf q}})]\, :a^\dagger_{{\bf p}-{\bf q},\sigma}a^\dagger_{{\bf p}'+{\bf q},\sigma'}a_{{\bf p}',\sigma'}a_{{\bf p},\sigma}:, \tag{13}$$

where we have used a normal product representation, and $a^\dagger_{{\bf p},\sigma}$ ($a_{{\bf p},\sigma}$) is a creation (destruction) operator of a quasiparticle of momentum **p** and spin $\sigma=\pm 1$, and $\epsilon_{\bf p}=p^2/2m$. In Eq. (13) $E_{\bf p}$, the quasiparticle energy, is given by

$$E_{\bf p}=\epsilon_{\bf p}-\sum_{\bf q}\left\{ n_{{\bf p}-{\bf q}}{\rm Re}[\Lambda_C({\bf q},\epsilon_{\bf p}-\epsilon_{{\bf p}-{\bf q}})]+\frac{1}{\pi}{\rm P}\int_0^\infty d\omega \frac{|{\rm Im}[\Lambda_C({\bf q},\omega)]|}{\omega-\epsilon_{\bf p}+\epsilon_{{\bf p}-{\bf q}}}\right\}, \tag{14}$$

where the symbol P mandates that the principal value of the integral must be taken. In Eqs. (13) and (14) $\Lambda_C({\bf q},\omega)$ is an effective potential which is defined in terms of $\chi_C({\bf q},\omega)$, the full momentum- and frequency-dependent charge response function of the system, as follows:

$$\Lambda_C({\bf q},\omega)=v({\bf q})[1+v({\bf q})\chi_C({\bf q},\omega)]. \tag{15}$$

Within the present approximation the function $\chi_C({\bf q},\omega)$ can be written as

$$\chi_C({\bf q},\omega)=\frac{\chi_0({\bf q},\omega)}{1-v({\bf q})\chi_0({\bf q},\omega)}, \tag{16}$$

where $\chi_0({\bf q},\omega)$ is the familiar Lindhard function for a 2D EG.[23] It should be emphasized here that the expression for the quasiparticle energy $E_{\bf p}$ of Eq. (14) is appropriate for an *unpolarized* electron system. Moreover, in Eq. (14) the second term is the dynamically screened exchange, whereas the third one represents the appropriate contribution of the corresponding Coulomb hole. As should be clear from Eqs. (14)–(16), the present approximation is equivalent to the familiar random-phase approximation (RPA).[24,25]

The next step consists of studying the response of the quasiparticle gas to an externally applied sinusoidal magnetic field $H_0\cos{\bf q}\cdot{\bf x}$. This can be achieved by adding to $H_{\rm qp}$ the suitably transformed coupling term $H_{\rm int}$ given by

$$H_{\rm int}=\tfrac{1}{2}\mu_B H_0\sum_{{\bf k},\sigma}(S_z)_{\sigma\sigma}\, :a^\dagger_{{\bf k}+{\bf q},\sigma}a_{{\bf k},\sigma}: +{\rm H.c.}, \tag{17}$$

where H.c. stands for Hermitian conjugate, $(S_z)_{\uparrow\uparrow(\downarrow\downarrow)}=\pm 1$. Then, to do the equivalent of solving the HF equation, the total Hamiltonian given by $H_{\rm qp}+H_{\rm int}$ is again canonically transformed so as to remove the off-diagonal terms in the single quasiparticle operators. The transformed Hamiltonian is given by

$$H_{\rm total}=e^{-T}(H_{\rm qp}+H_{\rm int})e^{T}, \tag{18}$$

where the appropriate form of the operator T can be surmised from the HF analysis to be

$$T=\frac{1}{2}\sum_{\mathbf{k},\sigma} C(\mathbf{k})(S_z)_{\sigma\sigma}\, :a^{\dagger}_{\mathbf{k}+\mathbf{q}/2,\sigma}a_{\mathbf{k}-\mathbf{q}/2,\sigma}: -\text{H.c.}\ . \tag{19}$$

Then, upon requiring that the off-diagonal one-particle terms vanish, and upon defining

$$u(\mathbf{k})=(E_{\mathbf{k}+\mathbf{q}/2}-E_{\mathbf{k}-\mathbf{q}/2})C(\mathbf{k}), \tag{20}$$

one can see that $\chi_S(\mathbf{q},0)$ is given again by the HF expression of Eq. (4). In this case, however, $u(\mathbf{k})$ is determined by the following integral equation:

$$u(\mathbf{k})=1+\frac{1}{2}\sum_{\mathbf{p}}\left\{\frac{n_{\mathbf{p}-\mathbf{q}/2}-n_{\mathbf{p}+\mathbf{q}/2}}{E_{\mathbf{p}+\mathbf{q}/2}-E_{\mathbf{p}-\mathbf{q}/2}}[\Lambda_C(\mathbf{k}-\mathbf{p},\epsilon_{\mathbf{p}+\mathbf{q}/2}-\epsilon_{\mathbf{k}+\mathbf{q}/2})+\Lambda_C(\mathbf{p}-\mathbf{k},\epsilon_{\mathbf{k}-\mathbf{q}/2}-\epsilon_{\mathbf{p}-\mathbf{q}/2})]u(\mathbf{p})\right\}. \tag{21}$$

It is easily seen that if $\Lambda_C(\mathbf{q},\omega)$ is replaced by the screened potential $v(\mathbf{q})$, then one simply recovers the HF susceptibilities of the previous section.

In the limit of small $\mathbf{q}$ we have found that, interestingly enough, the integral equation (21) for $u(\mathbf{p})$ can be solved exactly. In this limit the many-body enhancement of the susceptibility can then be cast in the following elegant and suggestive form:

$$\frac{\chi_S}{\chi_P}=\frac{\dfrac{m^*}{m}}{1-\dfrac{m^*}{m}f_{2,3}(r_s)}, \tag{22}$$

where the functions $f_{2,3}(r_s)$ depend only on the density parameter r_s and the dimensionality of the system. For the 3D EG case, $f_3(r_s)$ is given by the following single quadrature,

$$f_3(r_s)=\int_0^1 dz\frac{1}{1+(18\pi^4)^{1/3}r_s^{-1}z+\dfrac{1-z}{2z^{1/2}}\ln\left|\dfrac{1+z^{1/2}}{1-z^{1/2}}\right|}\,. \tag{23}$$

For a 2D EG, instead, $f_2(r_s)$ can be written in a closed analytic form and is identical to that obtained from Eq. (8) for the case of Thomas-Fermi screening, i.e.,

$$f_2(r_s)=\frac{r_s}{8^{1/2}\pi}\left[\frac{2\Theta(r_s^2-2)}{(r_s^2/2-1)^{1/2}}\left[\frac{\pi}{2}-\tan^{-1}\frac{1}{(r_s^2/2-1)^{1/2}}\right]-\frac{\Theta(2-r_s^2)}{(1-r_s^2/2)^{1/2}}\ln\left|\frac{1-(1-r_s^2/2)^{1/2}}{1+(1-r_s^2/2)^{1/2}}\right|\right]. \tag{24}$$

A plot of both $f_3(r_s)$ and $f_2(r_s)$ is provided in Fig. 4. Both these curves are proportional to $-r_s\ln r_s$ in the limit of small r_s. It should be noted that the r_s dependence of $f_2(r_s)$ is more pronounced. The resulting values for χ_S/χ_P in three dimensions can be readily shown to reproduce the results of Ref. 3, i.e., the "consensus curve". A plot of the susceptibility ratio χ_S/χ_P for the case of a 2D EG is shown in Fig. 3 by the curve labeled RPA.[25] This curve was obtained using Eqs. (22) and (23) and the appropriate value of the effective mass ratio m^*/m as calculated from Eqs. (5) and (14).[20,26] Notice the large enhancement over the result of the Thomas-Fermi screened potential. It should be stressed here that for large r_s the many-body corrections are comparatively significantly larger in two dimensions.

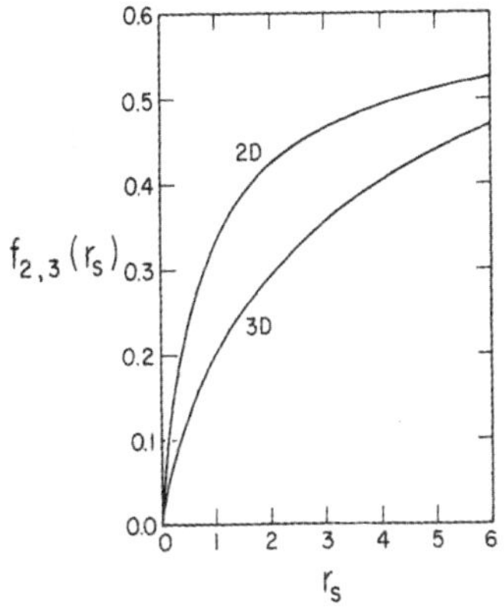

FIG. 4. Plot of $f_{2,3}$, Eqs. (20) and (21) in the text, vs the density parameter r_s. Notice that $f_2(r_s)$ is larger than its 3D counterpart and leads to a more pronounced many-body enhancement. Both functions behave as $-r_s\ln r_s$ for small r_s.

It is interesting to notice here that the perhaps surprising result of Eq. (24) (i.e., the fact that the RPA result basically displays the same structure as the Thomas-Fermi theory) is peculiar to the 2D situation and can be traced to the fact that in such a case the static Lindhard response function $\chi_0(\mathbf{q},0)$ is independent of the wave vector $\mathbf{q}$ for $q<2p_F$.

IV. HUBBARD VERTEX CORRECTIONS

In order to go beyond the dynamically screened exchange theory, we discuss here an approximate approach which allows one to account for the effects of charge- and spin-fluctuation-induced vertex corrections.

As originally suggested by Hubbard[19] and later exploited,[27] and generalized by several other authors,[28,29] it is possible to include some of the short-range effects of exchange and correlation by introducing suitable many-body local fields $G_+(\mathbf{q},\omega)$ and $G_-(\mathbf{q},\omega)$. For the sake of the present analysis it will suffice to define here these quantities through their relation to the full momentum- and frequency-dependent charge and spin susceptibilities of the system. A complete discussion of such a procedure can be found in Ref. 26. We have

$$\chi_C(\mathbf{q},\omega)=\frac{\chi_0(\mathbf{q},\omega)}{1-v(\mathbf{q})[1-G_+(\mathbf{q},\omega)]\chi_0(\mathbf{q},\omega)}, \qquad (25)$$

and

$$\chi_S(\mathbf{q},\omega)=-\mu_B^2\frac{\chi_0(\mathbf{q},\omega)}{1+v(\mathbf{q})G_-(\mathbf{q},\omega)\chi_0(\mathbf{q},\omega)}. \qquad (26)$$

Clearly the fields $G_+(\mathbf{q},\omega)$ and $G_-(\mathbf{q},\omega)$, respectively, account for charge- and spin-fluctuation-induced vertex corrections. In Eqs. (25) and (26) the quantity $\chi_0(\mathbf{q},\omega)$ differs from the usual Lindhard[30] function in that in its evaluation the expression for the momentum-space (bare) occupation number $n_\mathbf{p}$ appropriate to the interacting system must, in principle, be used.[26]

The many-body local fields not only enter the response functions, but can be shown also to modify in a significant fashion the effective potentials appearing in the theory.[31,26] For instance, in order to account for charge-fluctuation-induced vertex corrections, the expression for Λ_C of Eq. (15) must be modified as follows:

$$\Lambda_C(\mathbf{q},\omega)=v(\mathbf{q})\{1+v(\mathbf{q})[1-G_+(\mathbf{q},\omega)]^2\chi_C(\mathbf{q},\omega)\}. \qquad (27)$$

The physical processes associated with $G_-(\mathbf{q},\omega)$ necessitate here further discussion. The inclusion of G_- in the theory accounts for the effects of spin fluctuations and leads to extra terms in the quasiparticle energy. For instance, for an unpolarized state, Eq. (14) is in this case modified to read[26]

$$E_\mathbf{p}=\epsilon_\mathbf{p}-\sum_\mathbf{q}\left\{n_{\mathbf{p}-\mathbf{q}}\mathrm{Re}[\Lambda_C(\mathbf{q},\epsilon_\mathbf{p}-\epsilon_{\mathbf{p}-\mathbf{q}})+3\Lambda_S(\mathbf{q},\epsilon_\mathbf{p}-\epsilon_{\mathbf{p}-\mathbf{q}})]+\frac{1}{\pi}\mathrm{P}\int_0^\infty d\omega\frac{|\mathrm{Im}[\Lambda_C(\mathbf{q},\omega)]|+3|\mathrm{Im}[\Lambda_S(\mathbf{q},\omega)]|}{\omega-\epsilon_\mathbf{p}+\epsilon_{\mathbf{p}-\mathbf{q}}}\right\}, \qquad (28)$$

where Λ_C is defined in Eq. (27) and the new effective potential Λ_S is defined in terms of the full momentum- and frequency-dependent spin response $\chi_S(\mathbf{q},\omega)$ as follows:

$$\Lambda_S(\mathbf{q},\omega)=-\mu_B^{-2}[v(\mathbf{q})G_-(\mathbf{q},\omega)]^2\chi_S(\mathbf{q},\omega). \qquad (29)$$

The exact expressions for $G_+(\mathbf{q},\omega)$ and $G_-(\mathbf{q},\omega)$ are not known in general, so appeal must be made to approximate procedures. A possible way to tackle the problem is to investigate the exact asymptotic behaviors of these functions and then, as is customarily done, assume for them simple analytic formulas designed to interpolate between the known regimes. We find that suitable formulas for the 2D EG case are given by

$$G_\pm(\mathbf{q})=\frac{G_\pm(\infty)q}{\{q^2+[\beta_\pm G_\pm(\infty)p_F]^2\}^{1/2}}, \qquad (30)$$

where for the sake of simplicity we have neglected the frequency dependence of these functions. In Eq. (30) the quantities $G_\pm(\infty)$ and $\beta_\pm$ are density dependent and are related to the limiting values of the functions $G_+(\mathbf{q},\omega)$ and $G_-(\mathbf{q},\omega)$, respectively, for large and small wave vectors $\mathbf{q}$.

The exact large-wave-vector limits of the many-body local fields in a 2D EG have been analyzed in Ref. 32. There it was shown how the appropriate limiting values of $G_+(\mathbf{q},\omega)$ and $G_-(\mathbf{q},\omega)$ can be expressed in terms of $g(0)$, the value at the origin of the pair correlation function of the system. $g(0)$ is a function of the electronic density and its theoretical value can be approximately obtained via direct perturbative or numerical approaches.[33,34]

The coefficient β_+ can be simply obtained from the compressibility sum rule which relates the static charge susceptibility, Eq. (25), to E, the total ground-state energy of the electronic system, a quantity which has been the object of several detailed investigations and is therefore approximately known.[35–37] For a more detailed analysis the reader is referred to Ref. 26.

Finally, once β_+ is known β_- can be determined through a self-consistent procedure that will be discussed in detail in the next section.

V. SPIN SUSCEPTIBILITY

Making use of the results of the previous section, we can now proceed to the evaluation of the many-body enhancement of the spin susceptibility.

The first possible improvement upon the calculation contained in Sec. III is the inclusion of the effect of charge-fluctuation-induced vertex corrections described by the function G_+. The procedure employed to obtain the susceptibility in this case is identical to that of Sec. III in which these corrections were neglected. As is readily found, the susceptibility ratio in this case is still determined by the effective mass ratio m^*/m and the function $u(\mathbf{k})$ through Eqs. (4) and (21). In this situation, however, in Eq. (21) and (14) the modified expression for the effective potential $\Lambda_C(\mathbf{q},\omega)$ of Eq. (27) must be used. The results of such a calculation are displayed in Fig. 3 by the curve labeled G_+. It should be noticed that the inclusion of the many-body local field G_+ leads to a rather large enhancement of χ_S/χ_P as compared to the RPA calculation. An analysis of the effective mass for this case can be found in Ref. 26.

To carry the susceptibility analysis further, the effects of the processes associated with the spin-fluctuation-induced vertex corrections will be considered next. In this case the procedure employed above lands into difficulty in view of the fact that spin fluctuations in the electron gas will couple directly to any externally applied magnetic field, so that an alternative method of deriving the spin susceptibility must be used. A possible way to proceed is to make use of the Landau theory of the Fermi liquid.[38]

Within such a framework the static spin susceptibility χ_S can be obtained in terms of the quasiparticle interaction function as follows:

$$\frac{\chi_P}{\chi_S}=\frac{m}{m^*}+\frac{m}{\pi}\int_0^{2\pi}\frac{d\phi}{2\pi}f_a, \tag{31}$$

where the antisymmetrized interaction function f_a is given by

$$f_a\equiv\tfrac{1}{2}(f^{\uparrow\uparrow}-f^{\uparrow\downarrow}), \tag{32}$$

and $f^{\sigma\sigma'}$ can be obtained from the quasiparticle energy $E_{\mathbf{p}}^{\sigma}$ via a functional derivative with respect to the occupation number $n_{\mathbf{p}}^{\sigma}$ as follows:

$$f^{\sigma\sigma'}\equiv\frac{\delta E_{\mathbf{p}}^{\sigma}}{\delta n_{\mathbf{k}}^{\sigma'}}. \tag{33}$$

In order to perform the functional derivative of Eq. (33), the expression for the quasiparticle energy in a system with arbitrary polarization must be found. By following the procedure outlined in Refs. 3 and 26 a generalization of Eq. (14) to the polarized state can be obtained.[39] The enhancement of the spin susceptibility is then calculated by a straightforward application of Eqs. (31) and (33). The result can be cast in the following form:[39]

$$\frac{\chi_P}{\chi_S}=\frac{m}{m^*}-m\int_0^{2\pi}\frac{d\phi}{(2\pi)^2}[\Lambda_C(\mathbf{k}_F-\mathbf{p}_F,0)-\Lambda_S(\mathbf{k}_F-\mathbf{p}_F,0)]$$
$$+\frac{2}{\pi(a_0p_F)^2}\int_0^{\infty}dz\frac{1}{z^2}\int_0^{\infty}du[Q_-(\mathbf{q},i\omega)Q_+(\mathbf{q},i\omega)P_+(z,u)+Q_-(\mathbf{q},i\omega)^2P_-(z,u)], \tag{34}$$

where ϕ is the angle between a fixed vector $\mathbf{p}_F$ and a variable vector $\mathbf{k}_F$, both of which lie on the Fermi surface, and we have introduced the variables $z=q/2p_F$ and $u=\omega m/qp_F$. In Eq. (34) the functions $Q_\pm$ and $P_\pm$ are defined as follows:

$$Q_+(\mathbf{q},\omega)=\frac{1-G_+(\mathbf{q},\omega)}{1-v(\mathbf{q})\chi_0(\mathbf{q},\omega)[1-G_+(\mathbf{q},\omega)]}, \tag{35}$$

$$Q_-(\mathbf{q},\omega)=\frac{G_-(\mathbf{q},\omega)}{1+v(\mathbf{q})\chi_0(\mathbf{q},\omega)G_-(\mathbf{q},\omega)}, \tag{36}$$

and

$$P_\pm(z,u)=\frac{[(z^2-u^2-1)^2+(2zu)^2]^{1/2}\pm(z^2-u^2-1)}{[(z^2-u^2-1)^2+(2zu)^2]}. \tag{37}$$

It should be pointed out here that in formulating a complete theory for the spin susceptibility with the inclusion of the effects of spin fluctuations, particular care must be taken to also allow for *transverse* spin fluctuations. In deriving the expression of Eq. (34), this has been done by treating longitudinal and transverse spin fluctuations on the same footing, so that, for the sake of simplicity, only one many-body local field (i.e., G_-) is used here to describe the phenomenon.[39]

We have made use of Eqs. (30) and (34) – (37) to evaluate the spin susceptibility. The necessary effective-mass ratio has also been determined by using Eq. (28) for the quasiparticle energy. A crucial input of the present analysis is represented by the many-body local field $G_\pm(\mathbf{q},\omega)$ discussed in the previous section. As explained there, we have made use of the interpolation formulas of Eq. (30), which, in turn, depend on the choice of the density-dependent quantities $\beta_\pm$ and $g(0)$. We have chosen for $g(0)$ the theoretical value obtained in Ref. 34. As far as β_+ is concerned, as mentioned above, we have determined this parameter as a function of the electronic density from the total ground-state energy E via the compressibility sum rule. For E we have used the approximate interpolation formula proposed by Jonson,[35] which was obtained by implementing for the case of a 2D EG the classic numerical method of Singwi, Tosi, Land, and Sjolander.[40] Figure 5 displays the appropriate values of β_+ for a 2D EG as a function of the density parameter r_s.

Furthermore, and most importantly, we have determined the coefficient β_- via the following self-consistent procedure. Once the value of β_+ has been established, one starts with a trial value for β_- and proceeds to evaluate the corresponding m^* and χ_S/χ_P, respectively, from Eqs. (5) and (34). Then, by equating such a value to $\chi_S(\mathbf{q}\to 0,0)$, as given by Eq. (26), a new β_- is then determined from the relation

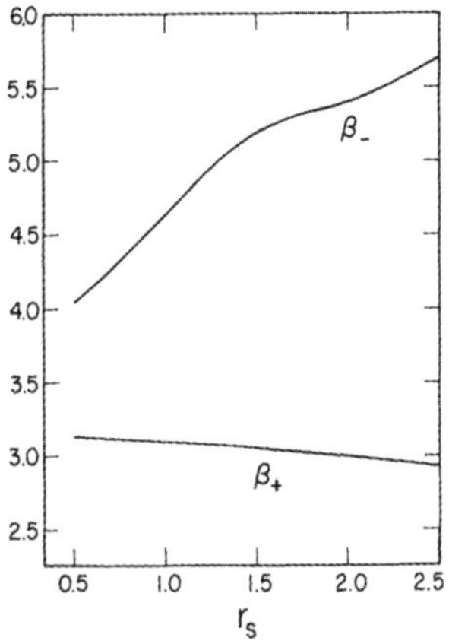

FIG. 5. Theoretical self-consistent results for the coefficients β_+ and β_-, defined in Eq. (30) in text, vs the density parameter

$$\frac{\chi_S(\mathbf{q}\to 0,0)}{\chi_P}=\frac{1}{1-\frac{2^{1/2}r_s}{\beta_-}}. \quad (38)$$

which is readily obtained from Eq. (26) if one makes use of the familiar Lindhard function for $\chi_0(\mathbf{q},\omega)$. This value for β_- becomes then the starting input for a new iteration. This procedure is repeated until convergence is reached. The appropriate self-consistent values for β_+ and β_- obtained in this way are plotted in Fig. 5 as a function of the density parameter r_s. It is important to realize that once β_+ and β_- are determined, our theory is free of arbitrary parameters. A plot of the susceptibility enhancement for this last case, representing our new full theory, is finally shown by the solid curve labeled G_+ & G_- in Fig. 3.

VI. DISCUSSION AND CONCLUSIONS

We have theoretically investigated the problem of the many-body enhancement of the paramagnetic spin susceptibility in a 2D EG. We have studied in detail the implications of the HF theory for the cases of local and screened interactions, in which case the problem has a simple analytic solution given by Eqs. (8) and (9) of Sec. I.

We have accounted for correlation effects beyond HF by a number of methods of increasing sophistication and physical significance. We have first evaluated the spin-response ratio χ_S/χ_P, by solving exactly in the long-wavelength limit an integral equation first introduced by Hamann and Overhauser,[3] and based on a generalization of the original Wolff HF theory formulation.[21] Our results, notably Eqs. (22)–(24), are extremely simple and allow a direct and straightforward calculation of χ_S. In fact, once the effective mass is known, in the 3D EG case we reduce the problem to a single quadrature and easily recover the established result of the "consensus curve".[3,4] We find that for the case of a 2D EG our result has a simple analytic form and formally coincides with that obtained within the HF approximation making use of the Thomas-Fermi screened potential, the only difference stemming from the different value attained by the quasiparticle effective mass. It is also interesting to notice in this respect that the result for χ_S/χ_P of Eqs. (22) and (24) has the same structure of the simple Thomas-Fermi formula of Eq. (8), so that it formally coincides with the result first obtained by Janak.[15]

We have analyzed the effect on the susceptibility of both charge- and spin-fluctuation-induced vertex corrections which we have accounted for by means of the many-body fields G_+ and G_-, which we have here approximated by suitable interpolation formulas in the spirit of Hubbard.

The results of our study are summarized in Fig. 3, where the curve labeled G_+ & G_- represents our new result for the many-body enhancement of the spin susceptibility in a 2D EG. An important conclusion which can be drawn from our study is that the final value of χ_S results from a subtle balance between various competing effects and that the simple RPA,[25,3] although providing a reasonable starting point, does not account for the full extent of the many-body physics inherent in the phenomenon at hand. It should also be stressed, however, that it is not enough to go beyond the RPA just by introducing, as is customary, the symmetric local field G_+ while altogether neglecting the effects of the spin fluctuations: in general, such a procedure tends to make things worse. We have arrived at the conclusion that the concomitant effects of both charge- and spin-fluctuations-induced vertex corrections must be accounted for in a satisfactory approach.

It must be stressed here that, in general, for large values of the density parameter r_s the many-body corrections are comparatively significantly larger in two dimensions than in three dimensions.

It must also be remarked that the results are sensibly dependent on the specific values used for $g(0)$ which enters the many-body fields, as well as the particular approximate interpolation formulas used for the latter. In particular, the choice of Eq. (30) as suitable expressions for $G_\pm$ was motivated only by natural requirements of simplicity and adherence to the spirit of Hubbard's original diagrammatic analysis of Ref. 19. In order to check the ultimate validity of the present theoretical approach, we have investigated the importance of our specific choice of Eq. (30) by making use of more complicated, yet still frequency-independent, reasonable forms of $G_\pm$. We have concluded that in spite of possible small changes in the actual numerical values, the results and conclusions reported above remain valid, although further rigorous studies on the importance of the frequency and wave-vector dependence of the many-body local fields in an electron gas are still needed.

Finally, although the present work is strictly concerned with the simple electron-gas model, we expect that a similar qualitative behavior will characterize the spin susceptibility of electrons and holes in layered electronic systems and superlattices. In particular, the present approach can be generalized to the more realistic case of electrons in quasi-two-dimensional semiconducting heterostructures and more specifically to inversion layers. As it turns out, the inclusion of the specific physical features and parameters related to the structure, such as the finite-thickness effects, the image potentials, the various background dielectric constants, the valley degeneracy, and band mass, is of crucial importance in such cases. Work on this particularly interesting problem is reported elsewhere.[20]

ACKNOWLEDGMENTS

We wish to thank A. W. Overhauser, and G. E. Santoro, and G. Vignale for useful discussions.

[1]For an illuminating review of the work prior to 1966, see C. Herring, *Magnetism*, edited by G. T. Rado and H. Suhl (Academic, New York, 1966), Vol. IV.
[2]A. W. Overhauser, Phys. Rev. **128**, 1437 (1962).
[3]D. R. Hamann and A. W. Overhauser, Phys. Rev. **143**, 183 (1966).
[4]A discussion of the experimental situation for the 3D EG can be found in Refs. 5–7. For a more recent review, see T. Kushida, J. C. Murphy, and M. Hanabusa, Phys. Rev. B **13**, 5136 (1976).
[5]R. Dupree and D. J. W. Geldhart, Solid State Commun. **9**, 145 (1971).
[6]G. Pizzimenti, M. P. Tosi, and A. Villari, Nuovo Cimento **2**, 81 (1971).
[7]T. M. Rice, Phys. Rev. **175**, 858 (1968).
[8]J. Lam, Phys. Rev. B **5**, 1254 (1972).
[9]P. Vashishta, and K. S. Singwi, Solid State Commun. **13**, 901 (1973).
[10]For an introduction to the work up to 1981, see T. Ando, A. B. Fowler, F. Stern, Rev. Mod. Phys. **54**, 437 (1982). For a more recent account of the field, see, for instance, *Proceedings of the Seventh International Conference on Electronic Properties of Two-Dimensional Systems*, Santa Fe, 1988, edited by J.M. Worlock [Surf. Sci. **196**, 1 (1988)].
[11]J. L. Smith and P. J. Stiles, Phys. Rev. Lett. **29**, 102 (1972).
[12]G. Abstreiter, J. P. Kotthaus, J. F. Koch, and G. Dorda, Phys. Rev. B **14**, 2480 (1976).
[13]F. F. Fang and P. J. Stiles, Phys. Rev **174**, 823 (1968).
[14]T. Neugebauer, K. von Klitzing, G. Landwehr, and G. Dorda, Solid State Commun. **17**, 295 (1975).
[15]J. F. Janak, Phys. Rev. **178**, 1416 (1969).
[16]K. Suzuki and Y. Kawamoto, J. Phys. Soc. Jpn. **35**, 1456 (1973).
[17]T. Ando and Y. Uemura, J. Phys. Soc. Jpn. **37**, 1044 (1974); these authors have used a formula for g^* in which the renormalization of the quasiparticle mass is neglected.
[18]C. S. Ting, T. K. Lee, and J. J. Quinn, Phys. Rev. Lett. **34**, 870 (1975).
[19]J. Hubbard, Proc. R. Soc. London, Ser. A **242**, 539 (1957); **243**, 336 (1957).
[20]S. Yarlagadda and G. F. Giuliani, Phys. Rev. B **38**, 10 966 (1988), and unpublished.
[21]P. A. Wolff, Phys. Rev. **120**, 814 (1960).
[22]G. F. Giuliani and J. J. Quinn, Phys. Rev. B **31**, 6228 (1985); Surf. Sci. **170**, 316 (1986).
[23]F. Stern, Phys. Rev. Lett. **30**, 278 (1973).
[24]A systematic study of the quasiparticle self-energy in a 2D EG can be found in G. E. Santoro and G. F. Giuliani, Solid State Commun. **67**, 681 (1988), and unpublished.
[25]It should be stressed here that, strictly speaking, the use made here of the notation RPA simply signifies that all the response functions appearing in this formulation have been calculated within such an approximation. In the 3D case our notation would refer to the calculation of Ref. 3 as a RPA calculation. More details can be found in Ref. 29.
[26]S. Yarlagadda and G. F. Giuliani, Solid State Commun. **69**, 677 (1989), and unpublished.
[27]T. M. Rice, Ann. Phys. (N.Y.) **31**, 100 (1965).
[28]G. Niklasson, Phys. Rev. B **10**, 3052 (1974).
[29]For a recent review of the field, see K. S. Singwi and M. P. Tosi, in *Solid State Physics*, edited by H. Ehrenreich, F. Seitz, and D. Turnbull (Academic , New York, 1981), Vol. 36, pp. 177–266.
[30]J. Lindhard, K. Dan. Vidensk. Selsk. Mat.-Fys. Medd. **28**, No. 8 (1954).
[31]C. A. Kukkonen and A. W. Overhauser, Phys. Rev. B **20**, 550 (1979).
[32]G. E. Santoro and G. F. Giuliani, Phys. Rev. B **37**, 4813 (1988).
[33]D. L. Freeman, J. Phys. C **16**, 711 (1983).
[34]S. Nagano, K. S. Singwi, and S. Ohnishi, Phys. Rev. B **29**, 1209 (1984).
[35]M. Jonson, J. Phys. C **9**, 3055 (1976).
[36]G. D. Mahan, Phys. Scr. **32**, 423 (1985).
[37]D. Ceperley, Phys. Rev. B **18**, 3126 (1978).
[38]L. D. Landau, Zh. EKsp. Teor. Fiz. **30**, 1058 (1956) [Sov. Phys.—JETP **3**, 920 (1956)].
[39]Details of this calculation will be given elsewhere.
[40]K. S. Singwi, M. P. Tosi, R. H. Land, and A. Sjolander, Phys. Rev. B **176**, 589 (1968).

Quasiparticle pseudo-Hamiltonian of an infinitesimally polarized Fermi liquid

Sudhakar Yarlagadda* and Gabriele F. Giuliani
Department of Physics, Purdue University, West Lafayette, Indiana 47907
(Received 24 May 1993)

We present the microscopic derivation of a quasiparticle pseudo-Hamiltonian for an infinitesimally polarized electron liquid. The Hamiltonian is expressed in terms of suitably defined quasiparticle operators. Our approach is based on a canonical transformation which allows one to replace the bare Coulombic coupling between the interacting electrons with an effective interaction between quasiparticles in which collective charge and spin fluctuations are explicitly accounted for. The relevant matrix elements of the charge and spin-density operators enter our theory via linear-response functions: the charge response, the longitudinal and transverse spin responses, and the mixed charge-spin response. These susceptibilities are in turn expressed in terms of the appropriate many-body local fields. As a consequence our method can be seen as an attempt to satisfactorily include in a self-consistent manner the effects of the vertex corrections associated with charge and spin-fluctuations of the electron liquid. As a result useful expressions for the quasiparticle energy and the effective interaction between two quasiparticles are determined. These can, in turn, be employed in a microscopic determination of the parameters of the Landau theory of the Fermi liquid. The generalization of our results to a multicomponent system is also discussed.

I. INTRODUCTION

Understanding the many-body aspects of an electron gas (EG) has been a subject of steady interest for the past few decades.[1-3] The EG, unlike a system of classical particles, behaves like a gas at high densities and like a solid at low densities.[4] In both these extreme limits, the ground-state energy of the EG has been evaluated exactly. In the high density limit, the ground-state energy was obtained as a series in terms of r_s, the average electronic separation in units of the Bohr radius, in three dimensions (3D) by Gell-Mann and Brueckner[5] and in two dimensions (2D) by Rajagopal and Kimball[6] and by Isihara and Toyoda.[7] Wigner showed that at sufficiently small densities the electrons localize to form a crystal lattice and hypothesized that in 3D a bcc structure is the most stable one.[4] Later on it was verified that, among the simple lattices, the bcc structure has the lowest energy and the ground-state energy for the bcc lattice was obtained as a power series in terms of $r_s^{-1/2}$.[8] In 2D, Bonsall and Maradudin[9] calculated the ground state energy for arbitrary electron lattices, and showed that the triangular one, as expected, has the lowest energy.

In the intermediate density regime, which is relevant in three dimensions for simple metals, and in two dimensions for systems like the electrons in an inversion layer in most density regimes, the usual perturbative techniques are not effective owing to the lack of an expansion parameter. Hence, one has to take recourse to approximate methods which are not completely rigorous but are physically justifiable. A review of a number of these techniques can be found in Ref. 1. Quite useful in this respect are the numerical techniques based on the quantum Monte Carlo methods.[10-12]

Among the various approximate methods designed to deal with the intermediate density regime of particular interest for its physical appeal and elegance is Landau's[13] original phenomenological theory of the Fermi liquids which treats accurately low-lying excitations. Landau called these excitations quasiparticles and postulated a one-to-one correspondence between them and the excited states of a noninteracting Fermi liquid. He wrote down the excitation energy of the system in terms of the energy of the quasiparticles and their effective interaction. The quasiparticle interaction function can be used in turn to obtain various physical properties of the system and can also be parametrized in terms of experimentally obtainable data. Within the framework of perturbative Green's function techniques it was shown by Luttinger and Nozières[14] that the Landau theory is valid in the limit of zero temperature, long wavelength, and zero frequency.

For an EG long-range screening of the Coulombic interaction is an important factor. The simplest approximation that takes this into account is the random phase approximation (RPA). In this approximation the screened charge response of the EG is assumed to be that of the noninteracting system.[15] Using a many-body local field, commonly named after him, Hubbard[16] improved upon this approximation of the screened charge susceptibility by including, in an approximate fashion, some exchange corrections.

More recently, Hubbard approach was generalized in such a way as to include the effects of vertex corrections due to both charge and spin fluctuations in an unpolarized EG.[17-21] In these papers, using formally different approaches, expressions for the quasiparticle self-energy and effective interaction were obtained. As it was pointed out in Ref. 20, these results are basically equivalent.

This body of work showed that the quasiparticle self-energy and effective interaction can be expressed in terms of suitable generalized Hubbard many-body local fields and the charge and spin susceptibilities of the system, which, in turn, can be also expressed in terms of the same quantities. As for the case of an infinitesimally polarized EG, work on the subject was carried out by Ng and Singwi by means of a diagrammatic approach.[21] This approach to the many-body theory of the EG has already found it applications.[22,23]

In this paper, we extend the results obtained previously for the unpolarized system by the present authors[20] and derive a quasiparticle pseudo-Hamiltonian for an infinitesimally polarized liquid in terms of the response functions and many-body local fields of the system. One of the motivating factors for the present work is to arrive at useful expressions that can then be employed within a Landau theory of the Fermi liquid to evaluate various physical quantities of interest.[23]

The basic idea leading to the definition of a pseudo-quasiparticle-Hamiltonian was previously developed by Hamann and Overhauser for the case of an unpolarized system.[24] These authors limited their analysis to the simple case of the RPA. Our treatment, on the other hand, is much more general, and is designed to incorporate the effects of vertex corrections associated with charge and spin-density fluctuations to account for exchange and correlation effects in the infinitesimally polarized electron liquid.

We begin by viewing the electron liquid as a system comprised of a few interacting "test" electrons and a screening dielectric medium characterized only by its collective charge and spin density excitations. The test electrons and the medium interact via effective potentials which we express in terms of (a priori unknown) appropriate local field factors $\tilde{G}$ so as to account for their deviations (due to exchange and correlation effects beyond RPA) from the bare Coulomb potential. By using a canonical transformation this interaction terms are then eliminated to first order, thereby generating an effective coupling between the test electrons. Upon averaging over the coordinates of the screening medium we then obtain the sought renormalization of the test electrons states. An important step in this procedure is represented by the identification of the various a priori unknown matrix elements in terms of appropriate response functions of the medium: the charge, the longitudinal and the transverse spin, and the mixed charge-spin response susceptibilities. These response functions are in turn expressed via the corresponding generalized Hubbard many-body local fields. Finally we show that in order to achieve a physically self-consistent description of the situation the factors $\tilde{G}$ do in fact coincide with the Hubbard many-body local fields appearing in the response functions of the medium.

With this purpose in mind it is important to be able to generalize our treatment of the many-body effects in the electron liquid to the case of a multicomponent system. This is in fact necessary for the case of the electronic system occurring in a silicon inversion layer. There the multicomponent nature of the electronic band structure leads to further interesting and important modifications.

Our paper is structured as follows. In Sec. II, we introduce the total Hamiltonian which we use to model the EG. In Sec. III the bulk of the renormalization procedure is presented and a quasiparticle pseudo-Hamiltonian is arrived at. Next, in Sec. IV we express our results in terms of the quasiparticle energy and the quasiparticle effective interaction. Section V contains a discussion of our results and their implications. The connection of our theory to previous work is also provided there. The paper contains two Appendices. In Appendix A we derive useful expressions for the various response functions that enter our quasiparticle pseudo-Hamiltonian for the case of a multi-component system. Finally, in Appendix B an exact expression for the mixed charge-spin response function is obtained.

II. TOTAL HAMILTONIAN

To describe the excitations of an electron liquid we employ the picture based on the concept of quasiparticle and similar to Landau's phenomenological theory of the normal Fermi liquid. We start by selecting a few electrons from the EG and call them test electrons. The remaining EG is treated as a screening dielectric medium. As the test electrons move through the dielectric medium they produce fluctuations in the density of spin up and spin down electrons. These fluctuations provide virtual clothing to the test electrons and also screen the interaction between them. Thus, the dielectric mimics the true processes in an average way. It is important to realize that in reality the test electrons and the electrons comprising the dielectric are physically the same. This must be taken into account when exchange effects are considered.

The goal is to derive a Hamiltonian containing only the degrees of freedom of the clothed test electrons or quasiparticles. To this end we proceed as follows. We write the total Hamiltonian of the system as

$$H=H_0^{(p)}+H_0^{(m)}+H_1\,, \tag{1}$$

where $H_0^{(p)}$ is the Hamiltonian of the test electrons and is given by

$$H_0^{(p)}=\sum_{\mathbf{p},\sigma}\epsilon_{\mathbf{p}}^{\sigma}a_{\mathbf{p},\sigma}^{+}a_{\mathbf{p},\sigma}+\frac{1}{2}\sum_{\substack{\mathbf{p},\mathbf{p}',\\ \mathbf{q},\sigma,\sigma'}}v(q)a_{\mathbf{p}-\mathbf{q},\sigma}^{+}a_{\mathbf{p}'+\mathbf{q},\sigma'}^{+}a_{\mathbf{p}',\sigma'}a_{\mathbf{p},\sigma}\,, \tag{2}$$

where $q\neq 0$, $a_{\mathbf{p},\sigma}^{+}$ ($a_{\mathbf{p},\sigma}$) creates (destroys) a quasiparticle with momentum $\mathbf{p}$ and spin index σ, ($\sigma=\pm1$), $v(q)$ is the Fourier transform of the bare Coulomb potential, with $\epsilon_{\mathbf{p}}^{\sigma}$ being the bare (band) energy of the test electron. In Eq. (1) $H_0^{(m)}$ is the Hamiltonian of the dielectric medium and is described by specifying its eigenstates $|n\rangle$ and its eigenvalues ω_n. Furthermore, H_1 is the part of the total Hamiltonian that takes into account the test electron-medium coupling and is given as follows:

$$H_1=\sum_{\mathbf{p},\mathbf{q},\alpha}v(q)[1-\tilde{G}_{+}(\mathbf{q},\epsilon_{\mathbf{p}+\mathbf{q}}^{\alpha}-\epsilon_{\mathbf{p}}^{\alpha})]\rho_{-\mathbf{q}}a_{\mathbf{p}+\mathbf{q},\alpha}^{+}a_{\mathbf{p},\alpha}-\sum_{\substack{\mathbf{p},\mathbf{q},\\ \alpha,\beta,\mu}}v(q)\tilde{G}_{-}^{\mu}(\mathbf{q},\epsilon_{\mathbf{p}+\mathbf{q}}^{\alpha}-\epsilon_{\mathbf{p}}^{\beta})\sigma_{\alpha\beta}^{\mu}S_{-\mathbf{q}}^{\mu}a_{\mathbf{p}+\mathbf{q},\alpha}^{+}a_{\mathbf{p},\beta}\,, \tag{3}$$

where $q\neq 0$, $\alpha,\beta=\pm 1$; $\mu=x,y,z$; and $\sigma^{\mu}_{\alpha\beta}$ is a Pauli matrix. Furthermore, $\rho_{\mathbf{q}}$ and $S^{\mu}_{\mathbf{q}}$ are the operators associated with charge- and spin-density fluctuations, respectively. Since at this stage the nature and strength of the potentials appearing in H_1 is unknown we have introduced the quantities $\tilde{G}_+$ and $\tilde{G}^{\mu}_-$ in order to account for exchange and correlation effects. The $\tilde{G}$'s are taken to be functions of the change in momentum and the change in energy of the test electron. In the above equation it has been assumed that $\tilde{G}_+$ and $\tilde{G}^{\mu}_-$ have reflection symmetry with respect to the plane perpendicular to the axis of polarization, namely the z axis. In general, for a polarized EG in its ground state one expects that $\tilde{G}^x_-=\tilde{G}^y_-=\tilde{G}^T_-\neq\tilde{G}^z_-=\tilde{G}_-$. Later on we will identify these $\tilde{G}_\pm$ and $\tilde{G}^T_-$ factors in terms of the true many-body local fields of the EG as a whole (see Sec. III). It is of interest to note that in real space the effective potential felt by a test electron of momentum p and spin σ as obtained from Eq. (3), can be expressed as (see Appendix A and Ref. 18 for a similar result)

$$\begin{aligned}\phi^{\mathbf{p}}_{\sigma}(q,\omega)=v(q)\{&[\Delta n_\uparrow(q,\omega)+\Delta n_\downarrow(q,\omega)]\\&\times[1-\tilde{G}_+(\mathbf{q},\epsilon^{\sigma}_{\mathbf{p+q}}-\epsilon^{\sigma}_{\mathbf{p}})]\\&-\sigma[\Delta n_\uparrow(q,\omega)-\Delta n_\downarrow(q,\omega)]\\&\times\tilde{G}^z_-(\mathbf{q},\epsilon^{\sigma}_{\mathbf{p+q}}-\epsilon^{\sigma}_{\mathbf{p}})\}\ ,\end{aligned}\qquad(4)$$

where z is the quantization axis for the spin σ and Δn_σ represents the density fluctuations of electrons with spin projection σ.

III. RENORMALIZATION PROCEDURE

To obtain the quasiparticle Hamiltonian of an infinitesimally polarized system we adopt the following renormalization procedure. We first perform a canonical transformation on the total Hamiltonian so as to eliminate the term H_1 up to first order. This, in turn, produces an effective coupling between the test electrons through their interaction with the charge and spin-density fluctuations. This is similar to the procedure employed in deriving the Frohlich phonon mediated electron-electron effective interaction.[25] The transformed Hamiltonian is given by $H'=e^{-T}He^{T}$, where the operator T is determined from the requirement

$$H_1+\left[\sum_{\mathbf{p}\sigma}\epsilon^{\sigma}_{\mathbf{p}}a^{+}_{\mathbf{p},\sigma}a_{\mathbf{p},\sigma}+H^{(m)}_0,T\right]=0\ .\qquad(5)$$

The above form involving only the Hamiltonians of the noninteracting test electrons and the dielectric medium in the commutator is chosen for the definition of T for the following reasons. First, this form enables us to determine the matrix elements of T with respect to the eigenstates of the dielectric medium. Second, as will be shown below, this definition yields the renormalization term as a combination of identifiable dynamic response functions of the dielectric medium. Last, this form correctly yields the RPA result for the pseudo-Hamiltonian when the exchange and correlation vertex corrections are neglected.

A. Averaging out the medium

The second step involves explicitly removing the degrees of freedom of the dielectric medium. This is done by averaging the transformed Hamiltonian H' over the *uniformly infinitesimally polarized state* $|0\rangle$ of the medium. We thus obtain the following quasiparticle pseudo-Hamiltonian H'_{QP}:

$$H'_{QP}=\langle 0|H'|0\rangle=H^{(p)}_0+\tfrac{1}{2}\langle 0|[H_1,T]|0\rangle\ ,\qquad(6)$$

which now contains only the test electron operators. In the pseudo-Hamiltonian given above, constants and higher order terms in $v(q)$ have been omitted. Later on, in Sec. IV we present arguments to show that the neglect of higher-order terms is consistent with the requirement that the correct pseudo-Hamiltonian must contain all the correlation effects. The expectation value with respect to the polarized state $|0\rangle$ on the right-hand side of Eq. (6) is precisely the term that leads to a renormalization of the bare interaction potential and also to the clothing of a bare electron. This term can be evaluated from the matrix elements $\langle 0|H_1|n\rangle$ and $\langle 0|T|n\rangle$. Then, from Eqs. (3) and (5) we obtain

$$\begin{aligned}\langle n|T|0\rangle=\sum_{\mathbf{p},\mathbf{q},\sigma}\{&\{[1-\tilde{G}^*_+[\mathbf{q},\Delta^{\sigma}_{\mathbf{p}}(\mathbf{q})]\}\langle n|\rho_{\mathbf{q}}|0\rangle\\&-\sigma\tilde{G}^*_-[\mathbf{q},\Delta^{\sigma}_{\mathbf{p}}(\mathbf{q})]\langle n|S^z_{\mathbf{q}}|0\rangle\}\frac{v(q)a^{+}_{\mathbf{p-q},\sigma}a_{\mathbf{p},\sigma}}{\Delta^{\sigma}_{\mathbf{p}}(\mathbf{q})-\omega_{n0}}\\&-\tilde{G}^{T*}_-[\mathbf{q},\Delta^{T\sigma}_{\mathbf{p}}(\mathbf{q})]\langle n|S^{\sigma}_{\mathbf{q}}|0\rangle\frac{v(q)a^{+}_{\mathbf{p-q},-\sigma}a_{\mathbf{p},\sigma}}{\Delta^{T\sigma}_{\mathbf{p}}(\mathbf{q})-\omega_{n0}}\ ,\end{aligned}\qquad(7)$$

where for the sake of brevity we have defined

$$\omega_{n0}=\omega_n-\omega_0\ ,\qquad(8)$$

$$\Delta^{\sigma}_{\mathbf{p}}(\mathbf{q})\equiv\epsilon^{\sigma}_{\mathbf{p}}-\epsilon^{\sigma}_{\mathbf{p-q}}\ ,\qquad(9)$$

and

$$\Delta^{T\sigma}_{\mathbf{p}}(\mathbf{q})\equiv\epsilon^{\sigma}_{\mathbf{p}}-\epsilon^{-\sigma}_{\mathbf{p-q}}\ .\qquad(10)$$

In obtaining Eq. (7), we used the relationship $\tilde{G}^{(T)}_{\pm}(\mathbf{q},\omega)=\tilde{G}^{(T)*}_{\pm}(-\mathbf{q},-\omega)$ for which the justification will become clear in Sec. III when the $\tilde{G}$'s are shown to coincide with the many-body local fields. We now define $S^{\pm}=S^x\pm iS^y$ and use the fact that the quantities $\langle 0|S^+|n\rangle\langle n|S^+|0\rangle$, $\langle 0|S^-|n\rangle\langle n|S^-|0\rangle$, $\langle 0|S^z|n\rangle\langle n|S^{\pm}|0\rangle$, and $\langle 0|\rho|n\rangle\langle n|S^{\pm}|0\rangle$ vanish. Furthermore, for $q\neq 0$, we also utilize the fact that $\langle 0|\rho_{\mathbf{q}}|0\rangle$ and $\langle 0|S^{\mu}_{\mathbf{q}}|0\rangle$ vanish whereas $\langle 0|\rho_{\mathbf{q}}|n\rangle\langle n|S^z_{-\mathbf{q}}|0\rangle$ has in general a nonzero value for a polarized system. Then from Eqs. (3) and (7) we obtain

$$\langle 0|H_1T|0\rangle = \sum_{\substack{\mathbf{p},\mathbf{p}',\mathbf{q},\\ \sigma,\sigma',n}} [\{1-\tilde{G}_+[\mathbf{q},-\Delta_{\mathbf{p}'}^{\sigma'}(-\mathbf{q})]\}\{1-\tilde{G}_+^*[\mathbf{q},\Delta_{\mathbf{p}}^{\sigma}(\mathbf{q})]\}|\langle n|\rho_{\mathbf{q}}|0\rangle|^2$$
$$+\sigma\sigma'\tilde{G}_-[\mathbf{q},-\Delta_{\mathbf{p}'}^{\sigma'}(-\mathbf{q})]\tilde{G}_-^*[\mathbf{q},\Delta_{\mathbf{p}}^{\sigma}(\mathbf{q})]|\langle n|S_{\mathbf{q}}^z|0\rangle|^2$$
$$-(\sigma\{1-\tilde{G}_+[\mathbf{q},-\Delta_{\mathbf{p}'}^{\sigma'}(-\mathbf{q})]\}\tilde{G}_-^*[\mathbf{q},\Delta_{\mathbf{p}}^{\sigma}(\mathbf{q})]\langle 0|\rho_{-\mathbf{q}}|n\rangle\langle n|S_{\mathbf{p}}^z|0\rangle$$
$$+\sigma'\{1-\tilde{G}_+^*[\mathbf{q},\Delta_{\mathbf{p}}^{\sigma}(\mathbf{q})]\}\tilde{G}_-[\mathbf{q},-\Delta_{\mathbf{p}'}^{\sigma'}(-\mathbf{q})]\langle 0|S_{-\mathbf{q}}^z|n\rangle\langle n|\rho_{\mathbf{q}}|0\rangle)]$$
$$\times\frac{v(q)^2 a_{\mathbf{p}'+\mathbf{q},\sigma'}^{+}a_{\mathbf{p}',\sigma'}a_{\mathbf{p}-\mathbf{q},\sigma}^{+}a_{\mathbf{p},\sigma}}{\Delta_{\mathbf{p}}^{\sigma}(\mathbf{q})-\omega_{n0}}+\tfrac{1}{2}(1-\sigma\sigma')\tilde{G}_-^T[\mathbf{q},-\Delta_{\mathbf{p}'}^{T\sigma'}(-\mathbf{q})]$$
$$\times\tilde{G}_-^{T*}[\mathbf{q},\Delta_{\mathbf{p}}^{T\sigma}(\mathbf{q})]|\langle n|S_{\mathbf{q}}^{\sigma}|0\rangle|^2$$
$$\times\frac{v(q)^2 a_{\mathbf{p}'+\mathbf{q},-\sigma'}^{+}a_{\mathbf{p}',\sigma'}a_{\mathbf{p}-\mathbf{q},-\sigma}^{+}a_{\mathbf{p},\sigma}}{\Delta_{\mathbf{p}}^{T\sigma}(\mathbf{q})-\omega_{n0}}\,. \tag{11}$$

Now, on recognizing that the operator T is anti-Hermitian we obtain from Eq. (11) the renormalization term $\frac{1}{2}\langle 0|[H_1,T]|0\rangle$ of the quasiparticle pseudo-Hamiltonian H'_{QP}. Then, in the renormalization term, upon identifying the various matrix elements of the charge and spin fluctuations in terms of the various dynamic response functions, to be defined below, we obtain the following compact expression:

$$\langle 0|[H_1,T]|0\rangle = 2\sum_{\mathbf{p},\sigma}\tilde{E}_{\mathrm{CH}}^{\sigma}(\mathbf{p})a_{\mathbf{p},\sigma}^{+}a_{\mathbf{p},\sigma}+\sum_{\substack{\mathbf{p},\mathbf{p}',\\ \mathbf{q},\sigma,\sigma'}}\{\tilde{V}_{\sigma,\sigma'}[\mathbf{q},-\Delta_{\mathbf{p}'}^{\sigma'}(-\mathbf{q}),\Delta_{\mathbf{p}}^{\sigma}(\mathbf{q}),\Delta_{\mathbf{p}}^{\sigma}(\mathbf{q})]a_{\mathbf{p}-\mathbf{q},\sigma}^{+}a_{\mathbf{p}'+\mathbf{q},\sigma'}^{+}a_{\mathbf{p}',\sigma'}a_{\mathbf{p},\sigma}$$
$$+\tilde{V}_{\sigma,\sigma'}^{T}[\mathbf{q},-\Delta_{\mathbf{p}'}^{T\sigma'}(-\mathbf{q}),\Delta_{\mathbf{p}}^{T\sigma}(\mathbf{q}),\Delta_{\mathbf{p}}^{T\sigma}(\mathbf{q})]a_{\mathbf{p}-\mathbf{q},-\sigma}^{+}a_{\mathbf{p}+\mathbf{q},-\sigma'}^{+}a_{\mathbf{p}',\sigma'}a_{\mathbf{p},\sigma}\}\,, \tag{12}$$

where the terms $\tilde{V}_{\sigma,\sigma'}$ and $\tilde{V}_{\sigma,\sigma'}^{T}$ are the longitudinal and transverse components of the renormalization part of the effective interaction between two quasiparticles and are given by

$$\tilde{V}_{\sigma,\sigma'}(\mathbf{q},\epsilon,\omega,\delta)\equiv v(q)^2([1-\tilde{G}_+(\mathbf{q},\epsilon)][1-\tilde{G}_+^*(\mathbf{q},\omega)]\mathrm{Re}\chi_C(\mathbf{q},\delta)$$
$$+\sigma\sigma'\tilde{G}_-(\mathbf{q},\epsilon)\tilde{G}_-^*(\mathbf{q},\omega)\mathrm{Re}\chi_S(\mathbf{q},\delta)/(-\mu_B^2)-\{\sigma[1-\tilde{G}_+(\mathbf{q},\epsilon)]\tilde{G}_-^*(\mathbf{q},\omega)$$
$$+\sigma'[1-\tilde{G}_+^*(\mathbf{q},\omega)]\tilde{G}_-(\mathbf{q},\epsilon)\}\mathrm{Re}\chi_{CS}(\mathbf{q},\delta))\,, \tag{13}$$

and

$$\tilde{V}_{\sigma,\sigma'}^{T}(\mathbf{q},\epsilon,\omega,\delta)\equiv 2(1-\sigma\sigma')v(1)^2\tilde{G}_-^T(\mathbf{q},\epsilon)\tilde{G}_-^{T*}(\mathbf{q},\omega)\mathrm{Re}\chi^{T\sigma}(\mathbf{q},\delta)/(-\mu_B^2)\,, \tag{14}$$

with μ_B being the Bohr magneton. In Eq. (12) E_{CH}^{σ} is the Coulomb hole part[26] of the renormalization term and is obtained upon rearranging the creation and destruction operators in the usual order of a Hamiltonian expressed in the second quantized form

$$\tilde{E}_{\mathrm{CH}}^{\sigma}(\mathbf{p})=-\sum_{\mathbf{q}}v(q)^2P\int_0^{\infty}\frac{d\omega}{\pi}\mathrm{Im}\left[\frac{|1-\tilde{G}_+|^2\chi_C+|\tilde{G}_-|^2\chi_S/(-\mu_B^2)}{\Delta_{\mathbf{p}}^{\sigma}(\mathbf{q})-\omega}-\frac{2\sigma\,\mathrm{Re}[(1-\tilde{G}_+^*)\tilde{G}_-]\chi_{\mathrm{CS}}}{\Delta_{\mathbf{p}}^{\sigma}(\mathbf{q})-\omega}+\frac{4|\tilde{G}_-^T|^2\chi^{T\sigma}/(-\mu_B^2)}{\Delta_{\mathbf{p}}^{T\sigma}(\mathbf{q})-\omega}\right]. \tag{15}$$

In the above equation it is understood that the factors $\tilde{G}_{\pm}$ are functions of $\mathbf{q}$ and $\Delta_{\mathbf{p}}^{\sigma}(\mathbf{q})$, while $\tilde{G}_-^T$ is a function of $\mathbf{q}$ and $\Delta_p^{T\sigma}(\mathbf{q})$. Moreover, the response functions depend on q and ω.

The various response functions appearing in the effective interaction terms and the Coulomb hole term [see Eqs. (13)–(15)] are χ_C the charge response, χ_S the spin response, $\chi^{T\sigma}$ the transverse spin response, and χ_{CS} the mixed charge-spin response.[27] Exact expressions for these response functions are as follows:

$$\chi_C(\mathbf{q},\nu)=\sum_n\left\{\frac{|\langle n|\rho_{\mathbf{q}}|0\rangle|^2}{\nu-\omega_{n0}+i\eta}-\frac{|\langle n|\rho_{-\mathbf{q}}|0\rangle|^2}{\nu+\omega_{n0}+i\eta}\right\}, \tag{16}$$

$$\chi_S(\mathbf{q},\nu)=-\mu_B^2\sum_n\left\{\frac{|\langle n|S_{\mathbf{q}}^z|0\rangle|^2}{\nu-\omega_{n0}+i\eta}-\frac{|\langle n|S_{-\mathbf{q}}^z|0\rangle|^2}{\nu+\omega_{n0}+i\eta}\right\}, \tag{17}$$

$$\chi^{T\sigma}(\mathbf{q},\nu)=\frac{-\mu_B^2}{4}\sum_n\left\{\frac{|\langle n|S_{\mathbf{q}}^{\sigma}|0\rangle|^2}{\nu-\omega_{n0}+i\eta}-\frac{|\langle n|S_{-\mathbf{q}}^{-\sigma}|0\rangle|^2}{n+\omega_{n0}+i\eta}\right\}, \tag{18}$$

and

$$\begin{aligned}\chi_{\mathrm{CS}}(\mathbf{q},\nu)&=\sum_n\left\{\frac{\langle 0|\rho_{-\mathbf{q}}|n\rangle\langle n|S_{\mathbf{q}}^z|0\rangle}{\nu-\omega_{n0}+i\eta}-\frac{\langle 0|S_{\mathbf{q}}^z|n\rangle\langle n|\rho_{-\mathbf{q}}|0\rangle}{\nu+\omega_{n0}+i\eta}\right\}\\&=\sum_n\left\{\frac{\langle 0|S_{-\mathbf{q}}^z|n\rangle\langle n|\rho_{\mathbf{q}}|0\rangle}{\nu-\omega_{n0}+i\eta}-\frac{\langle n|S_{-\mathbf{q}}^z|0\rangle\langle 0|\rho_{\mathbf{q}}|n\rangle}{\nu+\omega_{n0}+i\eta}\right\}.\end{aligned} \tag{19}$$

Details of the derivation of χ_{CS} are presented in Appendix B.

B. Self-consistent identification of the $\tilde{G}$'s

The last step in the renormalization procedure involves identification of the vertex correction factors $\tilde{G}$. With this goal in mind, we will first express the response functions appearing in the renormalization term [see Eqs. (12)–(15)] as functionals of the many-body local fields $G_{\pm}$ and G_{-}^{T}. Then, based on the formal similarity of the potentials given by Eqs. (4) and (A6) we make the physically reasonable ansatz that the vertex correction factors $\tilde{G}$ coincide identically with the corresponding many-body local fields G that enter the expression for the response functions.

For a single-component system the many-body local fields $G_{\pm}$ (Ref. 16) are commonly defined through the various response functions of the unpolarized medium as follows:

$$\chi_C\equiv\frac{\chi_0}{1-v(q)(1-G_+)\chi_0}, \tag{20}$$

and

$$\chi_S\equiv\mu_B^2\frac{\chi_0}{1+v(q)G_-\chi_0}. \tag{21}$$

The χ_0 appearing in the above equations differs from the Lindhard[28] polarizability for a noninteracting EG in that here it is defined in terms of exact occupation numbers if the local fields $G_{\pm}$ are taken to be consistent with Niklasson's definition (see below).[29]

In Appendix A, for an infinitesimally polarized multicomponent system, we have derived expressions for the various response functions in terms of the many-body local fields of a single component. Here, we merely present the results for a single-component system

$$\chi_C=\frac{\chi_0^{\uparrow}+\chi_0^{\downarrow}+4v(q)\chi_0^{\uparrow}\chi_0^{\downarrow}G_-}{\mathcal{D}}, \tag{22}$$

$$\chi_S=-\mu_B^2\frac{\chi_0^{\uparrow}+\chi_0^{\downarrow}-4v(q)\chi_0^{\uparrow}\chi_0^{\downarrow}(1-G_+)}{\mathcal{D}}, \tag{23}$$

$$\chi_{\mathrm{CS}}=\frac{\chi_0^{\uparrow}-\chi_0^{\downarrow}}{\mathcal{D}}, \tag{24}$$

where

$$\begin{aligned}\mathcal{D}\equiv\,&1-v(q)(\chi_0^{\uparrow}+\chi_0^{\downarrow})(1-G_+-G_-)\\&-4v(q)^2\chi_0^{\uparrow}\chi_0^{\downarrow}G_-(1-G_+).\end{aligned} \tag{25}$$

The χ_0^{σ} appearing in the above equations is the response of a free EG defined in terms of the exact occupation numbers $n_{\mathbf{p}}^{\sigma}$ as follows:

$$\chi_0^{\sigma}(\mathbf{q},\omega)\equiv\frac{1}{\Omega}\sum_{\mathbf{p}}\frac{n_{\mathbf{p}-\mathbf{q}}^{\sigma}-n_{\mathbf{p}}^{\sigma}}{\omega-\Delta_{\mathbf{p}}^{\sigma}(\mathbf{q})+i\eta}, \tag{26}$$

with Ω being the volume of the system. For the transverse spin response $\chi^{T\sigma}$, we only present its defining equation in terms of the local field G_{-}^{T} as follows:

$$\chi^{T\sigma}\equiv-\mu_B^2\frac{\chi_0^{T\sigma}}{1+2v(q)G_-^T\chi_0^{T\sigma}}, \tag{27}$$

where, similarly to χ_0^{σ}, the noninteracting transverse response $\chi_0^{T\sigma}$ too is defined in terms of the exact occupation numbers, i.e.,

$$\chi_0^{T\sigma}(\mathbf{q},\omega)\equiv\frac{1}{\Omega}\sum_{\mathbf{p}}\frac{n_{\mathbf{p}-\mathbf{q}}^{-\sigma}-n_{\mathbf{p}}^{\sigma}}{\omega-\Delta_{\mathbf{p}}^{T\sigma}(\mathbf{q})+i\eta}. \tag{28}$$

For an unpolarized system, the expressions for the charge and spin responses, given by Eqs. (22) and (23), reduce to the defining equations of $G_{\pm}$ as given by Eqs. (20) and (21). The mixed charge-spin response χ_{CS}, as expected, reduces to zero for the unpolarized case. As for the transverse spin response, the isotropy of the unpolarized system reduces it to $\frac{1}{2}\chi_S$ with the transverse $\chi_0^{T\sigma}$ and G_{-}^{T} coinciding with their unpolarized counterparts $\frac{1}{2}\chi_0$ and G_-.

Now, the $\tilde{G}$'s that enter the expressions for effective interactions given in Eqs. (13) and (14) correspond to the exchange and correlation corrections to the Hartree interaction between a test electron and the charge and spin density fluctuations in the dielectric medium. Similarly, the local fields $G_{\pm}$ and G_{-}^{T} too represent the corrections to the Hartree interaction between an electron and the density fluctuations in the EG. These facts are made explicit in the expressions for the potentials given by Eqs. (4) and (A6). Now, since the test electrons and the screening dielectric medium are one and the same, we expect the $\tilde{G}$'s and the G's to coincide. Hence, we postulate that the $\tilde{G}$'s coincide identically with the corresponding many-body local fields G's for all values of q and ω for an infinitesimally polarized system. Then the quasiparticle pseudo-Hamiltonian is given as follows:

$$H'_{QP}=\sum_{p,\sigma}[\epsilon_p^\sigma+E_{CH}^\sigma(p)]a_{p,\sigma}^+a_{p,\sigma}$$
$$+\frac{1}{2}\sum_{\substack{p,p'\\q,\sigma,\sigma'}}(\{v(q)+V_{\sigma,\sigma'}[q,-\Delta_{p'}^{\sigma'}(-q),\Delta_p^\sigma(q),\Delta_p^\sigma(q)]\}a_{p-q,\sigma}^+a_{p'+q,\sigma'}^+a_{p',\sigma'}a_{p,\sigma}$$
$$+V_{\sigma,\sigma'}^T[q,-\Delta_{p'}^{T\sigma'}(q),\Delta_p^{T\sigma}(q),\Delta_p^{T\sigma}(q)]a_{p-q,-\sigma}^+a_{p'+q,-\sigma'}^+a_{p',\sigma'}a_{p,\sigma})\ , \tag{29}$$

where E_{CH}^σ, $V_{\sigma,\sigma'}$, and $V_{\sigma,\sigma'}^T$ are formally given by Eqs. (13)–(15) with the $\tilde{G}$'s replaced by the corresponding local fields G.

IV. QUASIPARTICLE PSEUDO-HAMILTONIAN

For the sake of clarity, we now express the quasiparticle Hamiltonian in terms of normal products so that the exchange contribution from the effective interaction appears explicitly in the quasiparticle self-energy. We have

$$H_{QP}=\sum_{p,\sigma}E_p^\sigma:a_{p,\sigma}^+a_{p,\sigma}:$$
$$+\frac{1}{2}\sum_{\substack{p,p',\\q,\sigma,\sigma'}}(\{v(q)+V_{\sigma,\sigma'}[q,-\Delta_{p'}^{\sigma'}(-q),\Delta_p^\sigma(q),\Delta_p^\sigma(q)]\}:a_{p-q,\sigma}^+a_{p'+q,\sigma'}^+a_{p',\sigma'}a_{p,\sigma}:$$
$$+V_{\sigma,\sigma'}^T[q,-\Delta_{p'}^{T\sigma'}(-q),\Delta_p^{T\sigma}(q),\Delta_p^{T\sigma}(q)]:a_{p-q,-\sigma}^+a_{p'+q,-\sigma'}^+a_{p',\sigma'}a_{p,\sigma}:)\ . \tag{30}$$

Here constant terms have been omitted. In the above equation, the longitudinal component $V_{\sigma,\sigma'}$ corresponds to the case in which the spins of the quasiparticles are unchanged after interaction while the transverse component $V_{\sigma,\sigma'}^T$ corresponds to the case where opposite spin quasiparticles interact and flip their spins. For the sake of completeness we provide the expressions for these two as follows:

$$V_{\sigma,\sigma'}(q,\epsilon,\omega,\delta)\equiv v(q)^2([1-G_+(q,\epsilon)][1-G_+^*(q,\omega)]\mathrm{Re}\chi_C(q,\delta)+\sigma\sigma'G_-(q,\epsilon)G_-^*(q,\omega)\mathrm{Re}\chi_S(q,\delta)/(-\mu_B^2)$$
$$-\{\sigma[1-G_+(q,\epsilon)]G_-^*(q,\omega)+\sigma'[1-G_+^*(q,\omega)]G_-(q,\epsilon)\}\mathrm{Re}\chi_{CS}(q,\delta))\ , \tag{31}$$

and

$$V_{\sigma,\sigma'}^T(q,\epsilon,\omega,\delta)\equiv 2(1-\sigma\sigma')v(q)^2G_-^T(q,\epsilon)G_-^{T*}(q,\omega)\mathrm{Re}\chi^{T\sigma}(q,\delta)/(-\mu_B^2)\ . \tag{32}$$

The (fully renormalized) quasiparticle energy E_p^σ occurring in Eq. (30) contains both a dynamically screened exchange part E_{SX}^σ and a Coulomb hole part E_{CH}^σ

$$E_p^\sigma=\epsilon_p^\sigma+E_{SX}^\sigma(p)+E_{CH}^\sigma(p)\ , \tag{33}$$

where

$$E_{SX}^\sigma(p)=-\sum_q(n_{p-q}^\sigma\{v(q)+V_{\sigma,\sigma}[q,\Delta_p^\sigma(q),\Delta_p^\sigma(q),\Delta_p^\sigma(q)]\}+n_{p-q}^{-\sigma}V_{\sigma,-\sigma}^T[q,\Delta_p^{T\sigma}(q),\Delta_p^{T\sigma}(q),\Delta_p^{T\sigma}(q)])\ . \tag{34}$$

In the above expression the first term corresponds to the familiar Hartree-Fock exchange energy while the remaining terms represent exchange contribution from the dynamical screening produced by charge and spin-density fluctuations. In Eq. (33) the Coulomb hole term E_{CH}^σ is given by Eq. (15) with the factors $\tilde{G}$ replaced by their corresponding many-body local fields G.

The above derivation of the quasiparticle Hamiltonian is readily extended to the case of an EG with ν_v degenerate components. The problem is considerably simplified under the assumption that the density fluctuations are the same for all the components. Also for the case of a multivalley system where the relevant valleys are separated by a large momentum, an electron retains its valley after being scattered by other electrons. Then the electrons in different valleys can be regarded as different components with the component index v representing an additional quantum number. An additional complication is however represented by the fact that in the derivation of the quasiparticle Hamiltonian, the many-body local fields $G_\pm$, must be replaced by $G_\pm^v$, i.e., those appropriate for a multicomponent system (see Appendix A for further details). The final expression for the quasiparticle Hamiltonian is given as follows:

$$H_{\mathrm{QP}}=\sum_{\mathbf{p},\sigma,v} E_{\mathbf{p}}^{\sigma}:a_{\mathbf{p},\sigma,v}^{+}a_{\mathbf{p},\sigma,v}:+\frac{1}{2}\sum_{\substack{\mathbf{p},\mathbf{p}',\mathbf{q},\\ \sigma,\sigma',v,v'}}\{[v(q)+V_{\sigma,\sigma'}(\mathbf{q},-\Delta_{\mathbf{p}'}^{\sigma'}(-\mathbf{q}),\Delta_{\mathbf{p}}^{\sigma}(\mathbf{q}),\Delta_{\mathbf{p}}^{\sigma}(\mathbf{q}))]:a_{\mathbf{p}-\mathbf{q},\sigma,v}^{+}a_{\mathbf{p}'+\mathbf{q},\sigma',v'}^{+}a_{\mathbf{p}',\sigma',v'}a_{\mathbf{p},\sigma,v}:$$
$$+V_{\sigma,\sigma'}^{T}(\mathbf{q},-\Delta_{\mathbf{p}'}^{T\sigma'}(-\mathbf{q}),\Delta_{\mathbf{p}}^{T\sigma}(\mathbf{q}),\Delta_{\mathbf{p}}^{T\sigma}(\mathbf{q})):a_{\mathbf{p}-\mathbf{q},-\sigma,v}^{+}a_{\mathbf{p}'+\mathbf{q},-\sigma',v'}^{+}a_{\mathbf{p}',\sigma',v'}a_{\mathbf{p},\sigma,v}:\} \ . \qquad (35)$$

The quasiparticle energy and the effective interaction terms are still formally the same as those appearing in Eq. (15) and Eqs. (31)–(34). As for the response functions that enter these terms, expressions are derived in Appendix A.

V. DISCUSSION AND CONCLUSIONS

We have derived a quasiparticle pseudo-Hamiltonian for a multicomponent infinitesimally polarized Fermi liquid. This quasiparticle Hamiltonian is constructed in such a way as to properly account in an averaged way for the usually unwieldy effects of correlations beyond the popular RPA. This was achieved through the approximate use of Hubbard generalized many-body local fields associated with both charge and spin fluctuations. Our results can at this point be used to perform explicit calculations of many-body effects in the EG. To this purpose suitable approximations to the Hubbard local fields must be used. Such approximate expressions involve in turn the knowledge of the exact limits acquired by such quantities.[29–32] The alternative is to use for these functions the output of numerical work.[12] Lately a elegant self-consistent method for the evaluation of useful expressions of the Hubbard local fields has been devised.[23] The results of such an analysis will be presented elsewhere.[33]

Our results of Eqs. (15), (33), and (34) for the self-energy can be recast in the following transparent and useful form (see Ref. 33):

$$\Sigma^{\sigma}(\mathbf{p},\omega)=-\sum_{\mathbf{q}}\int_{-\infty}^{\infty}\frac{d\epsilon}{2\pi i}\{[v(q)+v(q)^2(|1-G_{+}|_{\chi C}^{2}+|G_{-}|_{\chi C}^{2}/(-\mu_B^2)$$
$$-2\sigma\,\mathrm{Re}[(1-G_{+}^{*})G_{-}]\chi_{\mathrm{CS}})]g^{\sigma}(\mathbf{p}-\mathbf{q},\omega-\epsilon)+4v(q)^2|G_{-}^{T}|^2\chi^{T\sigma}/(-\mu_B^2)g^{-\sigma}(\mathbf{p}-\mathbf{q},\omega-\epsilon)\} \ , \qquad (36)$$

where in order to recover our results one must set $\omega=\epsilon_{\mathbf{p}}^{\sigma}$. Here $g^{\sigma}(\mathbf{k},\omega)$ is the bare one electron Green's function and is defined as follows:

$$g^{\sigma}(\mathbf{p},\omega)\equiv\frac{n_{\mathbf{p}}^{\sigma}}{\omega-\epsilon_{\mathbf{p}}-i\eta}+\frac{1-n_{p}^{\sigma}}{\omega-\epsilon_{\mathbf{p}}+i\eta} \ . \qquad (37)$$

Furthermore, in Eq. (36) it is understood that the $G_{\pm}$ are functions of $\mathbf{q}$ and $\Delta_{\mathbf{p}}^{\sigma}(\mathbf{q})$, that the G_{-}^{T} is a function of $\mathbf{q}$ and $\Delta_{\mathbf{p}}^{T\sigma}(\mathbf{q})$, and that the response functions depend on $\mathbf{q}$ and ϵ. The expressions for the effective interaction terms in Eqs. (31) and (32) and that for the self-energy in Eq. (36) can be seen to be equivalent to the corresponding results of Ng and Singwi,[21] who however did not attempt to express their expressions in a transparent form (see also Ref. 33). There are also some differences mostly associated with the frequency dependence of the various quantities. The results of Ref. 21 can in fact be recovered if in our Eqs. (31), (32), and (36) in the many-body local fields that are prefactors to the response functions the complex conjugate forms are replaced by the corresponding complex forms, and the frequencies of all the local fields are replaced by the frequency appearing in the response functions.

It is of interest to note that for the special case of an unpolarized electron liquid, the charge and spin fluctuations are not coupled and also, owing to the isotropy, the effective potential due to the transverse spin fluctuations is just twice that due to the longitudinal spin fluctuations. Accordingly the quasiparticle self-energy of Eq. (36) simplifies to the following form for the unpolarized case:

$$\Sigma^{\sigma}(\mathbf{p},\omega)=-\sum_{\mathbf{q}}\int_{-\infty}^{\infty}\frac{d\epsilon}{2\pi i}\{v(q)+v(q)^2[|1-G_{+}|_{\chi_C}^{2}+3|G_{-}|_{\chi_S}^{2}/(-\mu_B^2)]\}g(\mathbf{p}-\mathbf{q},\omega-\epsilon) \ . \qquad (38)$$

As for the quasiparticle Hamiltonian, it is still given by Eq. (30) with $\Delta_{\mathbf{p}}^{T\sigma}(\mathbf{q})=\Delta_{\mathbf{p}}^{\sigma}(\mathbf{q})$ and the following simplifications for the screened exchange, Coulomb hole and the effective interaction terms:

$$E^{\sigma}_{SX}(\mathbf{p})=-\sum_{\mathbf{q}}[n_{\mathbf{p}-\mathbf{q}}(v(q)+v(q)^2\{|1-G_+[\mathbf{q},\Delta^{\sigma}_{\mathbf{p}}(\mathbf{q})]|^2\mathrm{Re}\chi_C[\mathbf{q},\Delta^{\sigma}_{\mathbf{p}}(\mathbf{q})]+3|G_-[\mathbf{q},\Delta^{\sigma}_{\mathbf{p}}(\mathbf{q})]|^2\mathrm{Re}\chi_S[\mathbf{q},\Delta^{\sigma}_{\mathbf{p}}(\mathbf{q})]/(-\mu_B^2)\})] , \tag{39}$$

$$E^{\sigma}_{\mathrm{CH}}(\mathbf{p})=-\sum_{\mathbf{q}}v(q)^2P\int_0^{\infty}\frac{d\omega}{\pi}\mathrm{Im}\left[\frac{|1-G_+|^2\chi_C+3|G_-|^2\chi_S/(-\mu_B^2)}{\Delta^{\sigma}_{\mathbf{p}}(\mathbf{q})-\omega}\right] , \tag{40}$$

$$V_{\sigma,\sigma'}(\mathbf{q},\epsilon,\omega,\delta)=v(q)^2\{[1-G_+(\mathbf{q},\epsilon)][1-G^*_+(\mathbf{q},\omega)]\mathrm{Re}\chi_C(\mathbf{q},\delta)+\sigma\sigma'G_-(\mathbf{q},\epsilon)G^*_-(\mathbf{q},\omega)\mathrm{Re}\chi_S(\mathbf{q},\delta)/(-\mu_B^2)\} , \tag{41}$$

and

$$V^T_{\sigma,\sigma'}(\mathbf{q},\epsilon,\omega,\delta)=(1-\sigma\sigma')v(q)^2G_-(\mathbf{q},\epsilon)G^*_-(\mathbf{q},\omega)\mathrm{Re}\chi_S(\mathbf{q},\delta)/(-\mu_B^2) . \tag{42}$$

For this case the quasiparticle self-energy and effective interaction terms of Ref. 21 are obtained form Eqs. (38), (41), and (42) by making the same modifications as in the polarized case. The above results for the effective interaction terms of an unpolarized system agree with those derived in Refs. 17, 19, and 20. In these papers however, the question of the proper frequency dependence of the various response functions and the corresponding local fields was not tackled.

It must be mentioned that our quasiparticle energy yields what in a diagrammatic analysis would amount to the on-shell value of the self-energy. Furthermore, and in connection with the above, it should be made clear that the quasiparticle Hamiltonian derived here should only be used for calculations carried out to first order. This prescription is the consequence of the fact that in our derivation the renormalization term is explicitly constructed so as to ignore higher-order terms in H'. On the other hand, we believe that such a procedure is appropriate in view of the fact that, as readily verified, our pseudo-Hamiltonian leads by construction to the expected RPA results when the many-body local-field corrections are neglected.[35]

For the case of a multi-component system our approach does not take into account the effects due to the difference in the density fluctuations between components. Future work that includes these effects is in progress.

ACKNOWLEDGMENTS

The authors would like to thank A. W. Overhauser, G. E. Santoro, and G. Vignale for useful discussions. This work was partially supported by DOE Grant No. DE-FG02-90ER45427 through MISCON.

APPENDIX A

In this appendix, using arguments based on linear response theory, we will derive the response functions for an infinitesimally polarized EG with ν_v degenerate components. We will refer here to the various components as valleys (as in band valleys). We will assume that the valleys are separated by large vectors in momentum space and it is therefore reasonable to assume that electrons retain their valley index after a scattering process.

When an external potential $\phi_{\mathrm{ext}}(\mathbf{q},\omega)$ is applied to the electronic system it sets up density fluctuations $\Delta n_\uparrow$ of spin-up and $\Delta n_\downarrow$ of spin-down electrons. Assuming that these density fluctuations are equal for all the valleys, the total effective potential felt by a spin up electron can be written by generalizing the procedure of Refs. 17 and 34 as follows:

$$\phi^{\uparrow}=\phi_{\mathrm{ext}}+v(q)\left\{[\Delta n_\uparrow+\Delta n_\downarrow]-[G^{\uparrow\uparrow}_{x,\mathrm{intra}}+G^{\uparrow;\uparrow}_{c,\mathrm{intra}}+G^{\uparrow;\uparrow}_{c,\mathrm{inter}}(\nu_v-1)]\frac{2\Delta n_\uparrow}{\nu_v}-[G^{\uparrow;\downarrow}_{c,\mathrm{intra}}+G^{\uparrow;\downarrow}_{c,\mathrm{inter}}(\nu_v-1)]\frac{2\Delta n_\downarrow}{\nu_v}\right\} , \tag{A1}$$

where for the sake of brevity the $\mathbf{q}$ and ω dependence of the potentials, the density fluctuations, and the many-body local fields has not been displayed. In the above equation the G's are assumed to be the same for each valley, the subscripts x and c denote exchange and correlation effects, and the labels *intra* and *inter* refer to intravalley and intervalley processes. Furthermore, among the terms containing the Coulombic potentials $v(q)$, the sum of the first two terms involving the density fluctuations represents the Hartree term, the next one is the exchange term, and the remaining ones are correlation terms. The total effective potential felt by a spin down electron can be similarly written as

$$\phi^{\downarrow}=\phi_{\mathrm{ext}}+v(q)\left\{[\Delta n_\uparrow+\Delta n_\downarrow]-[G^{\downarrow;\downarrow}_{x,\mathrm{intra}}+G^{\downarrow;\downarrow}_{c,\mathrm{intra}}+G^{\downarrow;\downarrow}_{c,\mathrm{inter}}(\nu_v-1)]\frac{2\Delta n_\downarrow}{\nu_v}-[G^{\downarrow;\uparrow}_{c,\mathrm{intra}}+G^{\downarrow;\uparrow}_{c,\mathrm{inter}}(\nu_v-1)]\frac{2\Delta n_\uparrow}{\nu_v}\right\} . \tag{A2}$$

Furthermore, for an unpolarized system, we note that for symmetry reasons

$$G^{\uparrow\uparrow}_{x,\text{intra}}=G^{\downarrow\downarrow}_{x,\text{intra}} , \tag{A3}$$

and

$$G^{\uparrow\uparrow(\downarrow)}_{c,\text{intra(inter)}}=G^{\downarrow\downarrow(\uparrow)}_{c,\text{intra(inter)}} . \tag{A4}$$

We assume that the above relations remain approximately valid for an infinitesimally polarized system. Also, when the electrons after interacting with each other scatter back to their original valleys, we can write

$$G^{\uparrow\uparrow}_{c,\text{inter}}=G^{\uparrow\downarrow}_{c,\text{inter}}=G^{\uparrow\downarrow}_{c,\text{intra}} . \tag{A5}$$

With these approximations the potential felt by an electron with spin $\sigma(=\pm 1)$ can be cast in the following compact form:

$$\phi_\sigma=\phi_{\text{ext}}+v(q)[\Delta n_\uparrow+\Delta n_\downarrow](1-G^v_+) - \sigma v(q)[\Delta n_\uparrow-\Delta n_\downarrow]G^n_- , \tag{A6}$$

where the quantities $G^v_\pm$ are defined as follows:

$$G^v_\pm\equiv\frac{1}{\nu_v}[G^{\uparrow\uparrow}_{x,\text{intra}}+G^{\uparrow\uparrow}_{c,\text{intra}}+(\nu_v\pm\nu\nu_v-1)G^{\uparrow\downarrow}_{c,\text{intra}}] . \tag{A7}$$

Then on defining the single valley local fields $G_\pm$ as

$$G_\pm\equiv G^{\uparrow\uparrow}_{x,\text{intra}}+G^{\uparrow\uparrow}_{c,\text{intra}}\pm G^{\uparrow\downarrow}_{c,\text{intra}} , \tag{A8}$$

we obtain the following useful form for the multivalley local fields $G^v_\pm$ defined in Eq. (A7):

$$G^v_{+(-)}=G_+-G_{-(+)}+\frac{1}{\nu_v}G_- . \tag{A9}$$

We will now derive the expression for the charge response of an infinitesimally polarized system. In the presence of a spin independent infinitesimal external potential ϕ_{ext}, the total effective potential ϕ_σ felt by an electron of spin σ is still given by Eq. (A6). The density fluctuation Δn_σ is related to ϕ_σ via the relation

$$\Delta n_\sigma=\nu_v\chi^\sigma_0\phi_\sigma , \tag{A10}$$

where χ^σ_0 is the response for a free EG as defined in Eq. (26) in the text. Then using Eqs. (A6) and (A10) we obtain the expression for charge response from its definition

$$\chi_C\equiv\frac{\Delta n_\uparrow+\Delta n_\downarrow}{\phi_{\text{ext}}} = \frac{\nu_v\chi^\uparrow_0+\nu_v\chi^\downarrow_0+4v(q)\nu^2_v\chi^\uparrow_0\chi^\downarrow_0 G^v_-}{\mathcal{D}_v} , \tag{A11}$$

where

$$\mathcal{D}_v\equiv 1-v(q)(\nu_v\chi^\uparrow_0+\nu_v\chi^\downarrow_0)(1-G^v_+-G^v_-) -4v(q)^2\nu^2_v\chi^\uparrow_0\chi^\downarrow_0 G^v_-(1-G^v_+) . \tag{A12}$$

For an unpolarized system, for which $\chi^\uparrow_0=\chi^\downarrow_0$, the expression for χ_C simplifies to the following

$$\chi_C=\frac{\nu_v\chi_0}{1-v(q)(1-G^v_+)\nu_v\chi_0} , \tag{A13}$$

where $\chi_0=\chi^\uparrow_0+\chi^\downarrow_0$.

To derive the spin response function, we consider the case of an infinitesimal external magnetic field H^z_{ext} setting up density fluctuations $\Delta n_\uparrow$ and $\Delta n_\downarrow$. Using Eq. (A6) we get the relevant effective potential to be

$$\phi_\sigma=\mu_B\sigma H^z_{\text{ext}}+v(q)[\Delta n_\uparrow+\Delta n_\downarrow](1-G^v_+) -\sigma v(q)[\Delta n_\uparrow-\Delta n_\downarrow]G^v_- . \tag{A14}$$

Then from the definition of spin response and from Eqs. (A10) and (A14) it follows that

$$\chi_S\equiv\mu_B\frac{\Delta n_\downarrow-\Delta n_\uparrow}{H^z_{\text{ext}}} =-\mu^2_B\frac{\nu_v\chi^\uparrow_0+\nu_v\chi^\downarrow_0-4v(q)\nu^2_v\chi^\uparrow_0\chi^\downarrow_0(1-G^v_+)}{\mathcal{D}_v} . \tag{A15}$$

For an unpolarized system this reduces to the familiar form

$$\chi_S=-\mu^2_B\frac{\nu_v\chi_0}{1-v(q)G^v_-\nu_v\chi_0} . \tag{A16}$$

The mixed charge-spin response function is obtained by using Eqs. (A6), (A10), and (A14)

$$\chi_{CS}\equiv\frac{\Delta n_\uparrow-\Delta n_\downarrow}{\phi_{\text{ext}}}\equiv\frac{\Delta n_\uparrow+\Delta n_\downarrow}{\mu_B H^z_{\text{ext}}}=\frac{\nu_v\chi^\uparrow_0-\nu_v\chi^\downarrow_0}{\mathcal{D}_v} . \tag{A17}$$

It should be noted that for an unpolarized system, as can be expected from symmetry considerations, $\chi_{CS}=0$.

The transverse spin response $\chi^{T\sigma}$ can be defined for a multivalley system as follows:

$$\chi^{T\sigma}\equiv-\mu^2_B\frac{\nu_v\chi^{T\sigma}_0}{1+2v(q)G^{Tv}_-\nu_v\chi^{T\sigma}_0} , \tag{A18}$$

where $\chi^{T\sigma}_0(\mathbf{q},\omega)$ is defined in Eq. (28) in the text. The only unknown quantity in Eq. (A18) is the transverse many-body local field G^{Tv}_- for which we propose the following ansatz:

$$G^{Tv}_-\equiv\frac{1}{2\nu_v}[G^{\uparrow\uparrow}_{x,\text{intra}}+G^{\uparrow\uparrow}_{c,\text{intra}}+G^{\downarrow\downarrow}_{x,\text{intra}} +G^{\downarrow\downarrow}_{c,\text{intra}}-2G^{\uparrow\downarrow}_{c,\text{intra}}] . \tag{A19}$$

Now, for an unpolarized system $\chi^{T\sigma}$ simplifies to $\frac{1}{2}\chi_S$ with $\chi^{T\sigma}_0$ and G^{Tv}_- reducing to their unpolarized forms χ^σ_0 and G^v_-, respectively. However, it should be noted that within the present approximation of $G^{\uparrow\uparrow}_{x(c),\text{intra}}=G^{\downarrow\downarrow}_{x(c),\text{intra}}$, and the transverse field G^{Tv}_- coincides with the longitudinal field G^v_-.

APPENDIX B

We will consider here a system with arbitrary polarization and derive the expression for the mixed charge-spin response function χ_{CS}. Let the initial state of the system be $|0\rangle$ and let the Hamiltonian H be characterized by eigenstates $|n\rangle$ having excitation energies ω_{n0}. Let the ground state of the system be $|G\rangle$. In order to obtain an expression of χ_{CS}, we begin by considering the spin response of the system due to an external spin symmetric potential $\phi(r,t)$. The Hamiltonian corresponding to the perturbation is (with standard notation) given by

$$H_e=[\rho_{\mathbf{q}}\phi(\mathbf{q},\omega)e^{-i\omega t}+\text{c.c.}]e^{\eta t} . \tag{B1}$$

Then the Schrödinger equation can be written as

$$i\frac{\partial}{\partial t}|\psi(t)\rangle=(H+H_e)|\psi(t)\rangle , \tag{B2}$$

where $|\psi(t)\rangle$ is an eigenstate of the total Hamiltonian and can be projected onto the states $|n\rangle$ as follows:

$$|\psi(t)\rangle=\sum_n a_n(t)e^{-i\omega_n t}|n\rangle . \tag{B3}$$

The boundary conditions are given by

$$a_n(-\infty)=\begin{cases}1 & \text{if } n=0\\ 0 & \text{otherwise.}\end{cases} \tag{B4}$$

Substituting Eq. (B3) in Eq. (B2) and retaining only the terms that are first order in $\phi(\mathbf{q},\omega)$, we obtain after integration

$$a_n(t)=\frac{\langle n|\rho_{\mathbf{q}}|0\rangle\phi(\mathbf{q},\omega)e^{(-i\omega+i\omega_{n0}+\eta)t}}{\omega-\omega_{n0}+i\eta}-\frac{\langle n|\rho_{-\mathbf{q}}|0\rangle\phi(\mathbf{q},\omega)^*e^{(i\omega+i\omega_{n0}+\eta)t}}{\omega+\omega_{n0}-i\eta} , \tag{B5}$$

where $n\neq 0$. Then on defining

$$\langle S^z_{-\mathbf{q}}(t)\rangle\equiv\langle\psi(t)|S^z_{-\mathbf{q}}|\psi(t)\rangle , \tag{B6}$$

with $S^z_{-\mathbf{q}}$ being the induced spin density fluctuation operator and using Eqs. (B3), (B5), and (B6), we obtain

$$\langle S^z_{-\mathbf{q}}(t)\rangle=\sum_n\left\{\frac{\langle 0|S^z_{-\mathbf{q}}|n\rangle\langle n|\rho_{\mathbf{q}}|0\rangle\phi(\mathbf{q},\omega)e^{(-i\omega+\eta)t}}{\omega-\omega_{n0}+i\eta}-\frac{\langle 0|S^z_{-\mathbf{q}}|n\rangle\langle n|\rho_{-\mathbf{q}}|0\rangle\phi(\mathbf{q},\omega)^*e^{(i\omega+\eta)t}}{\omega+\omega_{n0}-i\eta}\right.$$
$$\left.+\frac{\langle n|S^z_{-\mathbf{q}}|0\rangle\langle 0|\rho_{-\mathbf{q}}|n\rangle\phi(\mathbf{q},\omega)^*e^{(i\omega+\eta)t}}{\omega-\omega_{n0}-i\eta}-\frac{\langle n|S^z_{-\mathbf{q}}|0\rangle\langle 0|\rho_{\mathbf{q}}|n\rangle\phi(\mathbf{q},\omega)e^{(-i\omega+\eta)t}}{\omega+\omega_{n0}+i\eta}\right\} , \tag{B7}$$

where use has been made of the fact that

$$\langle 0|S^z_{\mathbf{q}}|0\rangle=0 . \tag{B8}$$

In Eq. (B7) the second and third terms on the right hand side vanish since $|n\rangle$ cannot be coupled to $|0\rangle$ by both $S^z_{\mathbf{q}}$ and $\rho_{-\mathbf{q}}$ since the former has momentum $\mathbf{q}$ whereas the latter has momentum $-\mathbf{q}$. Then we obtain the following expression for the spin response:

$$\frac{\langle S^z_{-\mathbf{q}}(\omega)\rangle}{\phi(\mathbf{q},\omega)}=e^{i\omega t-\eta t}\frac{\langle S^z_{-\mathbf{q}}(t)\rangle}{\phi(\mathbf{q},\omega)}=\sum_n\left\{\frac{\langle 0|S^z_{-\mathbf{q}}|n\rangle\langle n|\rho_{\mathbf{p}}|0\rangle}{\omega-\omega_{n0}+i\eta}-\frac{\langle n|S^z_{-\mathbf{q}}|0\rangle\langle 0|\rho_{\mathbf{q}}|n\rangle}{\omega+\omega_{n0}+i\eta}\right\} . \tag{B9}$$

Similarly, upon applying an external magnetic field $h^z_{\text{ext}}(r,t)$, the charge response is given by

$$\frac{\langle\rho_{-\mathbf{q}}(\omega)\rangle}{\mu_B h^z_{\text{ext}}(\mathbf{q},\omega)}=\sum_n\left\{\frac{\langle 0|\rho_{-\mathbf{q}}|n\rangle\langle n|S^z_{\mathbf{q}}|0\rangle}{\omega-\omega_{n0}+i\eta}-\frac{\langle 0|S^z_{\mathbf{q}}|n\rangle\langle n|\rho_{-\mathbf{q}}|0\rangle}{\omega+\omega_{n0}+i\eta}\right\} . \tag{B10}$$

Then on using the definition of χ_{CS} given in Eq. (A17) we obtain the following relationship for the mixed charge-spin response function:

$$\chi_{CS}(\mathbf{q},\nu)=\sum_n\left\{\frac{\langle 0|\rho_{-\mathbf{q}}|n\rangle\langle n|S^z_{\mathbf{q}}|0\rangle}{\nu-\omega_{no}+i\eta}-\frac{\langle 0|S^z_{\mathbf{q}}|n\rangle\langle n|\rho_{-\mathbf{q}}|0\rangle}{\nu+\omega_{no}+i\eta}\right\}=\sum_n\left\{\frac{\langle 0|S^z_{-\mathbf{q}}|n\rangle\langle n|\rho_{\mathbf{q}}|0\rangle}{\nu-\omega_{n0}+i\eta}-\frac{\langle n|S^z_{-\mathbf{q}}|0\rangle\langle 0|\rho_{\mathbf{q}}|n\rangle}{\nu+\omega_{n0}+i\eta}\right\} . \tag{B11}$$

*Present address: NTT Basic Research Labs, Musashino-shi, Tokyo 180, Japan.

[1]For a review on the subject, see K. S. Singwi and M. P. Tosi, in *Solid State Physics: Advances in Research and Applications,* edited by H. Ehrenreich, F. Seitz, and D. Turnbull (Academic, New York, 1980), Vol. 36.
[2]For a discussion of the physical properties of two-dimensional system, see T. Ando, A. B. Fowler, and F. Stern, Rev. Mod. Phys. **54**, 437 (1982).
[3]A. Isihara, in *Solid State Physics: Advances in Research and Applications* (Ref. 1), Vol. 42.
[4]E. P. Wigner, Phys. Rev. **46**, 1002 (1934).
[5]M. Gell-Mann and K. A. Brueckner, Phys. Rev. **106**, 364 (1957).
[6]A. K. Rajagopal and J. C. Kimball, Phys. Rev. B **15**, 2819 (1977).
[7]A. Isihara and T. Toyoda, Ann. Phys. (N.Y.) **106**, 394 (1977); A. Isihara and T. Toyoda, *ibid.* **114**, 497 (1978).
[8]W. J. Carr, Phys. Rev. **122** 1437 (1961).
[9]Lynn Bonsall and A. A. Maradudin, Phys. Rev. B **15**, 1959 (1977).
[10]D. M. Ceperley and B. J. Alder, Phys. Rev. Lett. **45**, 566 (1980).
[11]B. Tanatar and D. M. Ceperley, Phys. Rev. B **39**, 5005 (1989).
[12]S. Moroni and G. Senatore, Phys. Rev. B **44**, 9864 (1991); S. Moroni, D. M. Ceperley, and G. Senatore, Phys. Rev. Lett. **69**, 1837 (1992).
[13]L. D. Landau, Zh. Eksp. Teor. Fiz. **30**, 1058 (1956) [Sov. Phys. JETP **3**, 920 (1957)].
[14]J. M. Luttinger and P. Nozières, Phys. Rev. **127**, 1423 (1962); **127**, 1431 (1962).
[15]See, for instance, D. Pines and P. Nozières, *The Theory of Quantum Fluids* (Benjamin, New York, 1966), Vol. I.
[16]J. Hubbard, Proc. R. Soc. London Ser. A **242**, 539 (1957); **243**, 336 (1957).
[17]C. Z. Kukkonen and A. W. Overhauser, Phys. Rev. B **20**, 550 (1979).
[18]X. Zhu and A. W. Overhauser, Phys. Rev. B **33**, 925 (1986). These authors employ a single pole approximation which, in general, is expected to only provide qualitative results.
[19]G. Vignale and K. S. Singwi, Phys. Rev. B **32**, 2156 (1985); see also K. S. Singwi, Phys. Scr. **32**, 397 (1985).
[20]S. Yarlagadda and G. F. Giuliani, Solid State Commun. **69**, 677 (1989).
[21]T. K. Ng and K. S. Singwi, Phys. Rev. B **34**, 7738 (1986); **34**, 7743 (1986).
[22]G. E. Santoro and G. F. Giuliani, Solid State Commun. **67**, 681 (1988); also, Phys. Rev. B **39**, 12 818 (1989). See comment in Ref. 18.
[23]For a derivation of the effective mass, enhanced g-factor, and spin susceptibility using the present expression for the self-energy of an infinitesimally polarized EG, see S. Yarlagadda and G. F. Giuliani, Surf. Sci. **229**, 410 (1989); Phys. Rev. B **40**, 5432 (1989).
[24]D. R. Hamann and A. W. Overhauser, Phys. Rev. **143**, 183 (1966).
[25]See, for instance, *Quantum Theory of Solids,* edited by C. Kittel (Wiley, New York, 1963), Chap. 8.
[26]For a discussion of the Coulomb hole term see L. Hedin and S. Lundquist, in *Solid State Physics: Advances in Research and Applications* (Ref. 1), Vol. 23.
[27]A Hartree-Fock expression for the mixed charge-spin response was introduced earlier on by D. J. Kim, B. B. Schwartz, and H. C. Praddaude, Phys. Rev. B **7**, 205 (1973).
[28]J. Lindhard, Kgl. Danske Videnskab. Selsk. Mat. Fys. Medd. **28** (8), 1 (1954).
[29]G. Niklasson, Phys. Rev. B **10**, 3052 (1974).
[30]X. Zhu and A. W. Overhauser, Phys. Rev. B **30**, 3158 (1984).
[31]G. E. Santoro and G. F. Giuliani, Phys. Rev. B **37**, 4813 (1988).
[32]S. Yarlagadda and G. F. Giuliani, Phys. Rev. B **39**, 3386 (1989).
[33]S. Yarlagadda and G. F. Giuliani (unpublished); S. Yarlagadda, Ph. D. thesis, Purdue University, 1989.
[34]L. Hedin and S. Lundquist, J. Phys. C **4**, 2064 (1971).
[35]On these points our analysis is consistent with the conclusions reached by Y. Takada, Phys. Rev. A **28**, 2417 (1983).

PHYSICAL REVIEW B VOLUME 49, NUMBER 20 15 MAY 1994-II

Many-body local fields and Fermi-liquid parameters in a quasi-two-dimensional electron liquid

Sudhakar Yarlagadda* and Gabriele F. Giuliani
Department of Physics, Purdue University, West Lafayette, Indiana 47907

(Received 14 June 1993)

We present a quantitative theory of the quasiparticle properties in a Fermi liquid. Our approach uses as an input our previous result for the quasiparticle energy which incorporates the vertex corrections associated with charge and spin-density fluctuations through suitably defined many-body local fields. The method is explicitly applied to the case of the quasi-two-dimensional electron liquid occurring in silicon inversion layers. In particular, we discuss results for the effective mass m^* and the modified Landé factor g^* (Wilson ratio) that are in reasonable agreement with reported findings. Our calculations are performed by making use of a self-consistent static model for the many-body local fields and are consequently free of arbitrary parameters.

I. INTRODUCTION

The problem of an electron gas (EG) neutralized by a uniform positive background is still not well understood. Although several of the physical properties have long been qualitatively explained, meaningful quantitative comparisons with experimental data are still not feasible. This work is concerned with furthering the microscopic theoretical understanding of the quasiparticle properties of such a system in the Fermi-liquid regime. Quinn and Ferrell[1] provided a framework for the microscopic evaluation of the quasiparticle energy in an electron gas by implementing through diagrammatic means what is commonly referred to as the random-phase approximation (RPA). Next, Rice[2] incorporated the vertex corrections in the RPA form of the self-energy by including the Hubbard many-body local field.[3] Some problems in Rice's theory were subsequently resolved within the same general framework by Ting, and co-workers in their theory of the quasi-two-dimensional (Q2D) electron liquid.[4,5] A more detailed discussion of the merits and shortcomings of these important papers will be presented below and elsewhere.[6,7] All these papers considered only the charge-density fluctuations while neglecting the effect of the spin-density fluctuations. More recently, a more detailed analysis that accounts for the vertex corrections associated with both the charge-density fluctuations and the spin-density fluctuations was carried out by the authors of Refs. 8–11 for an unpolarized electron gas. By making use of different approaches, these authors obtained equivalent results for the effective interaction between two electrons. Moreover, in Refs. 9–11, general and complete expressions for the self-energy were also obtained. Calculations performed making use of the results of these theories have been carried out both for three-dimensional[9,12] and for two-dimensional systems.[13,14] As for the polarized case, Ng and Singwi[15] and the present authors[16,7] have obtained the self-energy and the effective interaction for a system with infinitesimal polarization again by incorporating the many-body local fields associated with the charge- and spin-density fluctuations. Once the self-energy of an infinitesimally polarized EG is known, one can carry out calculations within the framework of Landau's Fermi-liquid theory[17] and obtain various physical properties such as the effective mass m^*, the modified Landé factor g^*, the spin susceptibility, and the compressibility.

As regards the experimental determination of these physical properties, the density dependence of the cyclotron mass has been obtained by Smith and Stiles[18] from a study of the fluctuation of magnetoconductance in inversion layers by using Shubnikov–de Haas (SdH) type of experiments. To obtain the same information on the cyclotron effective mass, Abstreiter *et al.*[19] used cyclotron resonance measurements. Fang and Stiles[20] and Neugebauer *et al.*[21] performed a series of SdH type of experiments on silicon inversion layers also and obtained the dependence of g^* on density. These authors did not take the enhancement of the electron mass into account while deducing the value of g^*. Consequently, their reported values of g^* show a dramatic increase as the density is changed. Suzuki and Kawamoto[22] noted this fact and rescaled the data values of g^* of Fang and Stiles[20] using the values of m^* experimentally obtained by Smith and Stiles.[18]

Several authors have tried to explain the behavior of m^* and g^* as a function of density. Although each of these approaches has interesting insights, they do not properly consider the various subtleties of present-day many-body theories. The first step toward explaining the dependence of g^* on the carrier density was taken by Janak.[23] Janak's expression for the electronic self-energy takes into account only the Coulombic interaction simply screened by a static RPA dielectric function. As a result, the expression for g^* has the right functional form, whereas the expression for the effective mass includes only screened-exchange effects with the Coulomb hole contribution being left out. Furthermore, the contribution to the self-energy from intervalley scattering is taken to be the same as that due to intravalley scattering leading to a considerable, yet spurious, enhancement in the predicted g^* values.

The next step toward explaining the behavior of g^* was taken by Suzuki and Kawamoto,[22] who introduced the Hubbard modification factor[3] in a static RPA dielectric function and allowed for some valley degeneracy effects. These authors employed the Landau's Fermi-liquid-theory results using an *ad hoc* exchange type quasiparticle interaction function as a result of which the spin-symmetric and the spin-antisymmetric parts of the interaction function are equal. Their calculated values of g^* are close in magnitude, although weaker in density dependence, when compared to their rescaled version of the experimental values reported in Ref. 20. As for the m^* values, they obtain much smaller values than those observed by Smith and Stiles.[18]

Further work along these lines was done by Ando and Uemura,[24] who also attempted to account for many-body effects beyond RPA by a Hubbard many-body local field. The Hubbard function chosen by these authors takes into account the quasi-two-dimensional nature of the system, but the expression for g^* incorrectly contains the bare mass of the electron instead of its effective mass. These authors, too, choose an electronic self-energy involving only a static dielectric function. The values of g^* calculated by these authors show stronger density dependence when the electronic wave function is assumed to have a finite extent in the third dimension.

Ting, Lee, and Quinn[4] went beyond the above theories by considering a dynamic dielectric function. These authors adopt the microscopic approach to Fermi-liquid theory formulated by Rice.[2] They start with the ground-state energy which they obtain via the integration-over-the-coupling-constant algorithm. Unfortunately, the functional dependence of their dielectric function with respect to the occupation numbers contained only the effects due to the local fields associated with parallel spins and neglects those due to antiparallel spins. The major shortcoming of this theory is that the trend of the anomalous g factor as a function of the density is the opposite of what one would expect and is, in fact, observed experimentally (see Fig. 2 of Ref. 4).

Vinter[25] also developed a theory to explain the effective-mass behavior in inversion layers by taking the Q2D nature of the system into account. His self-energy involves dynamic screening within the RPA. His calculations, however, were carried out in the single-pole approximation which, in general, cannot be trusted from a quantitative viewpoint.[26] Moreover, the reported values are in reasonable agreement with the observed data only when the thickness of the inversion layer is neglected.

It should also be pointed out that all the theories to date consider an oversimplified model for the local-field correction that enters the formula for the dielectric function. Also, proper attention has not been paid to the effects of spin-density fluctuations. In particular, the Hubbard-type local fields used in these theories do not have the appropriate behavior in the limiting cases of small and large momentum transfer.

The present paper gives a detailed account of some of the results reported previously by the present authors.[27] In Ref. 27 the derivation of g^* is put on a more rigorous and satisfactory footing than that presented earlier on in Ref. 28. Finally, an alternative, yet physically less satisfactory procedure based on the integration-over-the-coupling-constant algorithm will be discussed elsewhere.[6]

The paper is organized as follows. In Sec. II we discuss Landau's theory of Fermi liquids so as to set up a framework for deriving and calculating m^* and g^*. Section III deals with a necessary input to Landau's phenomenological theory—the quasiparticle energy of an infinitesimally polarized electron liquid. The quasiparticle energy fully takes into account the many-body effects related to charge and spin fluctuations in the liquid. Here we also discuss the various response functions that enter the expression for the quasiparticle energy. Then, using the quasiparticle energy, we derive expressions for the quasiparticle interaction function and for the enhanced g factor. In Sec. IV, we set up a scheme to calculate m^* and g^* without involving any arbitrary parameters. The method involves approximating the many-body local fields by Hubbard-type static functions whose large momentum values are known exactly and whose small momentum values are determined self-consistently in the course of the calculation. Last, in Sec. V we discuss our results and compare them with previous work. In the Appendix, we derive a formula for the parameter that enters the expression for the static spin-symmetric local field $G_+^v(q,0)$.

II. LANDAU'S FERMI-LIQUID THEORY

The Landau theory of the Fermi liquids is based on the following phenomenological yet fundamental relation expressing the change in the total energy of a system as a *functional* of $\delta n_{\mathbf{k}}^{\sigma}$, the change of the quasiparticle occupation number $n_{\mathbf{k}}^{\sigma}$ from its ground-state value $n_{\mathbf{k}}^{0\sigma}$:

$$\delta E[n_{\mathbf{q}}^{\sigma''}]=\sum_{\mathbf{k},\sigma} E_{\mathbf{k}}^{\sigma}[n_{\mathbf{q}}^{0\sigma''}]\delta n_{\mathbf{k}}^{\sigma}+\frac{1}{2}\sum_{\mathbf{k},\mathbf{p},\sigma,\sigma'} f_{\mathbf{k},\mathbf{p}}^{\sigma,\sigma'}\delta n_{\mathbf{k}}^{\sigma}\delta n_{\mathbf{p}}^{\sigma'} . \tag{1}$$

In the above equation, $\delta n_{\mathbf{k}}^{\sigma}$ is assumed to be nonvanishingly only for states lying close to the Fermi surface, so that all the various quasiparticle momenta have a magnitude equal to p_F. This stipulation clearly indicates that such an approach is explicitly designed to treat the low-energy processes in the electron gas.

The quasiparticle energy $E_{\mathbf{k}}^{\sigma}$ and the Landau interaction function $f_{\mathbf{k},\mathbf{p}}^{\sigma\sigma'}$ appearing in Eq. (1) can be related to the total energy via appropriate functional derivatives in the standard way:

$$E_{\mathbf{k}}^{\sigma}[n_{\mathbf{q}}^{\sigma''}]=\frac{\delta E}{\delta n_{\mathbf{k}}^{\sigma}} \tag{2}$$

and

$$f_{\mathbf{k},\mathbf{p}}^{\sigma,\sigma'}=\frac{\delta^2 E}{\delta n_{\mathbf{k}}^{\sigma}\delta n_{\mathbf{p}}^{\sigma'}}=\frac{\delta E_{\mathbf{k}}^{\sigma}[n_{\mathbf{q}}^{\sigma''}]}{\delta n_{\mathbf{p}}^{\sigma'}} . \tag{3}$$

It should be noted here that since **k** and **p** both lie on the Fermi surface, in an isotropic system $f_{\mathbf{k},\mathbf{p}}^{\sigma\sigma'}$ depends only on the angle ϕ between these two momenta. Accordingly, in what follows we shall drop the subscripts of f. Various physical quantities of interest can then be obtained once the interaction function is known. For instance, in a 2D system we have the following relations for the effective mass m^*, the enhanced compressibility κ^*, the enhanced spin susceptibility χ_S and the modified Landé factor g^* (see Ref. 4):

$$\frac{m^*}{m}=1+\frac{m^*}{\pi}\int_0^{2\pi}\frac{d\phi}{2\pi}f_s(\phi)\cos\phi \ , \tag{4}$$

$$\frac{\kappa}{\kappa^*}=\frac{m}{m^*}+\frac{m}{\pi}\int_0^{2\pi}\frac{d\phi}{2\pi}f_s(\phi) \ , \tag{5}$$

$$\frac{\chi_s^0}{\chi_s}=\frac{m}{m^*}+\frac{m}{\pi}\int_0^{2\pi}\frac{d\phi}{2\pi}f_a(\phi) \ , \tag{6}$$

and

$$\frac{g}{g^*}=1+\frac{m^*}{\pi}\int_0^{2\pi}\frac{d\phi}{2\pi}f_a(\phi) \ , \tag{7}$$

where χ_s^0 is the free-electron Pauli value of the spin susceptibility. In Eqs. (4)–(7) the notations f_s and f_a indicate the spin-symmetric and spin-antisymmetric components of the Landau interaction function, which are defined as follows:

$$f_s\equiv\frac{f^{\uparrow,\uparrow}+f^{\uparrow,\downarrow}}{2} \tag{8}$$

and

$$f_a\equiv\frac{f^{\uparrow,\uparrow}-f^{\uparrow,\downarrow}}{2} \ . \tag{9}$$

Also, in Eq. (5) the compressibility does not include the contribution from the rigidity of the positive background. From Eqs. (4), (6), and (7) we note the following exact relationship:

$$\frac{\chi_s}{\chi_s^0}=\frac{m^*}{m}\frac{g^*}{g} \ , \tag{10}$$

which displays the relation between g^* and the Wilson ratio. It should be mentioned here that the Eqs. (4), (5), and (6) represent only phenomenological expressions for the corresponding physical quantities. On the other hand, the same quantities can be expressed through a series of rigorous relations as follows:

$$\frac{1}{m^*}\equiv\frac{1}{p_F}\left[\frac{\partial E_{\mathbf{p}}^{\sigma}}{\partial p}\right]_{p_F} \ , \tag{11}$$

and at zero temperature

$$\frac{1}{\kappa^*}\equiv A\frac{\partial^2 E}{\partial A^2} \tag{12}$$

and

$$\chi_s\equiv-\frac{1}{A}\frac{\partial^2 E}{\partial H^2} \ , \tag{13}$$

where A is the surface area and H is the magnetic field. In any case, we find that for computational reasons it is easier, for instance, to obtain m^* using Eq. (11) rather than Eq. (4). In principle, both approaches should lead to the same answer. In practice, however, this is not always the case and care must be taken to insure the consistency of the theory.

III. MICROSCOPIC DERIVATION OF THE FERMI-LIQUID PARAMETERS

As was made clear in the preceding section, an essential ingredient in the evaluation of the effective mass and the enhanced g factor is the quasiparticle self-energy as a functional of the occupation numbers $n_{\mathbf{k}}^{\sigma}$. To this end, we have developed a quasiparticle pseudo-Hamiltonian for an infinitesimally polarized electron gas.[16] The Hamiltonian takes into account the vertex corrections associated with both charge and spin fluctuations. The fully renormalized quasiparticle energy that appears in the Hamiltonian can be expressed in the following physically transparent form:

$$E_{\mathbf{p}}^{\sigma}[n_{\mathbf{k}}^{\sigma'}]=\epsilon_{\mathbf{p}}^{\sigma}+E_{\mathrm{SX}}^{\sigma}(\mathbf{p})+E_{\mathrm{CH}}^{\sigma}(\mathbf{p}) \ , \tag{14}$$

where $E_{\mathrm{SX}}^{\sigma}(\mathbf{p})$ is the screened-exchange term and $E_{\mathrm{CH}}^{\sigma}(\mathbf{p})$ is the Coulomb-hole term. The screened-exchange contribution is given by

$$E_{\mathrm{SX}}^{\sigma}(\mathbf{p})=-\sum_{\mathbf{q}}\mathrm{Re}[n_{\mathbf{p}-\mathbf{q}}^{\sigma}(V(q)+V(q)^2\{|1-G_+^{v}|^2\chi_C+|G_-^{v}|^2\chi_S/(-\mu_B^2)$$
$$-2\sigma\mathrm{Re}[(1-G_+^{v*})G_-^{v}]\chi_{CS}\})+4n_{\mathbf{p}-\mathbf{q}}^{-\sigma}V(q)^2|G_-^{Tv}|^2\chi^{T\sigma}/(-\mu_B^2)] \ , \tag{15}$$

where G_+^{v}, G_-^{v}, and G_-^{Tv} are the generalized Hubbard many-body local fields associated with the charge fluctuations, longitudinal spin fluctuations, and transverse spin fluctuations, respectively. These quantities will be defined below in terms of the corresponding response functions: the charge susceptibility χ_C, the longitudinal spin susceptibility χ_S, and the transverse spin susceptibility $\chi^{T\sigma}$, respectively. Furthermore, $V(q)$ is the bare Coulombic interaction and χ_{CS} is the mixed charge-spin response. Here, all the G's and the χ's are appropriate for a multivalley electronic system with ν_v degenerate valleys. In Eq. (15) it is understood that the $G_{\pm}^{v}$ and the $\chi_{C,S,CS}$ are functions of $\mathbf{q}$ and $\Delta_{\mathbf{p}}^{\sigma}(\mathbf{q})=\epsilon_{\mathbf{p}}^{\sigma}-\epsilon_{\mathbf{p}-\mathbf{q}}^{\sigma}$ while the G_-^{Tv} and $\chi^{T\sigma}$ are functions of $\mathbf{q}$ and $\Delta_{\mathbf{p}}^{T\sigma}(\mathbf{q})=\epsilon_{\mathbf{p}}^{\sigma}-\epsilon_{\mathbf{p}-\mathbf{q}}^{-\sigma}$. Here, the $\epsilon_{\mathbf{p}-\mathbf{q}}^{\sigma}$ are the free-electron band energies. As for the Coulomb hole contribution, we have

$$E_{CH}^{\sigma}(\mathbf{p})=-\sum_{\mathbf{q}} V(q)^2 P\int_0^{\infty}\frac{d\omega}{\pi}\times\mathrm{Im}\left[\frac{|1-G_+^{v}|^2\chi_C+|G_-^{v}|^2\chi_S/(-\mu_B^2)}{\epsilon_{\mathbf{p}}^{\sigma}-\epsilon_{\mathbf{p}-\mathbf{q}}^{\sigma}-\omega}-\frac{2\sigma\,\mathrm{Re}[(1-G_+^{v*})G_-^{v}]\chi_{CS}}{\epsilon_{\mathbf{p}}^{\sigma}-\epsilon_{\mathbf{p}-\mathbf{q}}^{\sigma}-\omega}+\frac{4|G_-^{Tv}|^2\chi^{T\sigma}/(-\mu_B^2)}{\epsilon_{\mathbf{p}}^{\sigma}-\epsilon_{\mathbf{p}-\mathbf{q}}^{-\sigma}-\omega}\right]. \tag{16}$$

In the above equation, it is understood that the factors $G_{\pm}^{v}$ are functions of $\mathbf{q}$ and $\Delta_{\mathbf{p}}^{\sigma}(\mathbf{q})$ while G_-^{Tv} is a function of $\mathbf{q}$ and $\Delta_{\mathbf{p}}^{T\sigma}(\mathbf{q})$. Moreover, the response functions depend on $\mathbf{q}$ and ω.

For an infinitesimally polarized system the expressions for the various response functions, assuming that $G_{\pm}^{v}$ retain the symmetry of an unpolarized system, are given by[16]

$$\chi_C=\frac{v_v\chi_0^{\uparrow}+v_v\chi_0^{\downarrow}+4V(q)v_v^2\chi_0^{\uparrow}\chi_0^{\downarrow}G_-^{v}}{\mathcal{D}_v}, \tag{17}$$

$$\chi_S=-\mu_B^2\frac{v_v\chi_0^{\uparrow}+v_v\chi_0^{\downarrow}-4V(q)v_v^2\chi_0^{\uparrow}\chi_0^{\downarrow}(1-G_+^{v})}{\mathcal{D}_v}, \tag{18}$$

and

$$\chi_{CS}=\frac{v_v\chi_0^{\uparrow}-v_v\chi_0^{\downarrow}}{\mathcal{D}_v}, \tag{19}$$

where the denominator $\mathcal{D}_v$ in the above equations is defined to be

$$\mathcal{D}_v\equiv 1-V(q)(v_v\chi_0^{\uparrow}+v_v\chi_0^{\downarrow})(1-G_+^{v}-G_-^{v})-4V(q)^2v_v^2\chi_0^{\uparrow}\chi_0^{\downarrow}G_-^{v}(1-G_+^{v}). \tag{20}$$

As for the transverse spin response, it is simply defined as follows:

$$\chi^{T\sigma}\equiv-\mu_B^2\frac{v_v\chi_0^{T\sigma}}{1+2V(q)G_0^{T}v_v\chi_0^{T\sigma}}. \tag{21}$$

The functions χ_0^{σ} and $\chi_0^{T\sigma}$ appearing in the above equations are the free-electron longitudinal and transverse responses and can be defined in terms of a one-electron Green's function $\mathcal{G}^{\sigma}(\mathbf{p},\omega)$ as follows:

$$\chi_0^{\sigma}(\mathbf{q},\omega)\equiv\sum_{\mathbf{p}}\int_{-\infty}^{\infty}\frac{d\epsilon}{2\pi i}\mathcal{G}^{\sigma}(\mathbf{p},\epsilon)\mathcal{G}^{\sigma}(\mathbf{p}+\mathbf{q},\epsilon+\omega) \tag{22}$$

and

$$\chi_0^{T\sigma}(\mathbf{q},\omega)\equiv\sum_{\mathbf{p}}\int_{-\infty}^{\infty}\frac{d\epsilon}{2\pi i}\mathcal{G}^{-\sigma}(\mathbf{p},\epsilon)\mathcal{G}^{\sigma}(\mathbf{p}+\mathbf{q},\epsilon+\omega), \tag{23}$$

where it should be made clear that $\mathcal{G}^{\sigma}(\mathbf{p},\omega)$ is, in turn, defined in terms of the exact occupation numbers $n_{\mathbf{p}}^{\sigma}$ as follows:

$$\mathcal{G}^{\sigma}(\mathbf{p},\omega)\equiv\frac{n_{\mathbf{p}}^{\sigma}}{\omega-\epsilon_{\mathbf{p}}-i\eta}+\frac{1-n_{\mathbf{p}}^{\sigma}}{\omega-\epsilon_{\mathbf{p}}+i\eta}. \tag{24}$$

For unpolarized systems we recover the following familiar forms for χ_C and χ_S, which can be treated as the defining equations for $G_{\pm}^{v}$:

$$\chi_C\equiv\frac{v_v\chi_0}{1-V(q)(1-G_+^{v})v_v\chi_0} \tag{25}$$

and

$$\chi_S\equiv-\mu_B^2\frac{v_v\chi_0}{1+V(q)G_-^{v}v_v\chi_0}. \tag{26}$$

Furthermore, for unpolarized systems due to symmetry $\chi^{T\sigma}=\chi_S/2$ and $\chi_{CS}^{\sigma}=0$.

In order to obtain the quasiparticle interaction function by taking the functional derivative of the quasiparticle energy $E_{\mathbf{p}}^{\sigma}$ with respect to the occupation number $n_{\mathbf{p}}^{\sigma}$, it is useful to recast the expression for $E_{\mathbf{p}}^{\sigma}$ given by Eqs. (14)–(16) in the following compact form:

$$E_{\mathbf{p}}^{\sigma}[n_{\mathbf{k}}^{\sigma'}]=\epsilon_{\mathbf{p}}^{\sigma}-\sum_{\mathbf{q}}\int_{-\infty}^{\infty}\frac{d\epsilon}{2\pi i}[(V(q)+V(q)^2\{|1-G_+^{v}|^2\chi_C+|G_-^{v}|^2\chi_S/(-\mu_B^2)-2\sigma\,\mathrm{Re}[(1-G_+^{v*})G_-^{v}]\chi_{CS}\})\mathcal{G}^{\sigma}(\mathbf{p}-\mathbf{q},\epsilon_{\mathbf{p}}^{\sigma}-\epsilon)+4V(q)^2|G_-^{vT}|^2\chi^{T\sigma}/(-\mu_B^2)\mathcal{G}^{-\sigma}(\mathbf{p}-\mathbf{q},\epsilon_{\mathbf{p}}^{\sigma}-\epsilon)]. \tag{27}$$

In Eq. (27) it is understood that the many-body local fields $G_{\pm}^{v}$ are functions of $\mathbf{q}$ and $\Delta_{\mathbf{p}}^{\sigma}(\mathbf{q})$, while G_-^{Tv} is a function of $\mathbf{q}$ and $\Delta_{\mathbf{p}}^{T\sigma}(\mathbf{q})$. Furthermore, all the response functions depends on q and ϵ. In order to take the functional derivative of the quasiparticle energy with respect to the occupation number $n_{\mathbf{k}}^{\sigma}$, we recognize that

$$\frac{\delta\chi_0^{\sigma'}(\mathbf{q},\omega)}{\delta n_{\mathbf{p}}^{\sigma}}=\delta_{\sigma\sigma'}[\mathcal{G}^{\sigma}(\mathbf{p}+\mathbf{q},\epsilon_{\mathbf{p}}^{\sigma}+\omega)+\mathcal{G}^{\sigma}(\mathbf{p}-\mathbf{q},\epsilon_{\mathbf{p}}^{\sigma}-\omega)]. \tag{28}$$

Then by assuming that the local fields are independent of $n_{\mathbf{p}}^{\sigma}$ one obtains from Eqs. (3) and (27) the following result for the Landau interaction function of a nonpolarized electron liquid:

$$f^{\uparrow,\uparrow}=\mathcal{F}^{\uparrow,\uparrow}-\left[V(q)+V(q)^2\left[[1-G_+^{v}(\mathbf{q},0)]^2\chi_C(\mathbf{q},0)+[G_-^{v}(\mathbf{q},0)]^2\frac{\chi_S(\mathbf{q},0)}{(-\mu_B^2)}\right]\right]_{\mathbf{q}=\mathbf{k}-\mathbf{p}}, \tag{29}$$

and

$$f^{\uparrow,\downarrow}=\mathcal{F}^{\uparrow,\downarrow}-\left[2V(q)^2[G_-^v(\mathbf{q},0)]^2\frac{\chi_S(\mathbf{q},0)}{(-\mu_B^2)}\right]_{\mathbf{q}=\mathbf{k}-\mathbf{p}}. \quad (30)$$

In the above equations the terms in the square brackets represent the contribution from the screened exchange while the terms $\mathcal{F}^{\uparrow,\sigma}$ are associated to higher-order correlation effects and are given by

$$\mathcal{F}^{\uparrow,\sigma}=-v_v\sum_{\mathbf{q}}\int_{-\infty}^{\infty}\frac{d\epsilon}{2\pi i}\left\{V(q)^2\left\{\left[\frac{|1-G_+^v|}{Q_+}\right]^2+\left[\frac{|G_-^v|}{Q_-}\right]^2-2\sigma\frac{\mathrm{Re}[(1-G_+^{v*})G_-^v]}{Q_+Q_-}\right\}\tilde{\mathcal{G}}\right.$$
$$\left.+4V(q)^2\left[\frac{|G_-^v|}{Q_-}\right]^2 g(\mathbf{k}-\sigma\mathbf{q},\epsilon_{\mathbf{k}}-\sigma\epsilon)g(\mathbf{p}-\mathbf{q},\epsilon_{\mathbf{p}}-\epsilon)\right\}. \quad (31)$$

In Eq. (31) it is understood that the $G_\pm^v$ are functions of $\mathbf{q}$ and $\Delta_{\mathbf{p}}^\sigma(\mathbf{q})$ while the $Q_\pm$ are functions of q and ϵ. Furthermore, in Eq. (31) we have defined

$$\tilde{\mathcal{G}}=[\mathcal{G}(\mathbf{k}+\mathbf{q},\epsilon_{\mathbf{k}}+\epsilon)+\mathcal{G}(\mathbf{k}-\mathbf{q},\epsilon_{\mathbf{k}}-\epsilon)]\mathcal{G}(\mathbf{p}-\mathbf{q},\epsilon_{\mathbf{p}}-\epsilon), \quad (32)$$

$$Q_+=1-V(q)v_v\chi_0(1-G_+^v), \quad (33)$$

and

$$Q_-=1+V(q)v_v\chi_0G_-^v. \quad (34)$$

At this point we perform a contour integration so as to switch the integration in Eq. (31) from over the real axis to over the imaginary axis. We then obtain from Eqs. (7), (9), and (29)–(31) the following expression for the enhanced Landé factor:[29]

$$\frac{g}{g^*}=1-m^*\int_0^{2\pi}\frac{d\phi}{(2\pi)^2}\left[V(q)+V(q)^2\left\{[1-G_+^v(\mathbf{q},0)]^2\chi_C(\mathbf{q},0)-[G_-^v(\mathbf{q},0)]^2\frac{\chi_S(\mathbf{q},0)}{(-\mu_B^2)}\right\}\right]_{\mathbf{q}=\mathbf{k}-\mathbf{p}}$$
$$+\frac{2v_vm^*m}{\pi^3}\int_0^\infty dz\int_0^\infty du\; V(q)^2\left\{\frac{\mathrm{Re}(\{1-G_+^{v*}[\mathbf{q},\Delta_{\mathbf{p}}^\sigma(\mathbf{q})]\}G_-^v[\mathbf{q},\Delta_{\mathbf{p}}^\sigma(\mathbf{q})])}{Q_-(q,i\omega)Q_+(q,i\omega)}P_+(z,u)\right.$$
$$\left.+\left[\frac{|G_-^v[\mathbf{q},\Delta_{\mathbf{p}}^\sigma(\mathbf{q})]|}{Q_-(q,i\omega)}\right]^2P_-(z,u)\right\}, \quad (35)$$

where we have defined the variables $z=q/2p_F$, $u=\omega m/qp_F$, and the function

$$P_\pm(z,u)=\frac{[(z^2-u^2-1)^2+(2zu)^2]^{1/2}\pm(z^2-u^2-1)}{[(z^2-u^2-1)^2+(2zu)^2]}. \quad (36)$$

IV. SELF-CONSISTENT CALCULATION OF m^*, g^*, and $G_\pm^v$ IN INVERSION LAYERS

In this section we develop a simplified yet practical self-consistent static model for the many-body local fields. We begin by discussing first the nature of the bare effective interaction $V(q)$ in a Q2D structure. The model used consists of a semiconductor-oxide-metal sandwich. The semiconductor is characterized by a static dielectric constant ϵ_s while the oxide layer has dielectric constant ϵ_0 and thickness D. Stern and Howard[30] proposed the following simple expression for the extent of the wave function in the direction perpendicular to the surface for electrons in the lowest subband:

$$\eta(z)=\left[\frac{b^3}{2}\right]^{1/2}z\exp\left[-\frac{bz}{2}\right], \quad (37)$$

where the parameter b is given by $3/\langle z\rangle$ with $\langle z\rangle$ being the average electronic distance from the oxide-semiconductor interface. The appropriate value for b can be obtained by an approximate minimization of the energy per particle.[30] The corresponding bare Coulomb interaction between two electrons in the lowest subband can be expressed as

$$V(q)=\frac{2\pi e^2}{\epsilon_s q}L(q), \quad (38)$$

where the form factor $L(q)$ is given by[5]

$$L(q)=(1+x)^{-6}\left\{\frac{x}{8}[33+54x+44x^2+18x^3+3x^4]\right.$$
$$\left.+2\epsilon_s[\epsilon_s+\epsilon_0\coth qD]^{-1}\right\}, \quad (39)$$

with $x=q/b$.

In order to calculate m^* and g^* for various densities we need an expression for the many-body local fields $G^v_\pm(\mathbf{q},\omega)$. Such a knowledge can only result from an exact solution of the many-body problem at hand. Accordingly, one has to employ approximate analytic expressions for these quantities. An alternative procedure is to follow a suitable numerical route.[31]

In general, the $G^v_\pm(\mathbf{q},\omega)$ are wave-vector and frequency dependent. Following the work of Niklasson,[32] it has been recently possible to derive exact expressions for the limiting values of the local fields for large and small q in the static regime.[33,34] For intermediate values of the momentum, however, only educated guesses, direct numerical work, or suitable physically judicious combinations of both currently represent viable methods of approach. In the present work we will therefore approximate the $G^v_\pm(\mathbf{q},\omega)$ by means of suitable static interpolating functions in the manner of Hubbard.[3] Our functions will be accordingly constructed so as to interpolate between $G^v_\pm(q\to 0,\omega=0)$ at small wave vectors and $G^v_\pm(\infty)=G^v_\pm(q\to\infty,\omega=q^2/2m)$ for large wave vectors. It should be noticed here that our selections of the values of ω are designed so as to remain approximately close to the single-particle excitation frequency regime of the noninteracting system.

Taking all the above considerations into account, we write for our many-body local fields the following simple interpolation formulas:

$$G^v_\pm(\mathbf{q})=G^v_\pm(\infty)\frac{V\{\sqrt{q^2+[\beta_\pm G^v_\pm(\infty)p_F]^2}\}}{V(q)}$$
$$\approx\frac{L(\sqrt{q^2+p_F^2})}{L(q)}\frac{G^v_\pm(\infty)q}{\sqrt{q^2+[\beta_\pm G^v_\pm(\infty)p_F]^2}}\ , \qquad (40)$$

where, as explained below, the coefficients $\beta_\pm$ are determined in a self-consistent way from the static long-wavelength limits of the charge and spin susceptibilities.

From Eqs. (39) and (40) we see that for large values of q the local fields are independent of the form factor $L(q)$ so that their functional form resembles that of the local fields in two dimensions. Clearly in this limit the physical properties of the electron liquid are determined by short distance processes. In fact, as shown in Ref. 34, the following relationships hold for the case of a single-valley 2DEG: $G_+(\infty)=1-g(0)$ and $G_-(\infty)=g(0)$. In these formulas $g(0)$ is the value at the origin of the probability of finding two electrons separated by a distance r.[35] The corresponding expressions for a multivalley system are[16,7]

$$G^v_+(\infty)=[1+(\nu_v^{-1}-2)g(0)] \qquad (41)$$

and

$$G^v_-(\infty)=\nu_v^{-1}g(0)\ . \qquad (42)$$

$g(0)$ is a density-dependent property of the electron liquid. Numerical calculations of $g(0)$ for 2D electron systems have been carried out by Jonson for the case of a Q2D system with no valley degeneracy.[36] Jonson's work was an application of the classic method of Singwi *et al.*[37] In our calculation we therefore elect to fix $G^v_\pm(\infty)$ by using for $g(0)$ the numerical values calculated by Jonson.[38]

We turn next to the discussion of the small-wave-vectors regime and will describe explicitly how the coefficients $\beta_\pm$ appearing in Eq. (40) can be obtained. We deal first with the straightforward case of β_+. This coefficient is directly related to the $q\to 0$ limit of the static-charge-response function $\chi_C(\mathbf{q},0)$ and is simply fixed by means of the compressibility sum rule.[35] This exact relationship stipulates that

$$n^2\kappa^*=\lim_{q\to 0}\frac{\epsilon(\mathbf{q},0)-1}{V(q)}=-\lim_{q\to 0}\frac{\chi_C(\mathbf{q},0)}{1+V(q)\chi_C(\mathbf{q},0)}\ , \qquad (43)$$

where n is the electronic density, $\epsilon(\mathbf{q},0)$ is the static dielectric constant, and κ^* is the compressibility as defined by Eq. (12).[39] By making use of Eqs. (25), (38), (40), and (43) we can write the following relationship between β_+ and κ^*:

$$\frac{1}{\kappa^*}=\frac{2}{\pi r_S^3 a_B^{*2}}\left\{\frac{1}{r_S\nu_v}-\frac{2L(p_F)\epsilon_{\rm av}}{(2/\nu_v)^{1/2}\beta_+\epsilon_s}\right\}{\rm Ry}^*\ , \qquad (44)$$

where ${\rm Ry}^*=me^4/2\epsilon_{\rm av}^2$ is the effective Rydberg, $a_B^*=\epsilon_{\rm av}/me^2$ is the effective Bohr radius, r_S is the average electronic separation in units of a_B^*, and $\epsilon_{\rm av}=(\epsilon_s+\epsilon_0)/2$. At this point κ^* can be evaluated from Eq. (12) using the ground-state energy as calculated by Jonson[36] for the case of a Q2D electron gas with two valleys. After some simple manipulations (see Appendix for details) we arrive at the following expression for β_+:

$$\beta_+=\frac{2\epsilon_{\rm av}L(p_F)}{\epsilon_S}\left[\frac{2A_1}{\nu_v\pi}+\frac{r_S^2}{8\nu_v}\left[\frac{2}{\nu_v}\right]^{1/2}\frac{B_1}{(B_2+r_S)^2}+\frac{r_S^3}{4\nu_v}\left[\frac{2}{\nu_v}\right]^{1/2}\frac{B_1}{(B_2+r_S)^3}\right]^{-1}\ , \qquad (45)$$

where $A_1=0.625$, $B_1=1.782$, and $B_2=6.25$.

In order to determine β_- we adopt instead the following self-consistent procedure. We start with a trial value for β_- and proceed to a first evaluation of m^* via Eqs. (11) and (14)–(16). Once m^* is known, g^* is obtained from Eq. (35). Then, from these the value of the long-wavelength static spin susceptibility can be evaluated from Eq. (10). Now, from the definition of the spin susceptibility given in Eq. (26) we obtain the following equation for $\chi_S(\mathbf{q}\to 0,0)$ in terms of β_-:

$$\frac{\chi_S(\mathbf{q}\to 0,0)}{\chi_S^0}=\left[1-\frac{2L(p_F)\epsilon_{\rm av}\nu_v}{\beta_-\epsilon_s a_B^* p_F}\right]^{-1}\ . \qquad (46)$$

On substituting into Eq. (46) the value of $\chi_S(\mathbf{q}\to 0,0)$ obtained from Eq. (10), a new value for β_- is determined. This iterative procedure is repeated until convergence is reached.[40] Once β_+ and β_- are determined, our theory is free of arbitrary parameters.

V. RESULTS AND DISCUSSION

We have carried out a number of self-consistent calculations of both m^* and g^*, as a function of the electronic density, for a scenario appropriate to the quasi-two-dimensional electron liquid occurring in a [100] silicon inversion layer. In particular, all the numerical results reported here have been obtained for the following values of the parameters: band mass $m=0.19$, valley degeneracy $\nu_v=2$, oxide dielectric constant $\epsilon_0=3.8$, semiconductor dielectric constant $\epsilon_s=11.8$, thickness of the oxide layer $D=5330$ Å, the average electronic distance from the oxide-semiconductor interface $3/b=32.5$ Å.

m^* and g^* have been calculated making use of Eqs. (11) and (35). Given the nature of our theory the present results are parameter free. Our self-consistent results for the corresponding coefficients $\beta_\pm$ are plotted in Fig. 1 as a function of r_S. There the curves numbered (1) and (2) correspond to the theory in this paper and that of Ref. 6, respectively. The number of iterations necessary to achieve convergence (usually within a few percent) depends on the quality of the initial guess, with five being a typical number.

One of the important questions we have addressed is that of understanding the overall relevance and the separate effects of the density-fluctuation-induced vertex corrections as accounted for by means of the many-body local fields G_+^v and G_-^v. The results of our study are summarized in Figs. 2 and 3, where the three cases (i) $G_+^v=G_-^v=0$, (RPA), (ii) $G_+^v\neq 0$, $G_-^v=0$ (no spin-density fluctuations), and (iii) $G_+^v\neq 0$, $G_-^v\neq 0$ (the full theory), are displayed. As before, in Fig. 1, the curves numbered (1) and (2) pertain to the theory developed in the present paper and to that of Ref. 6, respectively. We

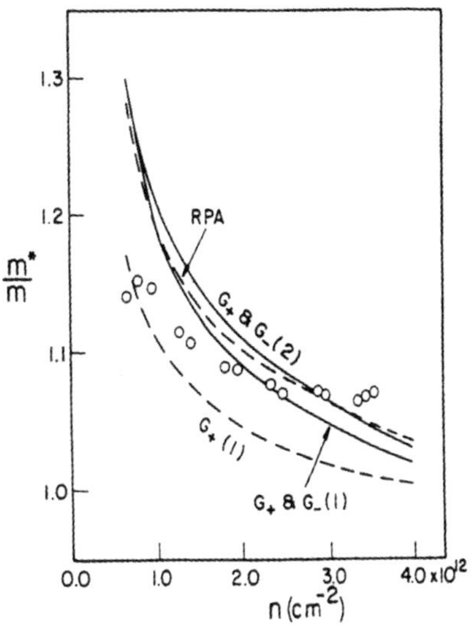

FIG. 2. Plot of the effective mass m^*/m vs the electronic areal density n. The open circles are the observed values reported in Ref. 18. The solid curves labeled $G_+ \& G_-(1)$ and $G_+ \& G_-(2)$ display, respectively, the result of the present theory and that of Ref. 6. The dashed curve labeled $G_+(1)$ is the result one obtains if spin-density fluctuations are neglected. The dashed curve labeled RPA is the result one obtains if all the many-body local fields were to be neglected. The parameters are the same as in Fig. 1.

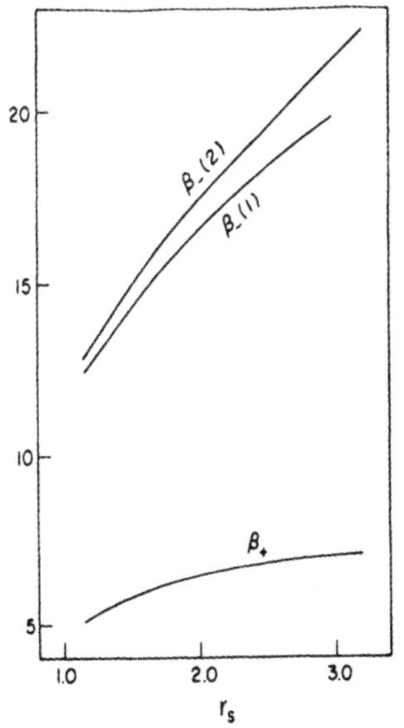

FIG. 1. Self-consistency values of the coefficients β_+ and β_-, defined in the text, vs the dimensionless average electronic separation r_S. The curves labeled $\beta_-(1)$ and $\beta_-(2)$ are, respectively, the result of the present theory and that of Ref. 6. The values of the parameters used in the calculations are given in the text at the beginning of Sec. V.

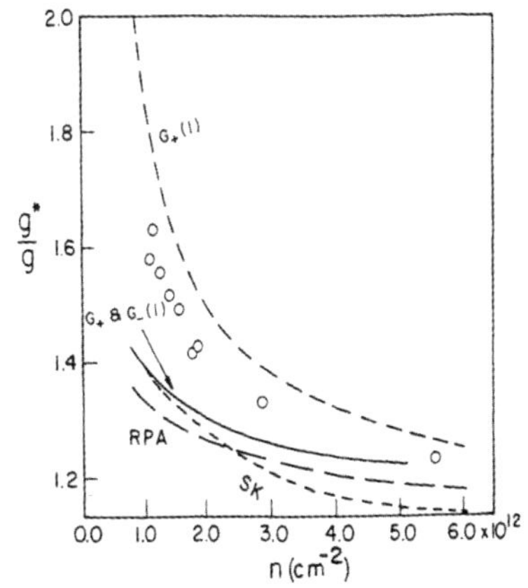

FIG. 3. Plot of the ratio g^*/g vs the electronic areal density n. The open circles are the original values reported by Fang and Stiles in Ref. 20. The dashed curve labeled SK represents the very same data as rescaled by Suzuki and Kawamoto in Ref. 22. The solid curve labeled $G_+ \& G_-(1)$ displays the results of the present theory. The dashed curve labeled $G_+(1)$ is the result one obtains if spin-density fluctuations are neglected. The dashed curve labeled RPA is the result one obtains if all the many-body local fields were to be neglected. The parameters are the same as in Fig. 1.

have also plotted in our figures the experimental values for m^* from Ref. 18, and the g^* data reported in Ref. 20. Furthermore, in Fig. 3 we have also displayed the very same g^* data as rescaled by Suzuki and Kawamoto [22]. This rescaling amounts to extracting the g^* values from the original data of Fang and Stiles[20] by making use not of the effective-mass values measured in that experiment (which showed no enhancement), but those obtained in the different experiment of Smith and Stiles.[18] The validity of such a procedure is still controversial.

It must first be noted that the agreement of the full theory with the experimental data is rather satisfactory. Also, interestingly, it is clear from Figs. 2 and 3 that including only the spin-symmetric local field G_+^v, as was done in the previous theories that tried to go beyond the simple RPA, leads to large deviation from the result of the full theory and the experimental data. In fact when the full theory is implemented a cancellation phenomenon seems to take place and the RPA result (except for the low-density region) appears to lead to a reasonable answer throughout most of the density regime analyzed. We therefore conclude that the effects due to the spin-density fluctuations should not be neglected and, in fact, one is likely to do worse than the simple RPA by only including spin-symmetric local fields, as is often done.

From the above discussion on the performance of the RPA theory one could naively surmise that the generalized Hubbard corrections can be safely ignored altogether. This is readily seen to be incorrect by recalling that, as discussed in Ref. 41, the RPA gives a sensibly smaller spin susceptibility enhancement than the present many-body theory in the case of a strictly two-dimensional electron liquid.

A few words must be spent to describe the limitations of the present approach and further developments of the present theory that should be pursued. It is obvious that the quantitative results of this theory are strongly dependent on the model chosen to describe the generalized Hubbard local fields. In particular, the popular static approximation used must be regarded with caution. A source of important numerical undeterminacy is also represented by the particular choice of the values of the density-dependent quantity $g(0)$. Since the evaluation of such an important physical property of the electron gas is by no means a trivial task, this tends to amplify the problem. The quantitative validity of our theory could be improved if more care is devoted to the effects of the valley degeneracy. In particular, a theory accounting for independent charge and spin fluctuations in different valleys should be developed.

ACKNOWLEDGMENTS

The authors would like to thank A. W. Overhauser, G. E. Santoro, and G. Vignale for useful discussions. This work was partially supported by DOE Grant No. DE-FG02-90ER45427 through MISCON.

APPENDIX

In this Appendix, we will derive an expression for the quantity β_+ by first obtaining the compressibility from the total ground-state energy and then using Eq. (44) of Sec. IV. The ground-state energy comprises three parts, namely, noninteracting kinetic energy, exchange energy, and correlation energy. The noninteracting kinetic energy per particle is simply given by

$$\epsilon_{\text{KIN}}=\frac{\text{Ry}^*}{v_v r_S^2}\,. \tag{A1}$$

The exchange energy for a strictly two-dimensional electronic system was first derived by Chaplik and it is given by[42]

$$\epsilon_x^{2D}=-\frac{8\text{Ry}^*}{3\pi r_S}\left[\frac{2}{v_v}\right]^{1/2}. \tag{A2}$$

Jonson[36] has calculated numerically the exchange energy for various r_S values for a Q2DEG system with two valleys. The system is a Si(100)-SiO_2 inversion layer that corresponds to a metal-oxide-semiconductor field-effect transistor with infinite oxide thickness. We curve fit his data to a simple extension of the formula given by the above equation and obtain

$$\epsilon_x^{Q2D}=-\frac{8\text{Ry}^*}{3\pi r_S}\left[\frac{2}{v_v}\right]^{1/2}(A_1+A_2 r_S)\,, \tag{A3}$$

where $A_1=0.625$ and $A_2=0.0428$. Here r_S ranges between 0.5 and 4.0. As for the correlation energy per particle, we assume a form first introduced by Wigner[43] as an interpolation between the high-density limit and the low-density Wigner lattice limit:[44]

$$\epsilon_c^{Q2D}=-\frac{\text{Ry}^*}{v_v}\frac{B_1}{B_2+r_S}\,. \tag{A4}$$

By curve fitting the above equation to the numerical correlation energy obtained by Jonson[36] employing the method of Singwi *et al.*,[37] we obtain for the Q2DEG system described above $B_1=1.782$ and $B_2=6.25$. For a large enough system we can write the total energy as follows:

$$E=Ah(r_S)\,, \tag{A5}$$

where A is the area of the system and from Eqs. (A1), (A3), and (A4) $h(r_S)$ is here given by

$$h(r_S)=\frac{1}{\pi r_S^2 a_B^{*2}}\left[\frac{1}{v_v r_S^2}-\frac{8}{3\pi r_S}\left[\frac{2}{v_v}\right]^{1/2}(A_1+A_2 r_S)-\frac{1}{v_v}\frac{B_1}{B_2+r_S}\right]\text{Ry}^*\,. \tag{A6}$$

At this point we make use of Eqs. (12) and (A5) to obtain the following formula for the compressibility κ^*:

$$\frac{1}{\kappa^*}=\frac{r_S^2}{4}h''(r_S)+\frac{3r_S}{4}h'(r_S)\,. \tag{A7}$$

Then, on substituting the expression for h of Eq. (A6), we obtain

$$\frac{1}{\kappa^*}=\frac{2\mathrm{Ry}^*}{\pi v_v r_S^4 a_B^{*2}}-\frac{2A_1\mathrm{Ry}^*}{\pi^2 r_S^3 a_B^{*2}}\left[\frac{2}{v_v}\right]^{1/2}-\frac{\mathrm{Ry}^*}{4\pi v_v r_S a_B^{*2}}\frac{B_1}{(B_2+r_S)^2}-\frac{\mathrm{Ry}^*}{2\pi v_v a_B^{*2}}\frac{B_1}{(B_2+r_S)^3}\ . \tag{A8}$$

Making use of Eqs. (44) and (A8) we finally arrive at the sought after expression for the coefficient β_+:

$$\beta_+=\frac{2\epsilon_{av}L(p_F)}{\epsilon_S}\left[\frac{2A_1}{v_v\pi}+\frac{r_S^2}{8v_v}\left[\frac{2}{v_v}\right]^{1/2}\frac{B_1}{(B_2+r_S)^2}+\frac{r_S^3}{4v_v}\left[\frac{2}{v_v}\right]^{1/2}\frac{B_1}{(B_2+r_S)^3}\right]^{-1}, \tag{A9}$$

where the form factor $L(q)$ is defined in Eq. (39) in the text.

---

*Present address: NTT Basic Research Labs, Musashino-shi, Tokyo 180, Japan.

[1] J. J. Quinn and R. A. Ferrell, Phys. Rev. **112**, 812 (1958).
[2] T. M. Rice, Ann. Phys. **31**, 100 (1965).
[3] J. Hubbard, Proc. R. Soc. London Ser. A **242**, 539 (1957); **243**, 336 (1957).
[4] C. S. Ting, T. K. Lee, and J. J. Quinn, Phys. Rev. Lett. **34**, 870 (1975).
[5] T. K. Lee, C. S. Ting, and J. J. Quinn, Solid State Commun. **16**, 1309 (1975).
[6] S. Yarlagadda and G. F. Giuliani, Phys. Rev. B **49**, 14 172 (1984). See also Ref. 7.
[7] S. Yarlagadda, Ph.D. thesis, Purdue University, 1989.
[8] C. A. Kukkonen and A. W. Overhauser, Phys. Rev. B **20**, 550 (1979).
[9] X. Zhu and A. W. Overhauser, Phys. Rev. B **33**, 925 (1986).
[10] G. Vignale and K. S. Singwi, Phys. Rev. B **32**, 2156 (1985); see also K. S. Singwi, Phys. Scr. **32**, 397 (1985).
[11] S. Yarlagadda and G. F. Giuliani, Solid State Commun. **69**, 677 (1989).
[12] G. E. Santoro and G. F. Giuliani, Phys. Rev. B **49**, 7887 (1994).
[13] G. E. Santoro and G. F. Giuliani, Phys. Rev. B **39**, 12 818 (1989).
[14] G. E. Santoro and G. F. Giuliani, Solid State Commun. **67**, 681 (1988).
[15] T. K. Ng and K. S. Singwi, Phys. Rev. B **34**, 7738 (1986); **34**, 7743 (1986).
[16] S. Yarlagadda and G. F. Giuliani, Phys. Rev. B **49**, 7887 (1994).
[17] L. D. Landau, Zh. Eksp. Teor. Fiz. **30**, 1058 (1956) [Sov. Phys. JETP **3**, 920 (1956)].
[18] J. L. Smith and P. J. Stiles, Phys. Rev. Lett. **29**, 102 (1972).
[19] G. Abstreiter, J. P. Kotthaus, J. F. Koch, and G. Dorda, Phys. Rev. B **14**, 2480 (1976).
[20] F. F. Fang and P. J. Stiles, Phys. Rev. **174**, 823 (1968).
[21] T. Neugebauer, K. von Klitzing, G. Landwehr, and G. Dorda, Solid State Commun. **17**, 295 (1975).
[22] K. Suzuki and Y. Kawamoto, J. Phys. Soc. Jpn. **35**, 1456 (1973).
[23] J. F. Janak, Phys. Rev. **178**, 1416 (1969).
[24] T. Ando and Y. Uemura, J. Phys. Soc. Jpn. **37**, 1044 (1974).
[25] B. Vinter, Phys. Rev. B **13**, 4447 (1976).
[26] S. Yarlagadda and G. F. Giuliani (unpublished).
[27] S. Yarlagadda and G. F. Giuliani, Surf. Sci. **229**, 410 (1990).
[28] S. Yarlagadda and G. F. Giuliani, Phys. Rev. B **38**, 10 966 (1988).
[29] It is important to mention here that in deriving Eq. (35) for g^* we have made use of the fact that for an unpolarized system $G_-^{T_v}=G_0^v$. Also m^* is here the effective mass appropriate to the unpolarized case as given by Eq. (11).
[30] F. Stern and W. E. Howard, Phys. Rev. **163**, 816 (1967).
[31] S. Moroni and G. Senatore, Phys. Rev. B **44**, 9864 (1991); S. Moroni, D. M. Ceperley, and G. Senatore, Phys. Rev. Lett. **69**, 1837 (1992).
[32] G. Niklasson, Phys. Rev. B **10**, 3052 (1974).
[33] X. Zhu and A. W. Overhauser, Phys. Rev. B **30**, 3158 (1984).
[34] G. E. Santoro and G. F. Giuliani, Phys. Rev. B **37**, 4813 (1988).
[35] See, for instance, D. Pines and P. Nozières, *The Theory of Quantum Fluids* (Benjamin, New York, 1966), Vol. 1.
[36] M. Jonson, J. Phys. C **9**, 3055 (1976).
[37] K. S. Singwi, M. P. Tosi, R. H. Land, and A. Sjolander, Phys. Rev. B **176**, 589 (1968).
[38] Since Jonson's work pertains to the simple case of no-valley degeneracy, his nominal densities must be scaled by a factor of v_v^{-1}.
[39] In this equation we have neglected the contribution to κ^* associated with the energy due to the capacitance of the semiconductor-oxide-metal sandwich.
[40] In principle it is possible to make use of a similar procedure to determine the parameter β_+ in a self-consistent manner. This can be accompanied by using Eqs. (5) and (44) and calculating κ^*.
[41] S. Yarlagadda and G. F. Giuliani, Phys. Rev. B **40**, 5432 (1989).
[42] A. V. Chaplik, Zh. Eksp. Teor. Fiz. **60**, 1845 (1971) [Sov. Phys. JETP **33**, 997 (1971)]; see also F. Stern, Phys. Rev. Lett. **30**, 278 (1973).
[43] E. P. Wigner, Phys. Rev. **46**, 1002 (1934).
[44] We are well aware that this way of proceeding has its shortcomings. See, for instance, P. W. Anderson, *Basic Notions in Condensed Matter Physics* (Addison-Wesley, Redwood City, 1983).

Charge-Density Excitations at the Surface of a Semiconductor Superlattice: A New Type of Surface Polariton

Gabriele F. Giuliani[(a)] and J. J. Quinn
Brown University, Providence, Rhode Island 02912

(Received 23 June 1983)

A new type of surface polariton, which can occur at the surface of a semiconducting superlattice, is introduced. Because of the quantization of the electron energy levels by the superlattice potential this new polariton mode has the remarkable property of being free from Landau damping.

PACS numbers: 71.36.+c, 71.45.-d, 73.40.Lq

Electromagnetic modes which propagate along the interface between two media with different dielectric properties and which involve photons coupled to dipole excitations are called surface polaritons.[1] In simple metals and degenerate semiconductors, which are well approximated by a jellium model, the surface plasmon-polariton occurs at a frequency ω_{sp} which is always smaller than the bulk plasmon frequency ω_p.[2] Because the presence of the surface destroys the lattice translational invariance, surface plasmons in these systems can always excite electron-hole pairs, and they are thus subject to Landau damping.[3] Both in semiconducting and metallic superlattices the bulk plasmon spectrum[4,5] is rich in structure. In the former the single-particle spectrum is characterized by quantized electronic minibands. In the quantum limit, in which only the lowest miniband is occupied, there can exist both intrasubband and intersubband collective modes. The frequency of these modes depends in different ways on $\vec{q}$ and k, the components of wave vector parallel and perpendicular to the superlattice layers, respectively. The existence of this rich structure in the bulk excitation spectrum gives rise to a novel set of surface polariton modes[6] with the remarkable property of being free of Landau damping. The surface polariton frequency for a given value of $\vec{q}$ can occur either above or below the bulk plasmon continuum, depending on the ratio of the background dielectric constants of the semiconductor and the bounding material. In the case of polar semiconductors the electric field of the excitations gives rise to plasmon-phonon interaction resulting in coupled surface plasmon-phonon polaritons. Because this new type of surface polariton has never been observed experimentally, the object of this note is to elucidate a few of its remarkable properties in the hope of stimulating experiment.

The simplest model of a superlattice which correctly describes the intrasubband plasma modes[7] consists of a periodic array of two-dimensional electron-gas layers imbedded in a material of background dielectric constant ϵ_s. In this model the miniband structure of the superlattice is neglected, and only the ground subband and the intrasubband collective modes are considered.[8] The bulk plasmon spectrum can most easily be obtained by writing the general solution of the wave equation in the regions between the electron layers, assuming that $\vec{E}(z+na)=\exp(ikna)\vec{E}(z)$, where a is the superlattice period, and imposing the standard electromagnetic boundary conditions at each of the two-dimensional electron layers. For $q \ll k_F$, k_F being the Fermi wave vector of a two-dimensional electron gas, and $\omega \ll qk_F\hbar/m$, the resulting dispersion relation is given by

$$\omega^2(q,k)=(1/2\epsilon_s)qa\,S(q,k)\omega_p{}^2. \quad (1)$$

Here ω_p is the effective three-dimensional plasma frequency, $\omega_p{}^2=4\pi n_s e^2/ma$, where n_s is the number of electrons per unit area in any layer. $S(q,k)=\sinh qa\{\cosh qa-\cos ka\}^{-1}$ is a structure factor. For small values of the parameter qa, corresponding to strong coupling between the layers, a band of plasma modes results with frequencies $\omega(q,k)$ between $\omega_+(q)=\omega(q,0)$ and $\omega_-(q)=\omega(q,\pi/a)$. This band appears as the upper shaded region in Fig. 1, a plot of frequency ω vs wave vector q.

In order to describe surface excitations we assume that the periodic array of two-dimensional electron layers described above fills the space $z>0$, while an insulator of dielectric constant ϵ_0 fills the space $z<0$. We follow exactly the same prescription of writing down solutions of the wave equation in each region and imposing standard boundary conditions at the planes $z=na$ for $n=0,1,2,\ldots$. However, in this case we are interested in the situation in which the electric

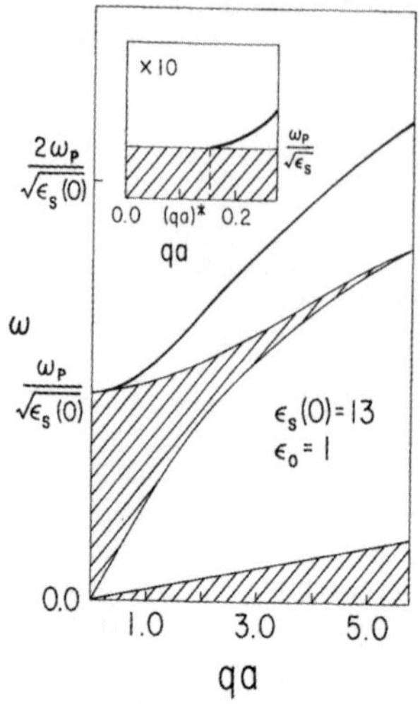

FIG. 1. A plot of frequency vs qa, the product of wave number parallel to the layers and superlattice spacing. The upper shaded region is the band of bulk intrasubband plasmons. The lower shaded region is the single-particle continuum. The surface polariton mode is the solid line which intersects the bulk plasmon continuum at $(qa)^* = 0.154$ as shown in the inset.

field in the region $z > 0$ satisfies the relation $\vec{E}(z+na) = e^{-\alpha na}\vec{E}(z)$, where α has a positive real part. In addition, the boundary condition at $z=0$ is different from those at $z=na$ for $n \geq 1$, because of the abrupt change in background dielectric constant from ϵ_s to ϵ_0, and because only the decaying-wave solution is allowed in the region $z<0$. In the electrostatic or nonretarded limit ($cq \gg \omega$) the resulting dispersion relation is

$$[1-e^{(\alpha+q)a}][\epsilon_+ - v_q\chi(q,\omega)] + [1-e^{(\alpha-q)a}][\epsilon_- + v_q\chi(q,\omega)] = 0. \qquad (2)$$

Here α, the inverse of the penetration depth, is the complex value of k for which the bulk dispersion relation [Eq. (1)] is satisfied for the given value of q and ω. We have introduced the symbols $\chi(q,\omega)$, the polarizability of the two-dimensional electron gas; $v_q = 2\pi e^2/q$, the Fourier transform of the two-dimensional Coulomb interaction; and $\epsilon_\pm = \frac{1}{2}(\epsilon_s \pm \epsilon_0)$. The solution of Eq. (2), which must be obtained numerically, depends in an important way on the ratio of ϵ_s to ϵ_0. For $\epsilon_s > \epsilon_0$ the parameter α is real, and the surface mode occurs at a frequency above the bulk plasmon continuum. For large values of qa, α is equal to q, but it decreases to zero as q decreases to the value q^*. Thus, for wavelengths short compared to superlattice period, the penetration depth is one wavelength, while for $q \simeq q^*$, the penetration depth becomes infinite. For $\epsilon_s < \epsilon_0$, α acquires an imaginary part (equal to π/a) and the surface mode falls below the continuum. The result is illustrated by the solid curve in Fig. 1 which corresponds to a semiconductor-vacuum interface with $\epsilon_s = 13$ and $\epsilon_0 = 1$. It is interesting to note that the frequency of the surface mode intersects the bulk plasmon continuum at a finite value of the wave vector q; an enlargement of the region of intersection is shown in the inset. The intersection with the continuum occurs at a value of $q = q^*$ given by

$$q^* = a^{-1}\ln|(\epsilon_s+\epsilon_0)/(\epsilon_s-\epsilon_0)|. \qquad (3)$$

For $q < q^*$ surface modes do not exist because the decay parameter α is purely imaginary. As ϵ_s approaches ϵ_0 the value of q^* increases logarithmically, so that the existence of surface modes depends quite critically on the difference in background dielectric constants of the semiconducting superlattice and the bounding medium. If the first two-dimensional electron layer occurs a small distance (compared to the superlattice period) from the interface, the value of q^* is increased as would be expected. We have obtained numerical results for the case in which $\epsilon_0 > \epsilon_s$; the surface mode lies below the continuum and intersects the lower edge at a value of q^* given by Eq. (3).

In a three-dimensional jellium model, a surface plasmon of wave vector $\vec{q}$, parallel to the surface, and frequency ω can always decay into an electron-hole pair conserving both energy and the parallel component of the wave vector. The reason for this is that the electronic energy spectrum is a continuous function of k, the normal wave number, and any change in k is allowed because the presence of the surface relaxes the condition of wave-vector conservation. In the semiconducting superlattice, however, the quantization of the electronic energy levels by the superlattice potential makes it impossible to conserve energy and parallel wave vector in the creation of an electron-hole pair by an elementary excitation lying outside the single-particle continuum. For the simple model used in this note, the single-particle continuum consists of that portion of the ω-q plane in which $\omega < \hbar q(k_F + q/2)2m$. The single-particle continuum appears as the lower shaded region in Fig. 1. For the model considered in Ref. 4, there are a number of "two-dimensional" electronic subbands separated by energy $\hbar\omega_{n0}$ from the ground subband, and there can exist a set of intersubband collective modes. In that case, there are additional regions of the single-particle continuum defined by $-\hbar q(k_F - q/2)/2m + \omega_{n0} < \omega < \hbar q(k_F + q/2)/2m + \omega_{n0}$ for each subband separation ω_{n0}. Collective surface excitations lying outside the single-particle continuum are unable to decay into a single electron-hole pair and are thus not subject to Landau damping. Therefore, in high-mobility semiconducting superlattices, these surface modes should have a very long lifetime.

A good candidate for possible observation of the surface polariton modes discussed here is the GaAs-$Al_xGa_{1-x}As$ superlattice system.[9] In this system the background dielectric function ϵ_s is not a constant, but it is a function of frequency: $\epsilon_s(\omega) = \epsilon_s(\infty)(\omega^2 - \omega_L^2)(\omega^2 - \omega_T^2)^{-1}$, where $\epsilon_s(\infty)$ is the high-frequency dielectric constant, and ω_L and ω_T are the longitudinal and transverse optical phonon frequencies respectively. By taking account of the frequency dependence of $\epsilon_s(\omega)$, we find a system of coupled bulk intrasubband-plasmon–optical-phonon bands, as shown by the two upper shaded regions in Fig. 2. The plasmon continuum is very similar to that in Fig. 1 in which coupling to phonons is neglected. The bulk longitudinal optical phonon mode becomes dispersive, and is broadened into a band by coupling to the plasmons. The parameters used in the numerical calculation are $\epsilon_s(\infty) = 10.9$, $\epsilon_s(0) = (\omega_L^2/\omega_T^2)\epsilon_s(\infty) = 13$, and $\epsilon_0 = 1$; the values of ω_L and ω_p are taken to be 5.5×10^{13} $\sec^{-1}$ and 3.13×10^{13} $\sec^{-1}$ respectively. The solid lines represent the coupled surface plasmon-phonon polariton modes. Again, we observe the plasmonlike mode above the bulk continuum for values of q larger than some critical value.[10]

The phononlike mode begins below the bulk phonon continuum, but the coupling to the plasmon forces it to merge with the continuum and eventually to reappear above it at a larger value of q.

Because the surface polaritons are nonradiative, they do not couple directly to light. In order to observe the modes in optical absorption or reflectance it will be necessary to destroy the translation invariance along the surface by, for example, producing a grating on the surface. The grating spacing l should satisfy the inequality $l < 2\pi q^{*-1}$; this is in the range of thousands of angstroms and should not be difficult to achieve. The relatively large values of q^* make attenuated total reflection seem an unlikely method of observation. However, resonant Raman scattering[11] and electron-energy-loss spectroscopy appear to be possible techniques for observing these surface polaritons. In these experiments large momentum transfer along the surface is possible, so that values of q greater than q^* can be attained.

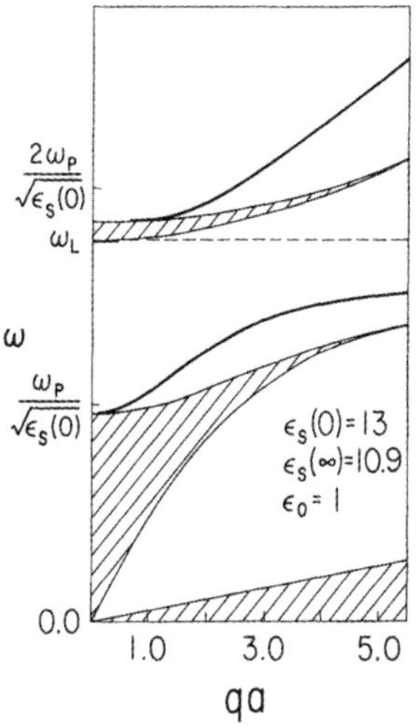

FIG. 2. Same plot as Fig. 1 when coupling to optical phonons is included. The two upper shaded regions are bulk intrasubband-plasmon–phonon modes. The two solid curves are the coupled surface polariton modes. The phononlike polariton intersects the continuum from above, and reappears below the continuum for very small values of wave vector.

The authors would like to thank Professor Guoyi Qin, Dr. G. Gonzalez de la Cruz, and Dr. A. C. Tselis for stimulating discussions. This work was supported in part by the National Science Foundation through Grant No. DMR-81-21-069 and by the U. S. Office of Naval Research.

---

[a]On leave from the Scuola Normale Supériore, Pisa, Italy.

[1]E. Burstein, in *Polaritons*, edited by E. Burstein and F. DeMartini (Pergamon, New York, 1974), p. 1.

[2]R. H. Ritchie, Phys. Rev. 106, 874 (1957); E. A. Stern and R. A. Ferrell, Phys. Rev. 120, 130 (1960).

[3]R. Fuchs and K. L. Kliewer, Phys. Rev. B 3, 2270 (1971); D. E. Beck, Phys. Rev. B 4, 1555 (1971).

[4]A. Tselis, G. Gonzalez de la Cruz, and J. J. Quinn, Solid State Commun. 46, 779 (1983); G. Gonzalez de

la Cruz, A. Tselis, and J. J. Quinn, J. Chem. Phys. Solids 44, 807 (1983).

[5]G. Giuliani, J. J. Quinn, and R. F. Wallis, Bull. Am. Phys. Soc. 28, 448 (1983); A. Caille, M. Banville, P. D. Loly, and M. J. Zuckerman, Solid State Commun. 41, 119 (1982); see also the review by E. Tosatti, in *Interaction of Radiation with Condensed Matter* (International Atomic Energy Agency, Vienna, 1977), Vol. 1, pp. 281–294.

[6]Surface plasma modes of a superlattice consisting of metallic layers, which can be described by a local three-dimensional dielectric function, separated by insulating layers have been considered recently by R. E. Camley and D. L. Mills, Bull. Am. Phys. Soc. 28, 408 (1983). Related problems of acoustic and magnetic excitations in semi-infinite periodic structures have been investigated by R. E. Camley, B. Djafari-Rouhani, L. Dobrzynski, and A. A. Maradudin, Phys. Rev. B 27, 7318 (1983); and R. E. Camley, T. Rahman, and D. L. Mills, Phys. Rev. B 27, 261 (1983).

[7]A. L. Fetter, Ann. Phys. (N.Y.) 81, 367 (1973); S. Das Sarma and J. J. Quinn, Phys. Rev. B 25, 7603 (1982).

[8]A more realistic model is introduced in Ref. 4; the surface modes associated with intersubband excitations will be considered in another publication.

[9]For a type-II superlattice like the GaSb-InAs system the surface polariton modes have been studied by G. Qin, G. Giuliani, and J. J. Quinn, to be published.

[10]The behavior of α, the inverse penetration depth, is more complicated in this situation. See G. Giuliani, G. Qin, and J. J. Quinn, in Proceedings of the Fifth International Conference on the Electronic Properties of Two-Dimensional Systems, Oxford, 1983 (to be published); and G. Qin, G. Giuliani, and J. J. Quinn, to be published.

[11]See, for example, D. Olego, A. Pinczuk, A. C. Gossard, and W. Weigmann, Phys. Rev. B 25, 7867 (1982); Z. J. Tien, J. M. Worlock, C. H. Perry, A. Pinczuk, R. L. Aggerwal, H. L. Stormer, A. C. Gossard, and W. Weigmann, Surf. Sci. 113, 89 (1982). Although values of qa up to approximately 0.6 were achieved, no surface modes were observed. We believe that this results from the last electronic layer being separated from the surface by a GaAlAs overlayer of unspecified thickness. An overlayer of background dielectric constant equal to the bulk value dramatically increases the value of q^* as mentioned in the text.

SELECTIONS

The volumes of this series arise from conferences and symposiums and focuse on a particular topic of current research in mathematics or physics.

Published volume

1. F. BELTRAM (editor), *Highlights in the quantum theory of condensed matter*, A symposium to honour Mario Tosi on his 72nd birthday, 2005. ISBN 88-7642-170-X
2. M. POLINI, G. VIGNALE, V. PELLEGRINI, J. K. JAIN (editors), *No-nonsense Physicist*, An overview of Gabriele Giuliani's work and life, 2016. ISBN 978-88-7642-535-6 eISBN 978-88-7642-536-3

Fotocomposizione "CompoMat" Loc. Braccone, 02040 Configni (RI) Italia
Finito di stampare nel mese di aprile 2016
dalla CSR, Via di Salone, 131/c, 00131 Roma